CDMA2000® EVOLUTION

CDMA2000® EVOLUTION
System Concepts and Design Principles

KAMRAN ETEMAD

WILEY-INTERSCIENCE

A JOHN WILEY & SONS, INC., PUBLICATION

CDMA2000® is a registered trademark of the
Telecommunications Industry Association (TIA-USA).

Library of Congress Cataloging-in-Publication Data
Etemad, Kamran.
    CDMA2000® evolution : system concepts and design principles / Kamran Etemad.
        p.   cm.
    "A Wiley-Interscience publication."
    Includes bibliographical references and index.
    ISBN 0-471-46125-3 (cloth)
    1. Code division multiple access.   I. Title.
    TK5103.452.E84   2005
    621.3845′6—dc22

                                                                                2004013846

Printed in the United States of America

10 9 8 7 6 5 4 3 2 1

To My Family:
Shiva and Soroush

CONTENTS

PREFACE

Wireless networks offering high-speed multimedia services to mobile users have finally become a reality and third generation (3G) wireless systems are being deployed and extended worldwide to meet the rising demand for these services.

Aiming to offer high rates of bearer services with appropriate Quality of Service (QoS) control in the mobile radio environment, the 3G standards have evolved from their original designs and gone through multiple revisions, each with much added flexibility, efficiency, and, of course, complexity.

The on-going evolution of wireless networks from legacy architectures toward next-generation platforms and the emergence of fundamentally new air interface technologies have created a significant need for a solid understanding of these networks and their design features and capabilities.

The radio access technology described by the cdma2000 standards is one of these systems that has been developed as part of the evolution of second-generation (2G) code division multiple access (CDMA) systems intended to achieve the new performance and service requirements of 3G systems. The cdma2000 family of standards includes the 1x-RTT or IS2000 as well as the High Rate Packet Data (HRPD) or IS856 standards.

With more than 190 million subscribers around the globe, cdma2000 is the second most widely deployed technology. Also, as some of the markets that are traditionally dominated by GSM systems are considering the adoption of this system, one can expect cdma2000's market share to increase over time.

The cdma2000 specifications developed by the 3rd Generation Partnership Project II (3GPP2) include all the technical details of this standard. However, while the ultimate sources for the various technical details of cdma2000 are the 3GPP2 specifications, reading and collating all these documents to gain an understanding of the key concepts and the design methodologies is not an easy task.

Considering the limited number of insightful, comprehensive, and up-to-date reference books about cdma2000, the author has developed the material in this volume with the aim of providing a clear presentation and a solid understanding of this radio access technology. The information contained in this book has been selectively gathered and enhanced using various sources ranging from the most recent cdma2000 standards and specifications to related technical papers and the author's practical insights which are integrated and documented for the first time. It should be mentioned that some of the figures in this book are reproduced from 3GPP2 specifications under written permission from Telecommunications Industry Association.

As a comprehensive single source about cdma2000 technology, this book will be very helpful to many engineers, students, and technical managers who are involved in the manufacturing, network design, and operation of these systems. It will also assist marketing managers and executives who are looking for a clear presentation of the key features of cdma2000 technology in general.

In addition to a clear presentation of cdma2000 standards the book also provides a solid foundation that explains some of the main enabling concepts and technologies commonly used by 3G systems as well as some network design and dimensioning methodologies.

The book is organized into 13 chapters, starting with the basic concepts of legacy systems and progressing to the more advanced concepts and features in the most recent releases of the cdma2000 standards.

Chapter 1 presents the evolution of cdma2000 standards from legacy cdmaONE systems to IS2000 release 0 (also known as 1x-RTT) and its subsequent revisions (Releases A to D). This evolution also includes a data-only or data-optimized air interface, called 1xEV-DO or High Rate Packet Data (HRPD), which can be deployed independently or as an overlay on existing integrated voice and low-rate data networks. This chapter also presents some discussions of the 3G standardization process and the main organizations contributing to cdma2000 specifications within 3GPP2.

Chapter 2 describes the key CDMA concepts and terminologies including spread spectrum communications and sequence spreading and dispreading, as well as spreading codes, rake receivers, and multipath diversity. This chapter also discusses the network aspects of CDMA such as universal channel reuse, soft cell capacity, power control, and soft hand off.

Chapter 3 is an overview of second generation (2G) or legacy CDMA systems based on cdmaONE or the IS95 air interface. Since cdma2000 is a backward-compatible evolution of IS95 networks, it includes all legacy 2G channels and protocols as a subset. Therefore a solid understanding of the IS95 system is not only needed as a preparation for studying more advanced features of cdma2000 technology, it is also an integral part of that study as well. This chapter describes the radio and physical channelization in IS95 as well as the coding, spreading, and modulation structures of forward and reverse link channels. Other topics of inertest in this chapter are system access and traffic channel power control and hand off procedures and parameters which are some of the distinct features of IS95 air interface.

Chapter 4 is one of the most important chapters in this book as it generically presents some of the advanced technologies and concepts used not only in cdma2000 but also in other 3G systems. These "enabling technologies" include both physical and MAC layer techniques introduced to maximize the system flexibility and efficiency. Link Adaptation is one of the main differentiating features of 3G systems in which higher-order modulation is combined with variable spreading and coding to allow variable data rate channels. Physical layer hybrid ARQ protocols are another means for link adaptation using incremental redundancy for high speed packet data channels. Adaptive scheduling using multiuser diversity and advanced turbo encoding are two other important techniques employed in 3G systems to maximize packet data efficiency. To provide better voice quality and/or higher voice capacity the 3G systems are also equipped with advanced multimode or multirate vocoders. The mode or rate of these encoders can be selected by the network to flexibly balance the voice quality and cell capacity under different traffic loads. This chapter also presents some optional techniques such as beam-forming and transmit diversity schemes, which provide further improvements in coverage and spectrum efficiency.

Chapters 5 through 8 are dedicated to a detailed description of thecdma2000 1x (or IS2000) air interface. The focus is on the features of the key physical, link, and signaling layers and the procedures related to voice and medium speed data bearer services. These features and procedures are mostly defined in Release A of the standard and have not changed in Releases B through D.

Chapter 5 presents the overall IS2000 airlink protocol layers and their respective functions and structures. It also provides an overview of all forward and reverse link physical channels and their role in the system's operation.

Chapter 6 provides a detailed description of physical layer structures for all forward and reverse physical channels, including their frame sizes, data rates, coding, spreading, and modulation characteristics.

Chapter 7 covers call processing and signaling procedures starting from system acquisition and mobile station initialization and continuing with the processes involved in the idle and access states. These procedures include the synchronization and monitoring of various broadcast and forward common control channels by a mobile station in order to obtain access and system parameters or updates and to respond to order messages when necessary. This chapter also presents the frame structure and usage for all broadcast and common channels used during the mobile station's initialization, idle, and access states. Another area of focus for this chapter is enhanced random and reservation access protocols as well as access state hand off procedures.

Chapter 8 is a continuation of the call processing and signaling procedures discussion of the previous chapter, focusing on the mobile station's control of traffic channel state where the mobile station is using a dedicated channel to exchange traffic frames with the network. The key issues discussed in this chapter are voice and data transmission structures on the traffic channels and channel assignment as well as call admission and radio resource control issues. Other areas of focus are the improved traffic channel soft hand off, inter-frequency hand off, and the enhanced traffic channel power control schemes.

Chapter 9 presents the cdma2000 network architecture and reference model. It also describes the function of the key network elements and their interfaces. One of the main subjects discussed in this chapter is network signaling and the procedures defined for mobility management and hand off support for packet data services. As the cdma2000 networks evolve to the so-called "All-IP" architectures system, there will be changes in the functional elements and interfaces. This chapter provides an overview of the cdma2000 All-IP model including a detailed discussion on QoS management.

Chapter 10 is an overview of the High Rate Packet Data (HRPD) or 1xEV-DO system. As some of the notions and concepts used in this system are introduced in Chapter 4 or they are common with IS2000, all discussions related to this standard are concisely presented. This chapter presents the main technical attributes of the HRPD system based on the original IS856 standard and it also includes recent enhancements in the system as part of its Release A or IS856-A. These enhancements are primarily related but not limited to uplink throughput and latencies.

Chapter 11 presents the enhancements in Releases C and D of IS2000, which are mainly related to the new high-rate packet data 1xEV-DV modes. This discussion is intentionally deferred to this chapter, as many design features of 1xEV-DV are similar to the HRPD standard which was discussed in Chapter 10. The high-speed packet data channel was introduced first for the forward link in Release C of IS2000 which was referred to as the 1xEV-DV mode. This mode was subsequently expanded to the uplink by defining a high-speed reverse packet data channel and its associated procedures and protocols in Release D. Using IS2000-D specifications as a reference, this chapter provides a clear description of the new high-speed packet channels and protocols.

Chapter 12 focuses on cdma2000 radio link performance from cell coverage and capacity perspectives and its impact on network planning. RF performance for both IS2000 and HRPD systems is covered in this chapter. The coverage performance is captured through forward and reverse link budget analysis for voice and different data bearer rates. The capacity performance is described for both voice and packet data services. This chapter also presents a practical and objective methodology for dimensioning the radio network based on cell coverage and capacity performance for mixed voice and data services. In this context both greenfield and expansion scenarios are discussed.

Chapter 13 is the last chapter and it provides an overview of other 3G radio access technologies, specifically focusing on WCDMA and TD-CDMA systems. The information in this chapter is provided to help readers to appreciate the similarities and differences between various 3G wireless technologies.

This book can be used as a reference for RF and network engineers, network operators, and technical and marketing managers as well as technologists with different backgrounds and objectives. It can also be used as a textbook or reference book for a graduate or senior-undergraduate level course specifically on cdma2000 or 3G wireless systems. For additional information, please see the book website at ftp://ftp.wiley.com/public/sci_tech_med/CDMA2000_evolution/

ACKNOWLEDGMENTS

Like any document covering a standard technology, this book has benefited the most from the work of many companies and individuals who have contributed to the standard development and specification. Therefore, the author's first and foremost acknowledgment is to all the entities contributing directly or indirectly to cdma2000 development within 3GPP2. Also many thanks are offered to the researchers whose original ideas and results are used as motivations for, and foundations of, the new generation of wireless networks and services.

The author also thanks many of his colleagues and students for their help and comments which have assisted in fine-tuning the scope and quality of this book. Specifically many thanks are due to Dr. Kamran Seyrafian-Pour for his review of and comments on various sections and to Dr. Sassan Ahmadi for sharing his insights on advanced vocoders. Also many thanks to Val Moliere and Kirsten Rohsted of John Wiley & Sons for their patience and cooperation throughout the extended development process of this book.

ACRONYMS

1xEV-DO	1x Evolution for Data Only (or Data Optimized)
1xEV-DV	1x Evolution for Data and Voice
2G	Second Generation Wireless Systems
2.5G	Wireless Systems in-between 2nd and 3rd generation
3G	Third Generation Wireless Systems
3GPP	Third Generation Partnership Project
3GPP2	Third Generation Partnership Project II
AAA	Authentication, Authorization, and Accounting
AAL	ATM Adaptation Layer
AAL2	ATM Adaptation Layer type 2
AC (AuC)	Authentication Center
ACH	Access Channel
ACK	Acknowledgment
ACN	Access Channel Number
ADA	Advertising Agent
AGCH	Access Grant Channel
A-GPS	Assisted GPS
AGW	Access Gateway
AH	Authentication Header
AMPS	Advanced Mobile Phone System
AN	Access Network
ANSI	American National Standards Institute
AP	Access Point
API	Application Program Interface
APN	Access Point Name
ARIB	Association of Radio Industries and Businesses
ARQ	Automatic Repeat reQuest
AS	Access Stratum
AT	Access Terminal
ATI	Access Terminal Identifier
ATM	Asynchronous Transfer Mode
AuC	Authentication Center
BCCH	Broadcast Control CHannel
BCH	Broadcast CHannel
BER	Bit Error Rate
BG	Border Gateway
BLOB	BLock Of Bits
BOD	Bandwidth on Demand

BPSK	Binary Phase Shift Keying
BR	Border Router
BS	Base Station
BSC	Base Station Controller
BSS	Base Station System
BTS	Base Station Transceiver System, Base Transceiver Station
CAC	Connection Admission Control
CALEA	Communications Assistance for Law Enforcement Act
CAM	Channel Assignment Message
CAMEL	Customized Applications for Mobile Enhanced Logic
CAVE	Cellular Authentication and Voice Encryption
CCCH	Common Control CHannel
CCH	Control CHannel
CCP	Compression Control Protocol
CCPCH	Common Control Physical CHannel
CCS	Common Channel Signaling
CD	Call Delivery
CDG	CDMA Development Group
CDMA	Code Division Multiple Access
CDPD	Cellular Digital Packet Data
CEPT	European Conference of Posts and Telecommunication Administrations
CGF	Charging Gateway Function
CGI	Common Gateway Interface
CHAP	Challenge Handshake Authentication Protocol
CID	Circuit ID
CN	Core Network
CNI	Calling Number Identification
CO	Central Office
CoA	Care-of Address
CPCH	Common Packet CHannel
CPI	Capability Preference Information
CPU	Central Processing Unit
CQM	Core Quality of Service Manager
CRC	Cyclic Redundancy Check
CS	Circuit Switched
CSMA/CA	Carrier Sense Multiple Access/Collision Avoidance
CSMA/CD	Carrier Sense Multiple Access/Collision Detect
CT	Cordless Telephony
CTCH	Common Traffic CHannel
CTIA	Cellular Telecommunications Industries Association
D-AMPS	Digital Advanced Mobile Phone Service
DB	DataBase
DCCH	Dedicated Control CHannel
DCH	Dedicated CHannel
DECT	Digital European Cordless Telephony
DHCP	Dynamic Host Configuration Protocol

DNS	Domain Name Server
DPCCH	Dedicated Physical Control CHannel
DPCH	Dedicated Physical CHannel
DPDCH	Dedicated Physical Data CHannel
DRC	Data Rate Control
DS	Direct Spread
DS-CDMA	Direct-Sequence Code Division Multiple Access
DSI	Dynamic Subscriber Information
DSL	Digital Subscriber Line
DSSS	Direct Sequence Spread Spectrum
DTCH	Dedicated Traffic CHannel
DTMF	Dual Tone Multi Frequency Signal
DTX	Discontinuous Transmission
E911	Enhanced 911
ECAM	Extended Channel Assignment Message
EDGE	Enhanced Data Rates for Global (GSM) Evolution
EIA	Electronic Industries Alliance
EIR	Equipment Identity Register
E-OTD	Enhanced Observed Time Difference
ESCAM	Extended Supplemental Channel Assignment Message
ESMR	Enhanced Specialized Mobile Radio
ESN	Electronic Serial Number
ETRI	Electronics and Telecommunication Research Institute
ETSI	European Telecommunication Standards Institute
EVRC	Enhanced Variable Rate CODEC
F-CCCH	Forward Common Control CHannel
F-PICH	Forward Pilot CHannel
FA	Foreign Agent
FAC	Foreign Agent Challenge
FACH	Forward Access Channel
F-APICH	Dedicated Auxiliary Pilot CHannel
F-ATDPICH	Auxiliary Transmit Diversity Pilot CHannel
F-BCCH	Broadcast Control CHannel
FBI	Feedback Information
F-CACH	Common Assignment CHannel
FCC	Federal Communications Commission
F-CCCH	Forward Common Control CHannel
FCH	Fundamental CHannel
F-CPCCH	Common Power Control CHannel
F-DCCH	Forward Dedicated Control CHannel
FDD	Frequency Division Duplex
FDD	Frequency Division Duplex
FDM	Frequency Division Multiplexing
FDMA	Frequency Division Multiple Access
FE	Functional Entities
FEC	Forward Error Correction

FER	Frame Error Rate
F-FCH	Forward/Reverse Fundamental CHannel
FHSS	Frequency Hopping Spread Spectrum
FM	Frequency Modulation
F-PCH	Paging CHannel
F-PDCCH	Forward Packet Data Control CHannel
F-PDCH	Forward Packet Date CHannel
F-QPCH	Quick Paging CHannel
FRAMES	Future Radio widebAnd Multiple accEss Systems
F-SCCH	Forward Supplemental Code CHannel
F-SCH	Forward Supplemental CHannel
FSK	Frame Shift Keying
F-SYNCH	Sync CHannel
FTAM	File Transfer and Access Management
FTC	Forward Traffic Channel
F-TDPICH	Transmit Diversity Pilot CHannel
FTP	File Transfer Protocol
GGSN	Gateway GPRS Support Node
GLR	Gateway Location Register
GMM	GPRS Mobility Management
GMSC	Gateway Mobile Switching Center
GMSK	Gaussian Minimum Shift Keying
GPRS	Generic Packet Radio Service
GPS	Global Positioning System
GRE	Generic Routing Encapsulation
GSM	Global Systems for Mobile communications
GSN	GPRS Support Node
GTP	GPRS Tunneling Protocol
HA	Home Agent
HCI	Host Controller Interface
HDLC	High-level Data Link Control
HDM	Hand Off Direction Message
HDR	High Data Rate
HLR	Home Location Register
HO	Handover
HRPD	High Rate Packet Data
HSCSD	High-Speed Circuit Switched Data
HSDPA	High Speed Downlink Packet Access
HS-DSCH	High Speed Downlink Shared CHannel
HTML	HyperText Markup Language
iDEN	Integrated Digital Enhanced Network
IEEE	Institute of Electrical and Electronics Engineers
IETF	Internet Engineering Task Force
IKE	Internet Key Exchange
IMEI	International Mobile Equipment Identity
i-mode	Information mode

IMSI	International Mobile Subscriber Identity
IMT-2000	International Mobile Telecommunications – 2000
IN/AIN	Intelligent Network/Advanced Intelligent Network
IOS	Inter Operability Specification
IP	Internet Protocol
IPSec	IP Security
IPv4	Internet Protocol Version 4
IPv6	Internet Protocol version 6
IS	Interim Standard
I-SCM	Interrogating SCM
ISDN	Integrated Services Digital Network
ISM	Industrial Scientific Medical
ISP	Internet Service Provider
ISUP	ISDN Signaling User Part
ITU	International Telecommunication Union
ITU-D	ITU-Development
ITU-R	ITU-Radio communication
ITU-T	ITU-Telecommunications
IWF	Inter Working Function
J2ME	Java 2 Micro Edition
JTC	Joint Technical Committee
kbps	kilo-bits per second
L1	Layer 1 (physical layer)
L2	Layer 2 (data link layer)
L3	Layer 3 (network layer)
LAC	Link Access Control
LAI	Location Area Identity
LAN	Local Area Network
LCP	Link Control Protocol
LEA	Law Enforcement Agencies
LEC	Local Exchange Carrier
LEO	Low Earth Orbit
LLC	Logical Link Control
LMDS	Local Multipoint Distribution Systems
LOC	Location Database
L-SCM	Local SCM
MAC	Medium Access Control
MAN	Metropolitan Area Network
MAP	Mobile Application Part
MC	Messaging Center
MCC	Mobile Country Code
MCF	Mobile Control Function
MCS	Modulation and Coding Scheme
ME	Mobile Equipment
MEO	Medium Earth Orbit
MGCF	Media Gateway Control Function

MGW	Media Gateway
MIB	Management Information Base
MIN	Mobile Identification Number
MIP	Mobile IP
MLP	Mobile Location Protocol
MM	Mobility Management
MM	Mobility Manager
MMDS	Multichannel Multipoint Distribution System
MN	Mobile Node
MOU	Minutes of Use
MPC	Mobile Positioning Center
MRF	Media Resource Function
MRFC	Media Resource Function Controller
MRFP	Media Resource Function Processor
MS	Mobile Station
MSC	Mobile Switching Center
MSID	Mobile Station ID
MT	Mobile Terminal
Mux	Multiplex
MWD	Mobile Wireless Data
NAI	Network Address Identifier
N-AMPS	Narrowband Advanced Mobile Phone System
NAS	Non-Access Stratum
NA-TDMA	North American TDMA, i.e., IS-136
NCGW	Network Capability Gateway
NID	Network Identifier
NMT	Nordic Mobile Telephone
NNI	Network to Network Interface
NOC	National Operations Centers
NSAPI	Network layer Service Access Point Identifier
NSS	Network SubSystem
OA&M	Operations, Administrations and Maintenance
OMC	Operations and Maintenance Centers
OSI	Open Systems Interconnection
OTAF	Over The air Activation Function
OTASP	Over The Air Service Provisioning
OTD	Orthogonal Transmit Diversity
OVSF	Orthogonal Variable Spreading Factor
PACA	Priority Access and Channel Assignment
PACCH	Packet Associated Control CHannel
PAGCH	Packet Access Grant CHannel
PAN	Personal Area Network
PAP	Password Authentication Protocol
PC	Power Control
PCB	Power Control Bit
PCF	Packet Control Function

PCH	Paging Channel
PCM	Pulse Code Modulation
PCMCIA	Personal Computer Memory Card International Association
PCS	Personal Communication Services
PCS	Personal Communication System
PCU	Packet Control Unit
PDA	Personal Digital Assistant
PDC	Personal Digital Cellular
PDCCH	Packet Dedicated Control CHannel
PDCH	Packet Data CHannel
PDCP	Personal Digital Cellular Packet
PDE	Position Determining Entity
PDN	Packet Data Node
PDP	Packet Data Protocol
PDSN	Packet Data Serving Node
PDTCH	Packet Data Traffic CHannel
PDU	Protocol Data Unit
PG	Processing Gain
PHB	Per Hop Behavior
PHS	Personal Handyphone System
PHY	Physical Layer
PI	Power Increase
PLMN	Public Land Mobile Network
PN	Pseudo-Noise
PPDN	Public Packet Data Network
PPP	Point-to-Point Protocol
PRAT	Paging channel data RATe
PS	Packet Switched
PSAP	Public Safety Answering Point
PS-CN	Packet Switched-Core Network
PSI	PCF Session ID
PSK	Phase Shift Keying
PSMM	Pilot Strength Measurement Message
PSTN	Public Switched Telephone Network
PTM	Point to Multipoint
PTP	Point to Point
PZID	Packet Zone ID
QAM	Quad Amplitude Modulation
QOF	Quasi Orthogonal Function
QoS	Quality of Service
QPCH	Quick Paging CHannel
QPSK	Quadrature Phase Shift Keying
RA	Routing Area
RAB	Reverse Activity Bit
RAC	Routing Area Code
R-ACH	Reverse Access CHannel

RACH	Random Access CHannel
R-ACKCH	Reverse Acknowledgement CHannel
RADIUS	Remote Authentication Dial In User Service
RAI	Routing Area Identity
RAN	Radio Access Network
RAND	Random Number
RAR	Resource Allocation Request
RBP	Radio Burst Protocol
RC	Radio Configuration
R-CCCH	Reverse Common Control CHannel
RCD	Resource Configuration Database
R-CQICH	Reverse Channel Quality Indicator CHannel
R-DCCH	Reverse Dedicated Control CHannel
RDF	Resource Description Framework
RDP	Remote Display Protocol
R-EACH	Enhanced Access CHannel
RF	Radio Frequency
RFC	Remote Feature Control
R-FCH	Reverse Fundamental CHannel
RLC	Radio Link Control
RLP	Radio Link Protocol
RN	Radio Network
RNC	Radio Network Controller
RNS	Radio Network Subsystem
R-P	RN-PDSN interface
RPC	Reverse Power Control
R-PICH	Reverse PIlot CHannel
RR	Radio Resource management
RRC	Radio Resource Control
RRI	Reverse Rate Indicator
RRP	Mobile IP Registration RePly
RRQ	Registration ReQuest (Mobile IP)
RSCAMM	Reverse Supplemental Channel Assignment Mini Message
R-SCCH	Reverse Supplemental Code CHannel
R-SCH	Reverse Supplemental CHannel
RSVP	Resource ReSerVation Protocol
RT	Random Time
RTC	Reverse Traffic Channel
RTP	Real Time Protocol
RTT	Radio Transmission Technology
SA	Security Association
SAP	Service Access Point
SAPI	Service Access Point Identifier
SAR	Segmentation and Reassembly sublayer
SCAM	Supplemental Channel Assignment Message
SCCH	Supplemental Code CHannel

SCF	Service Control Function
SCH	Supplemental CHannel
SCM	Session Control Manager
SCP	Service Control Point
SCRM	Supplemental Channel Request Message
SCRMM	Supplemental Channel Request Mini Message
SDB	Short Data Burst
SDU	Service Data Unit
SF	Spreading Factor
SGSN	Serving GPRS Support Node
SGW	Signaling GateWay
SIBB	Service Independent Building Block
SID	System Identifier
SIM	Subscriber Identity Module
SIP	Session Initiation Protocol
SLA	Service Level Agreement
SME	Short Message Entity
SMG	Special Mobile Group
SMR	Specialized Mobile Radio
SMS	Service Management System
SMS	Short Message Service
SMS-SC	Short Message Service-Service Center
SMTP	Simple Mail Transfer Protocol
SN	Service Node
SNMP	Simple Network Management Protocol
SNR	Signal to Noise Ratio
SONET	Synchronous Optical NETwork
SQM	Subscription Quality of service Manager
SR	Spreading Rate
SR_ID	Service Reference IDentifier
SRBP	Signaling Radio Burst Protocol
SS7	Signaling System 7
SSD	Shared Secret Data
STP	Signaling Transfer Point
STS	Space Time Spreading
TCP	Transmission Control Protocol
TD	Transmit Diversity
TDD	Time Division Duplex
TDMA	Time Division Multiple Access
TE	Terminal Equipment
TFCI	Transport Format Combination Indicator
TIA	Telecommunication Industry Association
TM	Traffic Mode
TMSI	Temporary Mobile Subscriber Identity
TOA	Time Of Arrival
TRAU	Transcoder and Rate Adaptor Unit

TT	Traffic Type
TTA	Telecommunication Technology Association
UDP	User Datagram Protocol
UDR	Usage Data Record
UE	User Equipment
UI	User Interface
UIM	User Identity Module
UMTS	Universal Mobile Telecommunications System
USB	Universal Serial Bus
UTRA	UMTS terrestrial Radio Access
UTRAN	UMTS terrestrial Radio Access Network
UWC	Universal Wireless Communication
UWCC	Universal Wireless Communication Consortium
VLR	Visitor Location Register
VMS	Voice Messaging System
VoIP	Voice over Internet Protocol
WAP	Wireless Application Protocol
WARC	World Administrative Radio Congress
WCDMA	Wideband Code Division Multiple Access
WIN	Wireless Intelligent Network
WLAN	Wireless Local Area Networks
WLL	Wireless Local Loop
WML	Wireless Markup Language
WNO	Wireless Network Operator
WNP	Wireless Number Portability
WWW	World Wide Web

CHAPTER 1

INTRODUCTION TO CDMA2000 STANDARDS EVOLUTION

1.1 INTRODUCTION

The ongoing growth in demand for high-speed packet data services and multimedia applications over mobile wireless networks has set new system requirements and objectives for the next generation of air interface protocols and network architectures.

Although the channelization, signaling, and access protocols of second-generation (2G) cellular systems were designed to efficiently support symmetric circuit switched data and voice traffic, most of the new data applications are IP based with highly asymmetric and packet-switch traffic. This asymmetric and bursty nature of multimedia packet data traffic along with the variability of data rates and packet sizes and complexity of quality of service (QoS) management makes conventional voice-oriented channelization and access protocols of 2G systems inefficient.

The third generations of radio access technologies, commonly known as 3G systems, are expected to use new physical and logical channelization schemes with enhanced media and link access control protocols. Also, to maximize the spectrum efficiency, the physical layer designs must utilize advanced coding, link adaptation, and diversity schemes as well as power and interference control mechanisms.

In the late 1990s, these observations and requirements motivated major efforts and studies in the International Telecommunication Union (ITU) and other regional standardization groups to define and harmonize a common set of specifications for new International Mobile Telecommunications standards referred to as IMT2000 systems. In Europe the IMT2000 is also referred to as Universal Mobile Telecommunication Services (UMTS).

ITU activities on IMT2000 are comprised of international standardization, including frequency spectrum and technical specifications for radio and network components, tariffs and billing, technical assistance, and studies on regulatory and policy aspects.

In this chapter we briefly present the overview of 3G evolution paths while we defer a more detailed description of technologies to later chapters.

The IMT2000 has defined a globally acceptable spectrum for the deployment of 3G systems, including uniband spectrum to support the time division duplex (TDD) mode as well as paired-band spectrum to allow the frequency division duplex

CDMA2000® Evolution: System Concepts and Design Principles, by Kamran Etemad
ISBN: 0-471-46125-3   Copyright © 2004 John Wiley & Sons, Inc.

(FDD) mode. In the FDD mode the system uses different frequency bands for the mobile station transmissions in the "uplink" and base station transmissions in the "downlink." In the TDD mode the uplink and downlink transmissions are on the same frequency channel but they are separated by time slots. Although most 3G deployments are expected to be in paired frequencies or in FDD mode, the TDD mode may also be used in unlicensed bands and when an FDD allocation is not feasible.

In early 1998, to expedite the process of IMT2000/3G standardization and the global acceptance of proposed radio transmission technologies (RTTs), a concept of a "Partnership Project" was proposed by the European Telecommunications Standards Institute (ETSI). This proposal initiated two Third-Generation Partnership Projects (3GPP and 3GPP2) with two different, but related, areas of focus. Each of the 3GPP and 3GPP2 projects involves a number of regional standardization bodies as organizational partners as shown in Figure 1.1.

For 3GPP the original scope was to produce globally applicable and acceptable technical specifications for a Third-Generation Mobile System based on evolved Global System for Mobile communication (GSM) core networks. Initially, the objective was to focus on the Universal Terrestrial Radio Access (UTRA) technologies with both FDD and TDD modes. This scope was subsequently amended to include the maintenance and development of Technical Specifications for GSM and its evolution to General Packet Radio Service (GPRS) and Enhanced Data rates for GSM Evolution (EDGE).

Similarly, the scope of the 3GPP2 work was to harmonize different variations of cdma2000® in a single family of standards that is based on the evolution of cdmaONE air interface. This scope was also expanded to include the development of a data-optimized air interface called the high-rate packet data (HRPD) system. In the development of cdma2000 systems the core network specifications are based on an evolved ANSI-41 and IP network; however, the specifications also include the necessary capabilities for operation with an evolved GSM-MAP-based core network. For more information on 3GPP and 3GPP2 the reader is referred to [1,2].

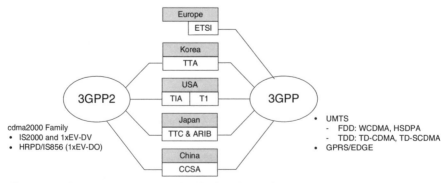

Figure 1.1   Organizational Partners in 3GPP and 3GPP2.

As a result of harmonization efforts in 3GPP and 3GPP2, the following three major technologies have been identified and included in the IMT2000 family of standards:

IMT2000 FDD Mode; Direct Spread: WCDMA (wideband code division multiple access) is one of the key radio access technologies adopted as an IMT2000 standard for global deployment in the FDD spectrum. WCDMA is based on direct spread spectrum technology in 5-MHz radio channels supporting mobile multimedia applications with up to 2 Mbps for local area access or 384 kbps for wide area access. The WCDMA standardization process has evolved from its first release in 1999 to an updated version (Release 5) in 2002, which contains major revisions and additions to the previous releases including a new high-speed data packet access (HSDPA) mode to allow high speed and low latency access for packet data applications.

IMT2000 FDD Mode; Multicarrier: The multicarrier CDMA, commonly referred to as the cdma2000 standards family, is the other FDD component of IMT2000 systems. The main member of the cdma2000 family of standards is the IS2000 air interface with the 1X and 3X components, corresponding to one and three 1.25-MHz carrier systems, respectively. The IS2000 is designed to provide a backward-compatible migration path for 2G-CDMA/IS95A(B) networks. The cdma2000 family also includes the IS856 standard, which was subsequently added as an optional and complementary radio access technology that is optimized HRPD access.

IMT2000 TDD Mode (UTRA-TDD/TD-CDMA): The TDD mode of IMT2000 standards involves a TDD variation of WCDMA, which uses a combination of time- and code division multiple access referred to as TD-CDMA. The TDD mode also has an optional spreading rate of 1.28 Mcps, which is based on synchronous code division multiple access called TD-SCDMA.

Figure 1.2 shows the timeline for the evolution of various 2G technologies toward IMT2000/3G systems, with emphasis on the commonly used FDD technologies [1,2].

Most GSM networks have evolved to include GPRS services and in some cases have been further enhanced to EDGE system for higher-speed packet data services. GPRS reuses GSM radio channels and frame structure and provides higher data rates to allow multislot traffic channels. The EDGE enhances GPRS spectral efficiency by using higher-order modulation with link adaptation but still maintaining GSM radio channels and frame structure.

Most GSM operators are planning or have begun deploying IMT2000/UMTS-based networks using WCDMA technology. Many IS136/TDMA-based networks have also joined the GSM group and have decided to migrate to WCDMA. In Japan the PDC-based networks were among the first to deploy the WCDMA system based on its 1999 release version [3].

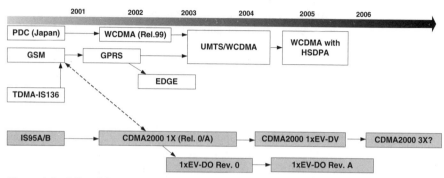

Figure 1.2   2G to 3G technology migration road map.

All 2G-CDMA (IS95/B)-based networks and some IS136/TDMA systems have migrated or are planning to migrate to cdma2000-1X technology, primarily based on the IS2000 Release 0 standard, and they are expected to transition their networks to either Release A or Release D (1xEV-DV) depending on the timing of their upgrade. Some operators who decided to devote separate carriers for high-speed packet data to complement their existing voice-based services have opted to use 1xEV-DO (or HRPD) carriers as an overlay to their existing 2G carriers. The deployment of multicarrier cdma2000 networks may happen in a later time frame.

We will revisit the overall CDMA technology evolution later in this chapter, after a brief overview of 3GPP2 and its standardization process.

1.2   3GPP2 AND CDMA2000 STANDARDIZATION

All standardization efforts related to cdma2000 are organized and managed under 3GPP2. There are five major regional standard organizations that contribute to 3GPP2 as Organizational Partners (see Fig. 1.1). These organizations are the following:

- ARIB: Association of Radio Industries and Businesses for Japan
- CCSA: China Communications Standards Association for China
- TIA: Telecommunications Industry Association for North America
- TTA: Telecommunications Technology Association for Korea
- TTC: Telecommunications Technology Committee for Japan

There are also a number of participating member companies, each required to be affiliated with at least one of the Organizational Partners.

The work of producing 3GPP2's specifications resides in four Technical Specification Groups (TSGs), comprised of representatives from Individual Member companies. The TSGs are:

- TSG-A (Access Network Interfaces)
- TSG-C (CDMA2000 Family of Standards)

- TSG-S (Services and Systems Aspects)
- TSG-X (Intersystem Operations)

There are different Working Groups within each TSG that focus on different areas within the main task. All 3GPP2 TSGs report to the Steering Committee, which is tasked with managing the overall work process and adopting the technical specifications forwarded by each of the TSGs. The following describes the function of each TSG.

TSG-A is responsible for the specifications of Interoperability Specifications (IOS) and interfaces between the radio access network and core network, as well as interfaces within the access network. One of the main IOS standards developed by TSG-A is specification for cdma2000 access network interfaces or IS-2001, which has gone through Revisions A–C. Meanwhile, the interworking function specification for the interface of 3GPP radio access technology to ANSI-41 core network is assigned to TSG-R.

TSG-C covers the radio layer 1–3 specification, mobile station MS and BS minimum performance specification, radio link protocols, support for enhanced privacy, authentication, and encryption. TSG-C is also responsible for developments related to digital speech and video codecs, data and other ancillary services support, conformance test plans, Removable User Identity Module (R-UIM), and location-based services support. TSG-C has so far developed cdma2000 air interface specifications IS-2000 Revisions 0 and A–D; air interface specification for High-Rate Packet Data IS-856; EVRC and SMV vocoders; and many other air interface-related specifications.

TSG-S is responsible for defining and developing system services and capabilities, stage 1 feature and service requirements definition, system reference model development and maintenance, as well as requirement definition for international roaming and operation, administration, management, and provisioning (OAM&P). Some of the working groups within TSG-S are focusing on a number of new features in the services and system aspects, such as enhanced messaging service, multimedia messaging service, broadcast/multicast, and multimedia streaming. Another key responsibility of TSG-S is coordinating and managing the working relationship among all TSGs.

TSG-X is responsible for the core network part of systems, including core network internal interfaces, IP support for packet data, voice, and multimedia services, and charging, accounting and billing specifications, etc. More specifically, TSG-X will address the following areas of work and the related technological developments:

- Evolution of core network to support interoperability and intersystem operations and International Roaming
- Network support for enhanced privacy, authentication, data integrity, and other security aspects, including User Identity Module (UIM) support
- Support for new supplemental services (including ISDN interworking)
- Wireless IP and multimedia services (e.g., voice over IP) and QoS support
- IP mobility management

To better capture and address the needs of operators and end users, 3GPP2 works with Market Representation Partners (MRPs), voiced by operators, who offer market advice and bring a consensus view of market requirements (e.g., services, features, and functionality) falling within the 3GPP2 scope. One of the main MRPs is the CDMA Development Group (CDG).

CDG is an international consortium of CDMA wireless equipment manufacturers and operators that have joined together to lead and help in the adoption and evolution of CDMA wireless systems around the world. One of the main tasks of CDG is to work with vendors and operators on CDMA-related technical requirements and deployment issues and to create consensus among the players and provide inputs to the standard organizations. CDG activities involve a number of technical teams with special interest areas ranging from interoperability specifications (IOS) and international roaming to applications and testing [4].

Another MRP of 3GPP2 is the IPv6 Forum. The IPv6 Forum is a worldwide consortium of leading Internet vendors and Research and Education Networks aimed at promoting IPv6-based solutions and interoperable implementations of IPv6 standards and also resolving issues that create barriers to IPv6 deployment.

1.3 CDMAONE EVOLUTION TO CDMA2000

The application of code division multiple access (CDMA) technology was introduced in cellular systems in the early 1990s with the development and commercialization of the IS-95 standard. Since then, the technology has been widely deployed throughout the world, reaching the 180 million subscriber mark in late 2003. Since its commercialization, the CDMA technology has evolved from IS-95 to cdma2000 and beyond with significant enhancements in voice capacity, data speed, and network features. Throughout this technology evolution each new standard or release is designed to maintain full backward compatibility with previous systems (see Fig. 1.3).

IS95 and its PCS version J-STD-008 appeared around 1994 as TIA standards, offering significant capacity improvement over existing TDMA-based networks. This CDMA technology uses direct sequence spread spectrum with frequency reuse of one and benefits from frequency and multipath diversity as well as statistical voice multiplexing to provide a high spectral efficiency.

After a few years of CDMA deployment experience, IS95-B was introduced in 1998, including the following enhancements:

- Support for medium-rate packet data service option, up to 64 kbps, using aggregation of code channels
- Improved soft handoff performance with dynamic thresholds
- Enhanced interfrequency handoff procedure to facilitate multicarrier network deployment
- Improved system access in handoff areas with access state handoff procedures
- Other improvements related to allowing position location and global roaming

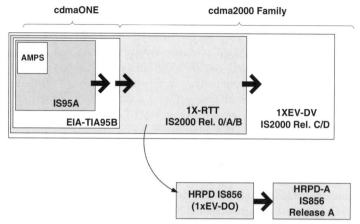

Figure 1.3   Backward compatibility in the evolution of cdma2000.

IS-95-A/B, which are subsequently called IS95A/B, along with some of their signaling standards form the basis of the 2G cellular technology known as cdmaONE.

Despite its improved feature sets and performance, IS95B was not widely deployed as most operators waited for the next generation of CDMA technology based on cdma2000 standards.

Following the IMT2000 efforts initiated by ITU the standardization of cdma2000 radio transmission technology (RTT) started in late the 1990s in TIA under the TR45.5 group, and it was subsequently continued more globally by 3GPP2.

The cdma2000 standardization aims at IMT2000 objectives on one hand and backward compatibility with existing cdmaONE networks and voice terminals on the other hand.

The preliminary release of cdma2000 that was proposed as an RTT to ITU was IS2000 Release 0, which is also referred to as the 1X-RTT system. Although this release was not a complete release, it was developed and built into the new CDMA chipsets and offered widely to the global market. Some of the key features of cdma2000 included in Release 0 are the following:

- Backward compatibility with IS95B including access and traffic state handoff enhancements
- Coherent uplink demodulation with reverse pilot
- Fast forward power control
- Variable-length Walsh spreading
- Data rates of up to 150 kbps or 300 kbps depending on the traffic channel radio configuration
- QPSK modulation on both forward and reverse links
- Enhanced channel coding with turbo encoders at higher data rates

- Optional support for transmit diversity
- Increased mobile terminal battery life with a new quick paging channel

The combination of all these enhancements provides a voice capacity that is twice that of the cdmaONE systems and a data rate of 153.6 kbps or 307.2 kbps depending on the radio configuration. The commercial deployment of cdma2000-based systems began as early as 2000 in South Korea and very soon in expanded in other countries including the US. Some have argued that as 1xRTT does not fully comply with IMT2000 requirements of 384 kbps data rate for pedestrian and 2 Mbps for fixed terminal it should be considered a "2.5G" system. The IMT2000 data rate requirements could only be met with the wideband direct spread or multicarrier 3X options of cdma2000, which have not yet been implemented.

The first complete cdma2000 release was Release A, which was published in the year 2000, including both the narrowband (1X) and the wideband (3X) multicarrier (MC) modes occupying 1.25 MHz and 3.75 MHz, respectively. As a result of harmonization efforts between cdma2000 and WCDMA, and to maintain only one direct spread (DS) mode for IMT2000 standards, the direct spread mode of cdms2000 with 3.68 Mcps was withdrawn from IS2000 release A [5].

The high data rate requirements of IMT2000 can be achieved in IS-2000A by aggregating three standard 1.25-MHz carriers in a multicarrier forward link signal. The reverse link of the 3X MC mode can optionally be transmitted using either a 1.2288 or a 3.6864 Mcps chip rate. The 3X forward/1X reverse mode leverages existing base station receiver and mobile transmitter designs while allowing higher forward link data rates for asymmetric packet data services.

Another important feature introduced in cdma2000 is the support of both IS-41 (native to IS-95) and the GSM's Mobile Application Part (MAP) network signaling. With the later option that is called cdma2000 MC-MAP, while the radio interface is handled according to the cdma2000 specification, the call control, mobility management, and other network signaling operate as per the GSM signaling protocol. This feature facilitates the worldwide adoption of cdma2000 radio technology and allows international roaming of cdma2000 terminals in the existing GSM-based networks.

Some of the key features of Release A are as follows:

- Backward compatibility to cdmaONE and Release 0
- Complete signaling support for MC 3X channels while DS mode is removed
- Signaling support for new common channels used for enhanced access and short data burst transmissions
- Enhanced signaling for concurrent services
- Flexible rates and frame formats
- QoS negotiation
- Enhanced encryption algorithm

Shortly after Release A, Release B of IS2000 was published with a few signaling protocol improvements such as rescue Channel Code Combining Soft Handoff, CDMA Off-Time Reporting, and Improved Traffic to Idle transition.

Following the widespread deployment of Release 0, and despite the significant improvements in the standard, Releases A and B of IS2000 did not motivate operators to upgrade their networks. Because Release 0 had already given operators enough voice capacity and because the pickup in demand for high-speed data was delayed by various factors, the costly upgrades to the new releases could not be justified.

Meanwhile, some operators who did want to introduce high-speed packet data services demanded much higher forward link data rates than the 150 kbps offered by cdma2000. Motivated by this demand 3GPP2 favorably considered a proposed data-only overlay radio technology based on a proprietary packet data optimized High Data Rate (HDR) air interface developed by Qualcomm Inc. 3GPP2 studied the original HDR specifications and further improved and published them as a new standard called High-Rate Packet Data (HRPD) in late 2000 [6]. This standard is also referred to as IS856 or 1xEV-DO because it is a Data-Optimized EVolution of, but not backward compatible to, the 1xRTT system.

1xEV-DO was designed to provide efficient HRPD services without the constraints of supporting legacy circuit switched channels in IS-95. The HRPD technology achieves very high spectral efficiency on the downlink by using high order modulation, fast rate adaptation, and scheduling on a single high-speed data channel that is time multiplexed among active users. The reverse link of HRPD, however, is very much like the 1xRTT system with much lower data rate and higher latencies than the forward link.

One of the main drawbacks of IS856 for some operators was that it would only provide packet data services on a best-effort basis and not applications with strict QoS requirements such as voice. Also, given the uncertainly in the revenue coming from data services, many operators at the time were not ready to dedicate some of their valuable spectrum resources and capital expenditure to data-only carriers. This concern motivated the 3GPP2 to work on a new release of cdma2000 specifically aimed at adding a high-speed packet data mode. The objective was to meet or exceed the HRPD performance without affecting the exiting backward-compatible framework for voice and low-rate data services. The result was the so-called 1xEV-DV system, which was published as Release C of IS2000, in 2002.

Some of the key features of Release C are as follows:

- Introduction of a new forward packet data channel mode and associated protocols
- High-order modulation and link adaptation on the new forward packet data channels supporting a peak data rate of 3.1 Mbps
- Short frames (1.25–5 ms) and fast scheduling to benefit from multiuser diversity
- Dynamic allocation of power and Walsh code resources among up to two packet data channels and the low-rate data/voice channels
- Fast call setup and enhanced authentication
- Data-only forward-link sector throughput comparable to 1xEV-DO

One can consider Release C as a combination of previous releases and the concepts used in 1xEV-DO with few additional improvements.

For some operators the imbalance between forward- and reverse-link through-put and latency performance was acceptable given the inherent traffic asymmetry of IP-based applications. However, for some applications with lower latency require-ment such as gaming, instant messaging, and Voice over IP (VoIP), the required QoS could not be provided with 1xEVDO or Release C of IS2000 unless some improve-ments were made in their uplink.

The 3GPP2 therefore started working on the uplink enhancement for IS2000 as part of Release D and in a parallel and somewhat controversial effort on Release A of HRPD/IS856. The proposed enhancements made in IS2000-D and IS856A were very similar in nature and would bring comparable performance gains for the uplink.

Release D of IS2000 was published in March 2004 by 3GPP2 and includes the following features mostly for uplink enhancements [5]:

- A New high-speed reverse packet data channel and associated protocols
- Link adaptation with improved rate selection and hybrid ARQ
- Shorter frames and lower latency for packet data channel
- A flexible MAC with multiple uplink rate control options and QoS control
- Support for peak data rate of 1.8 Mbps
- Maintenance of backward compatibility with cdmaONE and previous releases of IS2000
- Uplink throughput of more than 600 kbps.

Meanwhile, similar features have been added to the reverse link of 1xEV-DO as part of the IS856 Release A standard but in a design that is backward compatible with the original IS856. As a result, the uplink throughput, latency, and QoS control of HRPD have been significantly improved, providing a much more balanced design between the forward and reverse links.

In the next several chapters we will study the concepts, systems, and proto-cols features introduced in each of the major cdma2000 standard releases. As each system is backward compatible with the previous ones, we start with the most common denominator of all systems, namely, cdmaONE, and step by step expand our understanding of the cdma2000 family by focusing on new elements introduced in each release.

1.4   REFERENCES

1. The 3GPP2 website: www.3gpp2.org
2. The 3GPP website: www.3gpp.org
3. The CDG website: www.cdg.org
4. *WCDMA for UMTS*, H. Holma and A. Toskala Eds., John Wiley, 2001
5. C.S0001-A to D, Introduction to cdma2000 Standards for Spread Spectrum Systems
6. 3GPP2 C.S0024-0 v4.0 cdma2000 High Rate Packet Data Air Interface Specification, 2002

CHAPTER 2

CDMA CONCEPTS

2.1 INTRODUCTION

The first step toward introducing the cdma2000 radio access technologies should involve a basic presentation of code division multiple access (CDMA) concepts. This chapter provides a basic but technical introduction to CDMA concepts and the associated features and terminologies.

CDMA systems are based on spread spectrum communications principles that provide a flexible and efficient framework for coverage and capacity sharing. The spread spectrum schemes used in cdmaONE and cdma2000 provide these systems with frequency and multipath diversity gains, which increase the radio link's robustness against fading and interference.

In the following we will describe the concepts of spreading and de-spreading of a signal's spectrum from time and frequency perspectives. We will also look at the application of these concepts in channelization and multiple access with orthogonal and pseudo-orthogonal spreading codes. Other areas of our focus include CDMA system features such as multipath diversity or rake receivers, universal reuse, near-far and power control effects, and soft capacity.

2.2 SPREAD SPECTRUM CONCEPT

The basic idea of spread spectrum communications is based on transmitting information over channels much wider than required by the original signal bandwidth. The spread spectrum term is used to reflect the fact that the system spreads the energy of the information signal over a much wider band channel, allowing signal transmissions at very low power spectral densities.

In principle, the capacity of a radio channel is governed by Shannon's capacity equation:

$$C = W \times \log(1 + \text{SNR}) \tag{2.1}$$

where C is the system's capacity or throughput in bits/s, W is the channel bandwidth in Hz, and SNR is the effective signal-to-noise ratio [1]. This equation shows a theoretical trade-off between the signal bandwidth and the minimum acceptable SNR assuming a fixed capacity for the channel. Given this trade-off, a spread spectrum

CDMA2000® Evolution: System Concepts and Design Principles, by Kamran Etemad
ISBN: 0-471-46125-3   Copyright © 2004 John Wiley & Sons, Inc.

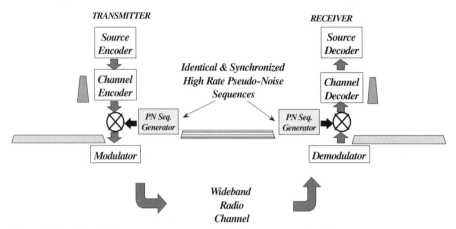

Figure 2.1   A basic direct sequence spread spectrum communications system.

communication is designed to operate at very low SNR using a very large bandwidth (W) compared with conventional non-spread spectrum systems.

Two methods of spectrum spreading are commonly used in communications systems:

- Direct sequence spreading, which is used in 2G and 3G CDMA systems, involves modulating or multiplying each information bit by a high-rate spreading sequence.

- Frequency hopping, which is used in 802.11 and several other broadband wireless radio access technologies, involves random hopping of narrowband subcarriers within a wide spectrum.

In this chapter, the focus is on direct sequence spread spectrum, as it is the foundation for the physical layer design in cdma2000 systems.

Figure 2.1 shows the basic block diagram for a direct sequence spread spectrum communications system. The diagram is similar to any conventional digital communication system with the exception of signal spreading and de-spreading processes.

In the following we describe the process of spreading and de-spreading. In this description the information and spreading signals are represented in the bipolar ("+1" and "−1") form rather than binary ("1" and "0") format.

At the transmitter, the coded information bits are multiplied by a high-rate pseudo-random noise-like sequence before modulation. Let us assume an information signal $x(t)$ of data rate R bits/s, with approximate bandwidth of about R Hz, and a spreading sequence $c(t)$ with a much higher rate and therefore a bandwidth of $W \gg R$ Hz (see Fig. 2.2).

By multiplying the signal $x(t)$ by $c(t)$ in the time domain, their spectral densities are convolved in the frequency domain, resulting in a signal $s(t)$ with bandwidth of $W + R$ or approximately W.

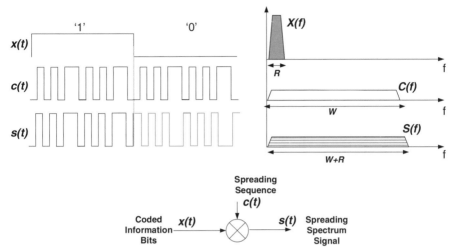

Figure 2.2   Spectrum spreading concept.

$$s(t) = x(t) \cdot c(t) \rightarrow S(f) = X(f) * C(f) \tag{2.2}$$

The transmitted wideband signal $S(t)$ therefore carries the information x but at a very low spectral density.

At the receiver the spread spectrum signal is correlated with the same spreading sequence $c(t)$, resulting in de-spreading to the original information signal $x(t)$. The correlation or de-spreading operation, which can also be represented as an inner product, consists of multiplication of the spread spectrum signal with a spreading sequence followed by integration over one bit period.

$$\hat{x}(t) = <s(t), c(t)> = \int_T s(\tau) \cdot c(\tau) d\tau = x(t) \cdot \int_T c_1(\tau) \cdot c_1(\tau) d\tau = x(t) \tag{2.3}$$

If the code used for de-spreading matches the one used for spreading the signal at the transmitter, as in, for example, the case of User 1 in Figure 2.3, the result of the correlation process will be the original bit value. In this case, all +1 and −1 chips of $c_1(t)$ consistently align and match with the corresponding chips on the $s(t)$. Thus the results of chip multiplications are either all +1 or all −1 and therefore the integration over the bit period also gives a +1 or −1 result reflecting the original coded information bit value.

In general, as a result of correlating the received spread spectrum signal with a specific code at the receiver:

- The signal from the intended transmitter is de-spread, and as a result its power spectral density is increased by a factor of W/R. The W/R factor is called the processing gain.
- The additive white Gaussian noise (AWGN) remains the same, that is, spreading and de-spreading have no effect on AWGN channels.
- Any narrowband interference is spread and appears as AWGN.

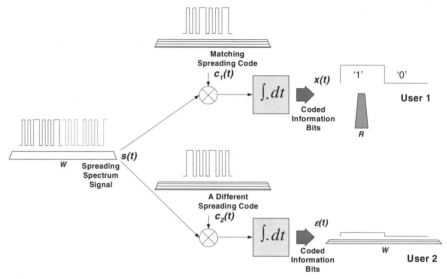

Figure 2.3   De-spreading concept.

- All signals from a unintended user, which have been spread with different codes, remain wideband and therefore, appear as AWGN, as in, for example, the case of User 2 in Figure 2.3.
- These effects are the underlying concepts behind the CDMA in a spread spectrum communications system.

A CDMA system provides a multiple-access framework by allowing users to share the same spectrum at all times as long as each user applies a different code to spread its signal over a wideband channel and the codes are selected such that cross-correlations between codes are zero or very small (see Fig. 2.4).

Although all cochannel transmissions from all transmitting users are present at the receiver, as part of the de-spreading process only the desired signal component that was spread with the matching code will be picked up by the correlator. Because other spreading codes have almost no correlation with the desired code, other users' signals will have almost no power at the output of the correlator and the receiver.

2.3   SPREADING CODES

The effectiveness of a CDMA scheme relies on the mutual orthogonality of spreading codes to minimize cross-interference among physical code channels. Remember that frequency and time division multiple access (FDMA and TDMA) schemes achieve this orthogonality by using nonoverlapping frequency channels or time slots, respectively. Therefore, spreading code should be consistently defined and generated at transmitters and receivers such that this orthogonality can be achieved and maintained across the radio link.

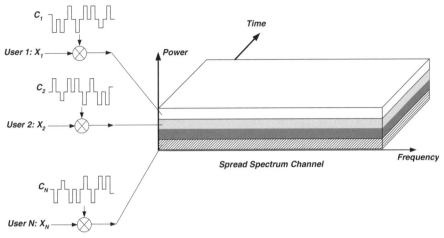

Figure 2.4   Multiple access using codes.

In this section, we introduce orthogonal Walsh codes and pseudo-noise (PN) codes, which are used in both 2G and 3G CDMA systems [2].

2.3.1   Walsh Codes

Walsh codes are a set of perfectly orthogonal codes typically used for channelization, that is, separating different physical channels transmitted by a node. The Walsh codes of length 2^n can be defined and generated as different rows of a $2^n \times 2^n$ Hadamard matrix. The following recursive equation can be used to generate higher-order Hadamard matrices from the lower ones.

$$H_{2^n} = \begin{bmatrix} H_{2^{n-1}} & H_{2^{n-1}} \\ H_{2^{n-1}} & \overline{H_{2^{n-1}}} \end{bmatrix} \tag{2.4}$$

For example, starting with a 1×1 matrix of $H_1 = [0]$, one can define Walsh codes of length 4 as follows:

$$H_1 = [0]$$

$$H_2 = \begin{bmatrix} 0 & 0 \\ 0 & 1 \end{bmatrix}$$

$$H_4 = \begin{bmatrix} H_2 & H_2 \\ H_2 & \overline{H_2} \end{bmatrix} = \begin{bmatrix} 0 & 0 & 0 & 0 \\ 0 & 1 & 0 & 1 \\ 0 & 0 & 1 & 1 \\ 0 & 1 & 1 & 0 \end{bmatrix} \rightarrow \begin{cases} W_0 = [-1,-1,-1,-1] \\ W_1 = [-1,+1,-1,+1] \\ W_2 = [-1,-1,+1,+1] \\ W_3 = [-1,+1,+1,-1] \end{cases} \tag{2.5}$$

where W_0 to W_3 are defined as different rows of a 4×4 Hadamard matrix and represented in binary or bipolar format. These rows are mutually orthogonal to each other.

For example, one can calculate the cross correlation between W_1 and W_3 as:

$$<W_1, W_3> = <[-1,+1,-1,+1],[-1,+1,+1,-1]>$$
$$= \frac{1}{4}[(-1 \times -1)+(+1 \times +1)+(-1 \times +1)+(+1 \times -1)]$$
$$= 0/4 = 0 \qquad (2.6)$$

which confirms that W_1 and W_3 are orthogonal. Similarly, any two rows of an $N \times N$ Hadamard matrix can be shown to be mutually orthogonal. In general, there always exist N orthogonal codes of length N that can be allocated to different physical channels to create an orthogonal channelization framework for multiple access.

In the TIA-EIA95 system, Walsh codes of length 64, which are the 64 rows of a 64×64 Hadamard matrix, are used for forward-link physical channelization. cdma2000 uses variable length Walsh codes, which are discussed later in this book.

2.3.2 PN Codes

The PN codes are a set of pseudo-random but deterministically generated sequences, which mimic certain properties of noise [3]. In principle, the desired properties for a spreading sequence include:

- Sharp Autocorrelation, so that any time-shifted version of a PN code has almost no correlation with the original sequence.

- Equal number of "1"s and "0"s in any long segment of the sequence, so that the signal has no bias.

- Random and independent appearance of "1"s and "0"s, so that it is difficult to reconstruct the sequence from any short segment.

Another desired property of PN codes is the possibility of generating a large number of codes with simple linear feedback shift registers.

The PN codes used in 2G and 3G CDMA systems discussed in this book are generated with maximal-length (ML) shift registers. The ML shift registers are linear feedback shift registers with specific feedback connections designed to ensure the maximum period of the output sequence. A sequence generated with an ML shift register is called an "m-sequence." Therefore, the output sequence of an ML shift register with N cells is an m-sequence with a period of $L = 2^N - 1$. By inserting different initial values or masks into the N cells of a ML shift register one can generate the same periodic m-sequence with different time offsets. The m-sequences have all the desired pseudo-noise properties described above.

Figure 2.5 shows a simple 5-cell ML register loaded with "1,0,0,0,0" as the initial condition. The output of this shift register $S_1(n)$ is an m-sequence of period $L = 2^5 - 1 = 31$.

By inserting another initial mask, in this shift register one can generate a sequence that is a shifted version of $S_1(n)$. For example, if "1,0,1,0,0" is used as the initial mask, the output sequence will be $S_2(n)$, which is equal to $S_1(n - 3)$, that is, the $S_1(n)$ shifted by three chips.

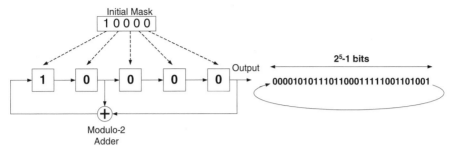

Figure 2.5   A 5-cell ML shift register.

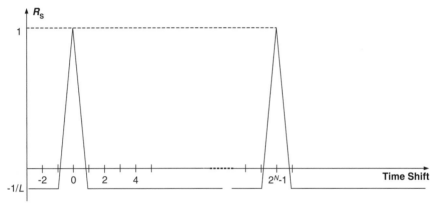

Figure 2.6   Autocorrelation function for a ML pseudo-noise code.

$$\text{Mask: } 1,0,0,0,0 \rightarrow S_1(n) = 00001010111011000111110011101001\ldots$$
$$\text{Mask: } 1,0,1,0,0 \rightarrow S_2(n) = 00101011101100011111100110100100\ldots$$
$$= S_1(n-3) \tag{2.7}$$

The correlation between an m-sequence and various nonzero shifts of that sequence is $-1/L$, where L is the period of the sequence. Therefore, as shown in Figure 2.6 for an m-sequence with a long period L, this cross-correlation is very small. Note that, by definition, the value of the autocorrelation at zero time shift is equal to 1, as each sequence is perfectly correlated with itself. However, any discrete time shift results in a very small autocorrelation function. Thus the autocorrelation function of PN codes has high peaks and sharp roll-offs that mimic the shape of an ideal impulse function in the autocorrelation of a noise sequence.

Going back to our previous example, one can see the correlation properties of m-sequences by calculating the cross-correlation between $S_1(n)$ and $S_2(n)$ as follows:

$$<S_1(n), S_2(n)> = \frac{1}{L} \times \sum_{i=1}^{L} S_1(i) S_2(i) \tag{2.8}$$

Note that the cross-correlation between two periodic sequences is defined as sum of their term-wise multiplication divided by their common period. In this case, because $S_2(n) = S_1(n - 3)$, the result is equal to the value of the autocorrelation function of $S_1(n)$, i.e. $[R_{S1}(\tau)]$, at point $\tau = 3$.

$$<S_1(n), S_2(n)> = <S_1(n), S_1(n-3)> = \frac{1}{L} \times \sum_{i=1}^{L} S_1(i)S_1(i-3) = R_{S1}(3) = -1/31 \quad (2.9)$$

For any m-sequence of period $L = 2^N - 1$, the number of 1s and 0s in the sequence is different only by 1. For example in $S_1(n)$ of Figure 2.5, there are 16 occurrence of 1s and 15 occurrence of 0s.

Also, there are $(L + 1)/2^{P+1}$ consecutive runs of all 0s or all 1s of length P in each period. This means that the long runs of all 1s and all 0s are infrequent in any long segments of the sequence. These two properties are also in line with the last two desired properties for the target PN codes.

Because of these characteristics, m-sequences are good candidates for spreading codes in CDMA systems.

In a typical cellular CDMA network both PN and Walsh codes are used to provide two levels of spreading. For example, in IS95A/B, the Walsh codes are used for physical channelization in the forward link, that is, to isolate the various traffic and control code channels transmitted by each base station. Each base station also uses a different time offset of a very long m-sequence as its specific signature PN code to spread the composite signal (containing all code channels) that is transmitted to the users in a cell.

This process is shown in the following example:

Let us assume that a base station has N users and has allocated one Walsh code C_i to each user i. Also assume that the symbol to be sent to the ith user is represented as a_i. Then one can write:

$$T_1 = a_1 \cdot C_1, \quad T_2 = a_2 \cdot C_2$$
$$\cdots \quad T_N = a_N \cdot C_N$$
$$T = T_1 + T_2 + \cdots + T_N$$
$$TX = T \otimes D_A \quad (2.10)$$

where T_i are users' symbols a_i spread with their assigned channelization code C_i and T is the composite signal including all code channels. Note that the user symbols can safely be added together, at the same time and in the same frequency, without causing any interference to each other because they have been "orthogonalized" by the Walsh codes. Also note that although a_i have bipolar (+/−1) values, the composite signals can take many different values depending on the number of users and their transmitted symbols.

The composite signal T is then spread by the base station's specific PN code D_A, which allows users to differentiate the composite signal TX_A from base station A from those received from other, perhaps unintended, base stations.

$$TX_A = T \otimes D_A \quad (2.11)$$

At the mobile station's receiver, the user's signal is recovered by de-spreading in two steps but in the reverse order of that used in the transmitter.

The received signal is first correlated with base station A's specific PN code (D_A). After the first correlation with D_A, the received signal from other base stations such as B, assumed to be the interferer, will be rejected, as D_B shows almost no correlation with D_A.

$$RX = (TX_A + TX_B)$$
$$\hat{T} = <RX, D_A>$$
$$\hat{T} = <(TX_A + TX_B), D_A> = <(T \otimes D_A), D_A> + <(T' \otimes D_B), D_A> = T + \delta_{AB} \quad (2.12)$$

Where $\delta_{AB} << 1$, assuming very small correlation between D_A and D_B. Therefore, the result is a close estimate of the combined transmitted signal T.

The receiver then applies the user's specific Walsh code, in this case C_1, to extract the first user's code channel from other channels transmitted by base station A.

$$r_1 = <T, C_1> = <a_1 \cdot C_1 + a_2 \cdot C_2 + \cdots + a_N \cdot C_N, C_1>$$
$$r_1 = <a_1 \cdot C_1, C_1> + <a_2 \cdot C_2 + \cdots + a_N \cdot C_N, C_1>$$
$$r_1 = a_1 + (\varepsilon_{21} + \cdots + \varepsilon_{N1}) = a_1 + \varepsilon \quad (2.13)$$

Ideally, the result of the second correlation should be the transmitted symbol a_1 only, because the Walsh codes are perfectly orthogonal to each other. However, in practice in the presence of fading and delay spread in the channel there may be some loss of orthogonality, captured by a very small intercode channel interference ε in the above equation. Similarly, other users can extract their signal by applying their own channel codes and as long as $\varepsilon\varepsilon << a_i$, each user's receiver can detect the symbols correctly.

2.4   MULTIPATH DIVERSITY AND THE RAKE RECEIVER

A CDMA system uses wideband signals and therefore narrow chip pulses as channel symbols for the transmission over the radio channel. With such narrow symbols the multipath components of the received signals will have small overlaps in the time domain and, therefore, they can be resolved and combined to achieve better receiver performance. The adaptive combination of multipath components, each experiencing different and sometime independent fading conditions, can provide some level of diversity gain that can be utilized by the receiver.

The rake receiver in a CDMA system is designed to take advantage of multipath diversity by estimating the relative timing and strength of the top multipath components and combining them to maximize the effective signal-to-noise ratio before the decoding process. Figure 2.7 shows a basic structure of a rake receiver with one searcher and three receiving correlators. The name rake receiver is based on the structure and function of correlators, which act like the fingers of a gardening rake, collecting and combining different portions of the received signal.

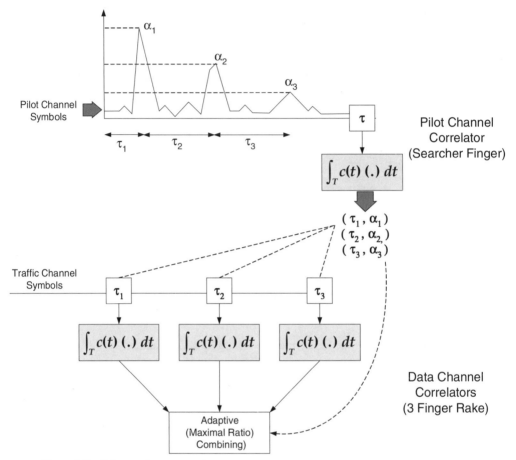

Figure 2.7   Rake receiver concepts.

The top portion of Figure 2.7 is the searcher or path delay estimator seeking to estimate the relative times τ_i and strengths α_i of the signal copies arriving from multiple paths. This is accomplished by cross-correlation of the received signal with a spreading code. The peaks of the correlation correspond to the arrival time of a multipath component. In a coherent demodulation case in which the transmitter along with the traffic channel sends a pilot channel as a phase reference, the searcher uses a pilot for capturing the multipath delay profile, namely, τ_i and α_i.

Once the arrival times τ_i are determined, the traffic channel correlators, that is, the fingers, can be tuned to capture the corresponding multipath copies of the traffic channels. Each finger of the rake receiver correlates the signal by a properly delayed spreading code. If the chip codes are correctly delayed, they hit the signal at the correct instance and all of the receiver fingers will have valid de-spread data. The output of the correlators can then be aligned and combined according to their relative strengths α_i to provide the maximum ratio combining gain.

This model of the rake receiver is very simple but effectively illustrates the concept of multipath reception diversity. However, the exact design of a rake receiver is implementation specific and varies in complexity and processing details.

2.5   UNIVERSAL FREQUENCY REUSE

In a typical TDMA-, FDMA-based cellular network the frequency channels must be carefully assigned to cells or sectors in a cluster such that the level of cochannel interference is maintained above the minimum carrier-to-interference ratio (C/I) required by the system's coding and modulation schemes. The C/I requirement typically limits the channel reuse in a network and reduces the effective spectrum availability per cell. For example, a network with the reuse factor of K divides all available radio channels into K groups and allocates one group to each cell in a K-cell cluster. The same frequency allocation pattern repeats in each cluster in the network. Figure 2.8a shows such a frequency reuse pattern for a reuse factor of 4, where each cell can use 1/4th of the spectrum.

A CDMA network can take advantage of the processing gain associated with de-spreading to operate at very low C/I. With a large processing gain, the network can be designed to allow successful demodulation and detection of signals at levels much below the cochannel interference.

Because of this very low C/I requirement, a typical CDMA system can be designed with frequency reuse of 1, or a so-called universal frequency reuse.

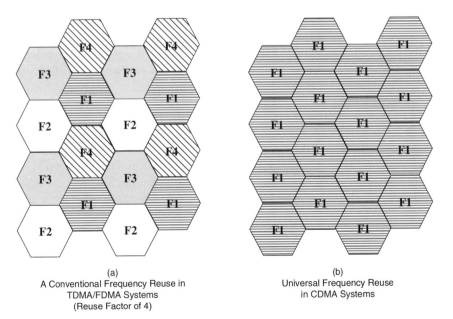

(a)
A Conventional Frequency Reuse in
TDMA/FDMA Systems
(Reuse Factor of 4)

(b)
Universal Frequency Reuse
in CDMA Systems

Figure 2.8   Conventional vs. universal channel reuse.

Universal frequency reuse implies that the same carrier frequency can be used by every cell and sector (see Fig. 2.8b) and that, in principle, the entire spectrum can be utilized in each cell to maximize cell capacity and overall spectrum efficiency. Using such a tight frequency reuse, the effective C/I in significant parts of the network area will be lower than 1, that is, negative in decibel scale. However, the interference rejection achieved as part of de-spreading can bring the E_b/N_o values to acceptable levels for the decoding process.

This feature of CDMA-based networks is one of the main reasons for their capacity advantage over conventional TDMA-, FDMA-based systems.

A CDMA system can be deployed with a single wideband direct spectrum carrier covering the entire spectrum or as a multicarrier system in which some or all channels can be allocated to each cell based on its capacity requirements. In either case no frequency planning would be required.

2.6   SOFT HANDOFF

Another important feature of CDMA-based systems is soft handoff (SHO). Handoff in a cellular network generally is a form of macrospace diversity that allows the mobile terminal to seamlessly switch between cells and maintain their connection with the strongest serving base station. In conventional TDMA-based systems, the handoff follows a break-before-make (BBM) process, because the mobile must disconnect from the original cell before it can be fully connected with the target cell.

In a CDMA system, because all base stations operate in the same frequency a mobile terminal with a single receiver can monitor and receive traffic channels from multiple base stations simultaneously. This feature allows the network to support a make-before-break (MBB) handoff process, which means that the mobile can establish a new link with the target base station while it continues receiving traffic from the orthogonal cell. During this MBB handoff the mobile is served by multiple base stations, resulting in a more robust handoff process with a smaller chance of failure or drop calls compared with conventional BBM procedures.

During the SHO the system may also allow simultaneous power control of mobiles by the base stations connected to the mobile. In this case, each mobile in the SHO will transmit power needed to reliably maintain only one of multiple links and not necessarily the weakest link. This joint power control strategy reduces the average transmit power of users in the handoff areas, who are the main sources of interference in the system. The reduced uplink interference has a direct impact on the cell capacity. Therefore, SHO not only improves the coverage performance as a macrospace diversity technique, it will also increase uplink capacity by reducing average interference received by the base station.

There is a diversity gain associated with SHO in both reverse and forward links. However, the type of diversity combining and the cost associated with SHO is different between the uplink and the downlink.

In the uplink, the mobile terminal transmits only one signal picked up by all base stations involved in SHO. At each base station, the received frame from all of its sectors involved in SHO is soft-combined before sending the frame to the switch.

The major diversity gain is obtained by the frame selection combining at the mobile switch center (MSC) or base station controller (BSC).

In the forward link, each base station involved in SHO needs to separately allocate power, channel elements, and Walsh code channels to the mobile station. Therefore, the network must allocate almost twice the resources to a user in SHO with two base stations. This resource redundancy requirement in the forward link to support handoff has impacts on cost as well as forward link capacity.

In general, although a proper level of SHO can help in both capacity and coverage enhancement, putting too many users in SHO can cause unnecessary interference and reduced forward link capacity as well as higher network complexity and cost.

2.7 POWER CONTROL AND SOFT CAPACITY

Another important aspect of CDMA systems in a mobile fading channel is the significant impact of and the need for power control. In a CDMA system, power is the main resource that is shared by the multiple access scheme, and thus the performance relies greatly on fast and accurate adaptive power control for both mobile and base stations.

In fact, a mobile station without power control can create enough uplink interference level at the base station to block any other user from accessing the network. Figure 2.9 shows this problem, typically referred to as the near-far problem. In this example, the user (U_1) is transmitting without power control, creating a large interference rise at the base station receiver and thus making any other user's signal undetectable by the base station.

In a CDMA system, the cell capacity and coverage are interdependent. This section describes the soft capacity feature of CDMA-based networks and the effect of cell loading on coverage and capacity. The capacity analysis presented in this section is primarily for voice services and for the reverse link, which is classically

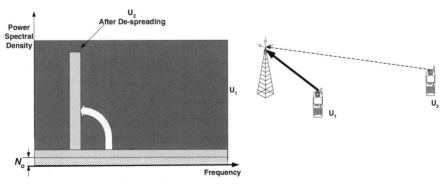

Figure 2.9   Near-far problem and power control.

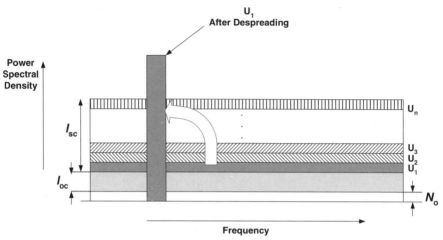

Figure 2.10   Uplink interference and reverse link capacity.

considered as the limiting link. The forward link capacity will be discussed later in this section.

We first assume a perfect power control on the uplink such that all received signals from N users in a cell arrive at the base station at the same power level S and therefore the same power spectral density. Figure 2.10 shows the received signal at the base station comprised of the desired signal from user U_1 as well as all cochannel interference and thermal noise.

As the number of users in a cell N increases the total interference and noise level faced by the demodulation and decoding system is proportionally raised up to a point where the decoding and de-spreading gains can no longer provide the bit error rate (BER) or frame erasure rate (FER) required by the voice application. In this case, the de-spread signal, despite its increased power spectral density, is "buried" in cochannel interference and noise or does not have enough energy to overcome the effective "noise floor". At this point, the ratio of the received energy per bit to total interference and noise spectral density or $E_b/(I + N_o)$ becomes smaller than $(E_b/N_o)_{min}$ that is required by the demodulator and decoder for acceptable quality of service, for example, BER or FER of 1%. Therefore, the capacity limit is reached when the following inequality turns to equality:

$$\left(\frac{E_b}{N_o}\right)_{min} < \frac{E_b}{I_t + N_o} = \frac{S/R}{I_{sc} + I_{oc} + N_o} \tag{2.14}$$

In this equation, S is the received signal power (in watts) from the desired user on the traffic channel and R is the data rate (in bits per second). Also, I_t represents the received total interference power spectral density, that is, received power divided by the bandwidth W, and it is assumed to be the cochannel interference only.

The thermal noise is characterized as additive white Gaussian noise (AWGN) with power spectral density of $N_o/2$ and in-band power of N_oW, where N_o is a function of temperature. The value of N_o is estimated as $N_o = kT$, where $k = 1.38 \times 10^{-23}$ J/K is Boltzmann's constant and T is the absolute temperature of the base station in Kelvins, typically $T = 290$ K.

The total interference I_t spectral density is the summation of same-cell interference I_{sc} and other-cell interference I_{oc} spectral densities. The I_{sc} component reflects the uplink interference power density received from users served by the same cell that is serving the intended user, and I_{oc} is the effective uplink interference power density received from users in other cells.

The same-cell interference power, $I_{sc} \times W$, is equal to total power received from $N - 1$ interfering users within the same cell, which, based on perfect power control assumption, can be estimated to be equal to $(N - 1) \times S$.

A CDMA system, however, takes advantage of voice activity and lowers transmit power significantly during periods of nonactive speech, namely, pauses and silences. Therefore, the effective average transmit power of each voice user is reduced based on the duty cycle of the speech waveform. This duty cycle, usually referred to as the voice activity factor, is measured to be about 40–50% in most cases. As a result, the average received power for each of the $N - 1$ users is reduced proportionally by the voice activity factor. Therefore, the same-cell interference I_{sc} can be estimated to be:

$$I_{sc}W = (N-1) \times S \times \bar{v} \tag{2.15}$$

The other-cell interference I_{oc} is typically modeled as a percentage of the same-cell interference by defining the *interference factor* (f) as the ratio of other-cell interference to same-cell interference, $f = I_{oc}/I_{sc}$. Alternatively, one can define the *frequency reuse efficiency*, F, as the ratio of the same-cell interference and total interference:

$$F = \frac{I_{sc}}{I_{sc} + I_{oc}} = \frac{1}{1 + \dfrac{I_{oc}}{I_{sc}}} = \frac{1}{1 + f} \tag{2.16}$$

The frequency reuse efficiency shows the inevitable capacity reduction as a result of the universal frequency reuse of 1 in the CDMA network.

Given the estimates for I_{sc} and I_{oc}, we can show the relationship between the number of users N and the effective E_b/N_o.

$$\left(\frac{E_b}{N_o} \right)_{min} \leq \frac{S/R}{I_{sc}(1+f) + N_o} = \frac{S \times W/R}{I_{sc}W(1+f) + N_oW}$$

$$= \frac{S \times W/R}{[(N-1) \times S \times \bar{v}(1+f)] + N_oW} \tag{2.17}$$

The capacity N_c is reached when the above inequality holds as an equation.

$$N_c = \frac{S \times W/R - N_oW}{(E_b/N_o)_{min} \times S \times \bar{v}(1+f)} + 1 \tag{2.18}$$

This equation can be simplified under some assumptions. If the noise level can be neglected, implying that $S/R \gg N_o$, the equation can be simplified to

$$N_{\text{pole}} = \frac{W/R}{(E_b/N_o)_{\min} \times \bar{\nu}(1+f)} + 1 \approx \frac{W/R}{(E_b/N_o)_{\min} \times \bar{\nu}(1+f)} \qquad (2.19)$$

where N_{pole} is used to represent the theoretical upper bound for reverse link cell capacity that is called the "pole capacity." As will be discussed later, a system operating at pole capacity will require mobiles to transmit at infinite power and this results in unstable cell coverage and interference, especially in mobile networks. Therefore, in practice, a percentage of pole capacity called the loading factor L is used to balance capacity and coverage in a stable interference environment. The capacity loading factor L can be defined as

$$L = \frac{N}{N_{\text{pole}}} \qquad (2.20)$$

Also, going back to the assumption used in derivation of the pole capacity, one can easily see the relation between loading and the noise-to-interference ratio.

$$L = \frac{N}{N_{\text{pole}}} = \frac{I_{\text{sc}} + I_{\text{oc}}}{I_{\text{sc}} + I_{\text{oc}} + N_o} = \frac{I_t}{I_t + N_o} = \frac{1}{1 + N_o/I_t} \qquad (2.21)$$

where again neglecting noise results in $L = 1$ or 100% loading, corresponding to the theoretical pole capacity. In practice, L is typically chosen to be 50–70%.

The reverse link CDMA cell capacity can therefore be simplified to the following:

$$N_{\text{rev}} = \frac{W/R}{(E_b/N_o)_{\min} \times \bar{\nu}(1+f)} \times L \qquad (2.22)$$

This equation, which is most commonly used and referenced for calculating a CDMA cell capacity, can provide some insights into the effect of different variables on the system capacity. For example:

- The cell capacity is proportional to the processing or spreading gain W/R, and the lower the information data rate or the vocoder rate, with same bandwidth W, the higher the capacity.

- The capacity is inversely proportional to the minimum E_b/N_o required by the physical layer for acceptable quality of service, that is, BER or FER. Higher capacities can be achieved by better demodulation and decoding performance or by accepting lower quality, that is, higher BER or FER.

- The lower the other-cell interference, the higher the capacity. This may be achieved by using more effective power control and/or some beam forming.

- Utilizing voice activity directly increases the capacity by lowering effective interference.

- A higher loading factor can be used to increase capacity as long as the cell coverage stability and handoffs can be managed. This option may be used for wireless local loop and other systems with fixed users, where there is no handoff in and out of cells, and thus dynamic range and rate of traffic load changes are smaller.

When sectorization is used, the interference received by the base station is reduced and, therefore, the capacity is increased proportionally. For example, with n_s sectors, each with $360/n_s$ degree beamwidth and with perfect isolation among sectors, one would expect the cell capacity for the combination of all sectors to be n_s times that of an omnidirectional cell.

The effect of reduced interference could be captured in CDMA capacity equation with a sectorization gain of $g_s = n_s$. In reality, because there is always some overlap between adjacent sectors on a site, the interference is reduced by a factor smaller than but close to n_s. For example, for three sectors g_s may be 2.4–2.7, depending on the actual sectors' beamwidth.

In the following, we use the EIA-TIA95 CDMA parameters as an example for CDMA cell capacity calculation.

- The channel bandwidth for the spread spectrum signal is equal to 1.2288 MHz.
- The vocoder bit rate is $R = 9.6$ kbps.
- A typical value for the voice activity factor is $vv = 0.4$.
- The minimum E_b/N_o required at the base station receiver is different for different categories of users. Its value is a function of mobile subscriber speed, multipath conditions, and the desired FER. We assume a typically used value of $(E_b/N_o)_{min} = 7$ dB, that is, 5 in linear scale.
- For a CDMA system consisting of omnidirectional cells, the interference factor of $f = 0.6$ is assumed, resulting in the frequency reuse efficiency of $F = 0.625$.
- The resulting pole capacity is:

$$N_{pole} = \frac{1.228 \times 10^6 / 9.6 \times 10^3}{5 \times 0.4 \times (1 + 0.6)} \approx 40 \qquad (2.23)$$

- With loading of $L = 50\%$ for an omnidirectional cell the capacity is 20 simultaneous voice channels.
- With loading of $L = 50\%$ for a three-sector site, assuming sectorization gain of $g_s = 2.6$, the site capacity is $2.6 \times 20 = 52$ voice users and the sector capacity is about $52/3 = 17$ voice users.

The forward link capacity in a CDMA system is usually limited by the maximum total transmission power of the base station or number of code channels. For voice-based system the power limitation is the key limiting factor that determines base stations' ability to support all users supported and covered based on the reverse link.

As the portion of total power allocated to a user depends on its channel condition and thus distance from the base station, the number of supported users would depend on their distribution within the cell area. Any objective estimate of the forward link capacity would require simulations based on the realistic user distribution. Even for uniform user distribution there is not a closed-form equation that can be used to calculate the forward link capacity.

In most deployment scenarios the CDMA cell capacity has been shown to be reverse link limited. However, in some cases, especially where a lot of traffic comes from a location far from the serving base stations, CDMA cells have also shown forward link limitations.

2.8 REFERENCES

1. *CDMA*, Andrew J. Viterbi, Addison-Wesley, 1995
2. *Applications of CDMA in Wireless/Personal Communications*, Vijay K. Garg, Kenneth Smolik, and Joseph E. Wilkes, Prentice-Hall, 1997.
3. *CDMA System Engineering Handbook*, Jhong S. Lee and Leonard E. Miller, Artech House Publishers, 1998.

CHAPTER *3*

OVERVIEW OF IS95A

3.1 INTRODUCTION

cdma2000 Systems are backward compatible with second-generation CDMA networks based on cdmaONE standards. In fact, the IS2000 specifications include all
legacy traffic and control channels and protocols defined in IS95A and -B. Therefore, an overview of cdmaONE not only helps in better understanding cdma2000
advanced features and capabilities but is also necessary because cdmaONE captures
a significant part of the IS2000 specifications.

In this chapter we focus on channelization, physical layer processing ,and basic
protocols and messaging used in IS95A [1], which is the basis for almost all legacy
CDMA systems deployed around the globe.

Note that throughout this chapter and some of the following chapters, we use
IS95A to refer to cdmaONE or the EIA-TIA95A/B standards.

3.2 RADIO AND PHYSICAL CHANNELIZATION

Each radio channel in IS95A has 1.25 MHz of bandwidth. The 1.25-MHz channel
separation is used to include the 3-dB bandwidth CDMA channel that is 1.23 MHz
plus some guard band (see Fig. 3.1).

The CDMA radio channels, sometimes referred to as frequency assignments,
can be defined in various spectrum allocations, including the North American cellular and PCS bands.

The 850-MHz cellular band consists of 25-MHz paired spectrum to allow frequency division duplexing of uplink and downlink transmissions, as shown in Figure
3.2. Each pair of uplink-downlink channels has 45-MHz separation. The CDMA
channels are arranged based on a 30-kHz grid to be consistent with the legacy analog
(AMPS) channel numbers, to allow CDMA and 30-KHz analog channel planning in
the same band.

The 1900-MHz PCS band consists of six blocks, A–D. The A, B, and C blocks
are paired 15-MHz channels, and the D, E, and F blocks are 5 MHz each way. Table
3.1 shows the 1900-MHz PCS blocks, and Figure 3.3 shows the CDMA channel
numbering in this band.

CDMA2000® Evolution: System Concepts and Design Principles, by Kamran Etemad
ISBN: 0-471-46125-3   Copyright © 2004 John Wiley & Sons, Inc.

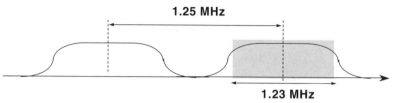

Figure 3.1   IS95A channel spacing.

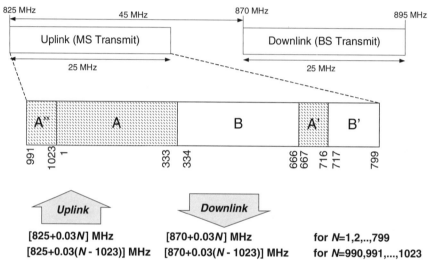

Figure 3.2   CDMA channel numbering in 850-MHz cellular band.

TABLE 3.1   **1900 PCS Blocks**

Block Designator	Transmit Frequency Band (MHz)	
	Mobile Station	Base Station
A	1850–1865	1930–1945
D	1865–1870	1945–1950
B	1870–1885	1950–1965
E	1885–1890	1965–1970
F	1890–1895	1970–1975
C	1895–1910	1975–1990

The IS95A air interface defines four types of physical channels in the forward link. These channels are the pilot, synch, paging, and traffic channels, each mapping to the corresponding logical channel. The reverse link physical channels include traffic and access channels (see Fig. 3.4).

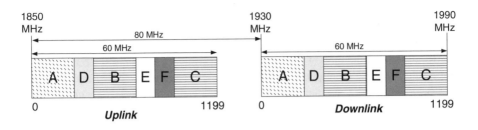

[1850+0.05*N*] MHz   *N*=0,1,..,1199   **[1930+0.05*N*] MHz   *N*=0,1,..,1199**

Figure 3.3   CDMA channel numbering in 1900-MHz PCS band.

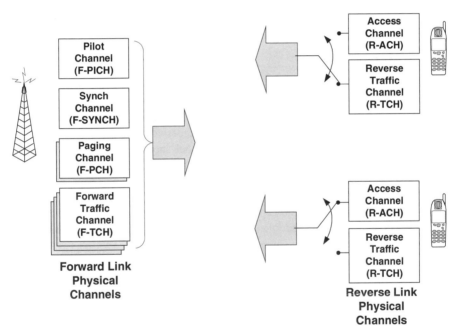

Figure 3.4   IS95A physical channels.

The forward link physical channels are code multiplexed with orthogonal Walsh spreading codes of length 64 chips. All orthogonally spread physical channels are added together to form the composite spread spectrum signal, which is further spread by a quadrature pair of short PN sequences and modulated before transmission over the air. The forward channel therefore involves concurrent transmission of all forward physical channels spread by base station-specific PN code. Although the Walsh codes provide orthogonal physical channelization within a cell, the base station-specific short codes serve as the base stations' signatures, allowing differentiation and de-spreading of desired base stations' signals in the presence of other-cell interference.

The reverse link physical channels are also code multiplexed, consisting of access and traffic channels. The mobile transmits only one channel at a time depending on the call processing state. Each reverse-link channel is spread and therefore can be identified by a unique long PN sequence. Each uplink transmission also includes a second spreading with base station-specific short codes to allow a base station to differentiate and de-spread its users' signals from other-cell interference.

The next two sections describe the functions of, and frame structures for, physical channels in the reverse and forward links.

3.3   REVERSE LINK PHYSICAL CHANNELS

In this section we describe the key features of reverse physical channels in IS95A.

3.3.1   Access Channel

The *reverse access channel* (R-ACH or ACH) is a contention-based random access channel used by the mobile to send uplink signaling messages to the base station when no traffic channel is assigned to the mobile. ACH can be used for registration, call origination, and page response messaging. Access channels are uniquely identified by their long codes. For each paging channel supported in the forward link, there may be up to 32 access channels numbered 0 through 31; also, each access channel is associated with a single paging channel.

Any R-ACH transmission consists of a preamble and a message capsule (see Fig. 3.5). The R-ACH message capsule consists of a Layer 2 encapsulated packet data unit (PDU) plus some padding bits added so that the R-ACH message capsule ends on a frame boundary.

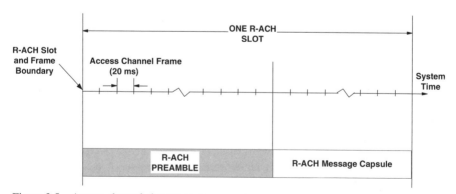

Figure 3.5   Access channel slot structure.

The preamble does not carry any massages and it is transmitted to help base station capture the phase and timing of user's transmission in the uplink. Once the preamble is detected the base station can demodulate the message capsule and process MS's request. Thus, the preamble length needs to be long enough to ensure the successful capture of the signal but not too long to cause unnecessary uplink interference.

Each R-ACH transmission needs to be contained within an R-ACH slot. The base station defines the total size of an R-ACH slot by specifying the preamble length to be used as well as and maximum size of the message capsule. The actual length of the capsule however, depends on the Layer 2 message size that is transmitted.

The following are some of main messages carried by the access channel.

- *Registration Message*: Every time the mobile terminal is turned, on the mobile sends this message to inform the base station. This message provides the mobile's location status, identification, and other parameters required for registration so that the mobile can initiate or receive calls.

- *Origination Message*: This message is sent by the mobile to initiate communication with the base station as part of a mobile-originated call setup.

- *Page Response Message*: This message is sent by the mobile to respond to an incoming page from the base station as part of a mobile-terminated call setup.

- *Order Message*: This includes typical order messages such as shared secret data (SSD) update confirmation, mobile station acknowledgments, or rejects.

- *Data Burst Message*: This is a user-generated data message sent by the mobile station to the base station.

- *Authentication Challenge Response Message*: This message is a response to the base station's authentication test sent before any channel assignment. The mobile may be granted access to the system only if this message provides a valid response.

3.3.2   Reverse Traffic Channels

The traffic channels in IS95A are bidirectional and symmetric. For each traffic channel assigned in the forward link, there is a corresponding reverse traffic channel at the same data rates. The traffic channels are designed to efficiently transport variable-rate user voice and data as well as in-band associated signaling messages.

Therefore, the forward link traffic channel in IS95A is designed to support two steams of data, one as primary and one as secondary traffic. Whereas the primary traffic is typically used for voice, the secondary traffic may carry signaling messages or user data.

Each 20-ms traffic channel frame starts with a mixed mode (MM) indicator to show whether any traffic and signaling multiplexing is used in that frame. An MM = 0 indicates no multiplexing, that is, primary voice traffic only. An MM = 1 implies a multiplexing mode that take be one of several forms indicated by the following TT and TM bits (see Fig. 3.6 and Table 3.2).

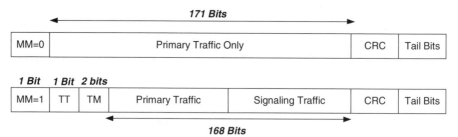

Figure 3.6   Multiplexing primary and secondary traffic on TCH.

TABLE 3.2   **Traffic and Signaling Multiplex Options for Traffic Channel**

Traffic/Signaling Multiplex Options	MM	TT	TM	Primary Traffic (bits)	Secondary Traffic (bits)
Blank and Burst	1	0	11	0	168
Dim and Burst	1	0	00	80	88
	1	0	01	40	128
	1	0	10	16	152
Primary Traffic Only	0		NA	171/80/40/16	0

On the basis of this structure the traffic channel frame can be used in one of the following modes:

Blank and Burst: In this mode, the entire traffic channel frame, that is, 168 bits, is used to send only secondary data or signaling messages. This is a fast mechanism to transport message at the expense of interrupting the voice frame transmission for few frames. Therefore, the blank and burst mode is used mainly for delay-sensitive and urgent messages.

Dim and Burst: This mode allows time multiplexing of primary and secondary traffic in the same frame. With this mode the signaling bits can be sent at a lower rate over several frames, without interrupting the voice transmission. The number of voice bits in this case should be reduced, that is, voice quality slightly degraded, to leave some available bits for signaling. Table 3.2 shows various multiplex formats available on the traffic channel.

Primary Traffic Only: In this mode the traffic channel frame typically carries the vocoder's output only.

In IS95A, the speech signal is encoded with a variable rate vocoder, the rate of which changes according to the voice activity. Therefore, the output rate of the vocoder varies dynamically during the conversation. IS95A traffic channels support two types of variable-rate vocoders, each with four possible data rates.

The 8-kbps vocoder generates traffic channel rates of 9.6, 4.8, 2.4, and 1.2 kbps, or the full, 1/2, 1/4, and 1/8 rates, respectively, which are collectively referred to as *Rate Set I*.

The 13-kbps vocoder similarly generates 14.4-, 7.2-, 3.6-, and 1.8-kbps traffic channels rates that correspond to full, 1/2, 1/4, and 1/8 rates of *Rate Set II*. Support

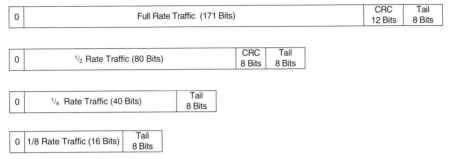

Figure 3.7   Traffic channel bits for Rate Set I.

for this rate set is optional. Figure 3.7 shows the number of bits used in for each vocoder rate with Rate Set I.

The lower-rate outputs correspond to higher compression and lower quality, which are used during nonactive periods of speech signal. For example, the 1/8 rate that corresponds to highest compression and lowest output quality is used during silences and pauses between words and sentences, when only the background noise is present. During the active speech the full rate is used, and 1/2 or 1/4 rates are used during the transition between active voice and silences. When lower rates are used, the mobile's transmission duty cycle is reduced accordingly. For example, when the vocoder is in 1/2 rate mode, the mobile only transmits half the time. The reduced transmission power during silence periods directly reduces the interference and thereby increases the uplink voice capacity.

The following is a brief list of important associated signaling messages carried by the reverse traffic channel:

- Pilot strength measurement message (PSMM) sends information about the strength of serving and other neighbors' pilot signals.

- Power measurement report message (PMRM) sends frame error rate statistics to the base station, either periodically or when a threshold is reached.

- Handoff completion message is the mobile response to a hand-off direction message in the forward link.

- Data burst is a user-generated data message sent by the mobile to the base station.

- Order messages include confirmations, updates, acknowledgements, and reject indications by the mobile while on the call, for example, SSD update confirmation or rejection, parameter update confirmation, mobile station acknowledgment, service option control with request or response, call connect and release indication, and authentication challenge response.

- Dual tone multifrequency (DTMF) burst message uses two tones, one low- and one high frequency, to represent a dialed digit to allow "touch tone" capability.

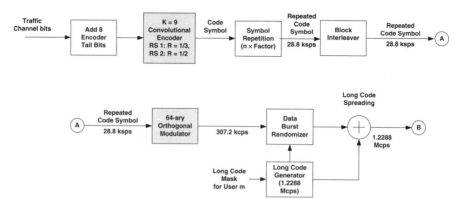

Figure 3.8   Reverse traffic channel coding and long code spreading. Reproduced under written permission from TIA.

- Parameter response message is the mobile response to the base station's retrieve parameters message.

3.3.3   Reverse Link Physical Layer Processing

This section describes the physical layer processing for the reverse traffic and access channels, including coding, spreading, and modulation features.

Figure 3.8 shows the physical layer processing of the reverse traffic channel. The baseband traffic channel bits for Rate Set I, including the primary and secondary traffic, are encoded using a 1/3-rate convolutional encoder followed by symbol repetition and block interleaving. The symbol repetition is used for lower vocoder rates such that the code symbol rate is fixed at 28.8 ksps, regardless of voice activity. For 1/2, 1/4, and 1/8 rates each symbol is repeated 2, 4, and 8 times, respectively.

For Rate Set II a similar process is followed with a 1/2-rate convolutional encoder, so that the final channel symbol rate for both RS-1 and RS-2 are the same, that is, 28.8 ksps.

After interleaving, the coded symbols are further modulated with 64-ary orthogonal Walsh functions. This orthogonal modulation, which maps every 6 input binary symbols to one of $2^6 = 64$ orthogonal Walsh codes of length 64, provides additional coding gain and simplifies the noncoherent detection of symbols in the reverse link. The resulting symbol rate after 64-ary Walsh modulation is $28.8 \times 64/6 = 307.2$ ksps.

The Walsh-modulated symbols are then fed into the data burst randomizer. The data burst randomizer takes advantage of the symbols' repetition on the reverse link to adjust the transmission duty cycle of the traffic channel according to the vocoder rate.

Each 20-ms traffic channel frame is divided into 16 power control groups (PCGs), each 1.25 ms. The data burst randomizer pseudo-randomly masks out individual power control groups, based on a masking pattern of 0 s and 1 s derived from

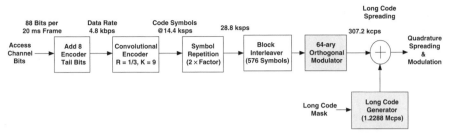

Figure 3.9   Access channel coding and long code spreading. Reproduced under written permission from TIA.

the long code sequence. The number of masked-out PCGs depends on the vocoder rate. For example, at full rate all 16 PCG are transmitted but at 1/2 , 1/4, and 1/8 rates 8, 12, and 14 PCGs, respectively, are removed from transmission mask. The resulting effect is proportionally lower average transmission power at lower vocoder rates, which reduces the average uplink interference and thereby increases the cell capacity.

After the burst randomization the symbols are spread by a long code generated based on a mask assigned to the specific user's reverse traffic channel. Because the long code sequence runs at 1.2288 Mcps, the resulting spread spectrum signal also has a rate of 1.2288 Mcps.

The baseband processing for reverse access channel is similar to the reverse traffic channel. The difference is that because the access channel has a fixed rate, the symbol repetition factor is fixed at 2 times and no data burst randomizer is utilized (see Fig. 3.9). Also, the long code spreading used for access channel is based on distinct access channel long code masks that are separate from those that can be assigned to users' traffic channels.

Note that for the access channel, this long code spreading operation involves modulo-2 addition of the 64-ary orthogonal modulator output stream and the long code. For the reverse traffic channel, spreading operation involves modulo-2 addition of the data burst randomizer output stream and the long code.

The long code is a periodic maximum-length shift register sequence of period $2^{42} - 1$ chips. The shift register structure is based on a linear recursion specified by the following characteristic polynomial:

$$p(x) = x^{42} + x^{35} + x^{33} + x^{31} + x^{27} + x^{26} + x^{25} + x^{22} + x^{21} + x^{19}$$
$$+ x^{18} + x^{17} + x^{16} + x^{10} + x^7 + x^6 + x^5 + x^3 + x^2 + x + 1. \qquad (3.1)$$

After long code spreading the symbols are fed into the quadrature spreading and modulation structure. The access channel and the reverse traffic channel are spread in quadrature as shown in Figure 3.10. The direct sequence spreading output is modulo-2 added to an in-phase and a quadrature-phase sequence. The in-phase and quadrature-phase components of this PN spreading sequence are periodic with a period of 2^{15} chips. These PN codes are the same short codes that the base station uses as its "signature" in the forward link transmission. After quadrature spreading,

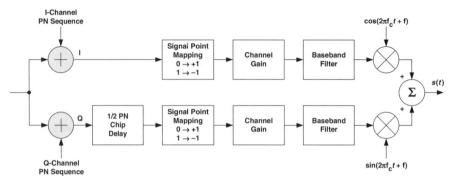

Figure 3.10   Quadrature spreading and modulation on the reverse link. Reproduced under written permission from TIA.

the Q-channel data are delayed by half a PN chip time (406.901 ns) with respect to the I-channel data.

Each of the I and Q channel data are converted to bipolar and multiplied by a digital gain before baseband filtering and quadrature modulation.

3.4   FORWARD LINK CHANNELS

This section describes the functions and frame structures for the forward physical channels.

3.4.1   Pilot Channel and PN Offsets

The pilot channel (PICH) is an unmodulated direct sequence spread spectrum signal continuously transmitted at a fixed power by each base station or sector in the network. The pilot is the first broadcast physical channel that is searched and acquired by the mobile stations immediately after the mobile is powered on. The PICH is used as a power, timing, and phase reference for the mobile stations and a physical layer-based identifier for the base station's signal.

The mobile uses the measured E_c/I_o, or the signal-to-noise ratio, of the pilot channels as a reference to select the strongest serving base station or sector and also to determine the need for soft handoff as well as the candidate sectors for soft handoff. The soft handoff triggers are therefore defined based on pilot strengths of base stations measured by the mobile. The pilot strength is measured in terms of the ratio of energy per chip to total interference spectral density, or E_c/I_o. The E_c/I_o, typically represented in decibel scale, is equivalent to the signal-to-noise ratio. Because of the de-spreading gain in a CDMA system, the value of E_c/I_o can be negative.

As a timing and phase reference the pilot is used for chip synchronization and coherent demodulation of symbols at the mobile's receiver, resulting in a more efficient and less complex receiver design. The pilot is used by the rake receiver's searching "finger" or correlator to identify the timing, phase, and relative strength of multipath received signals to be used as part of the rake receiver's maximal ratio combining algorithms.

The pilot channel does not have any frame structure. The baseband part of the pilot channel is a stream of all 0s spread by Walsh function 0, which is also a sequence of all 0s. The all-zero baseband stream is then multiplied by a pair of quadrature PN sequences. The PN sequence, with a specific offset, forms a short code, which uniquely identifies the pilot and thereby the particular sector that is transmitting that pilot signal. As a result, the pilot is an easy-to-capture signal, which mainly reflects the phase of the PN sequence used as the base station's short code.

The short codes, which are uniquely allocated to each base station in a large cluster, are defined based on different time offset of the same PN sequence. Therefore, all short codes can be generated with the same type of shift registers, initialized with different bit patterns. The period of this PN sequence is $2^{15} = 32768$ chips, which at the rate of 1.2288 Mcps corresponds to 26.67 ms.

Because the main differentiating factor for pilots assigned in a cluster of base stations is their time offset, it is important to allocate the PN offset with enough separation. This separation should be large enough such that the pilot received from a nearby, but undesired, base station cannot be confused with the time-delayed multipath pilot signals from the desired base station. To provide a time-offset margin, PN offsets are allocated in increments of 64 chips, specified with a PN offset index; thus there are $2^{15}/64 = 512$ defined PN offsets.

Typically, the PN offset is a power of 2. If PN offset index is 2, the minimum offset increment is $64 \times 2 = 128$ chips and there are $2^{15}/128 = 256$ possible PN offsets, which can be allocated uniquely to each base station in the network. For networks with more base stations than the number of available offsets, PN offsets can be reused as long as the base stations using the same offset are far from each other. In this case, the task of PN offset planning becomes similar to a simple frequency planning in a conventional 2G TDMA-based system. In a multicarrier base station, the same pilot PN sequence offset shall be used on all CDMA frequency assignments for a given base station.

3.4.2 Sync Channel

The sync channel (SYNCH) is the next channel acquired by the mobile right after pilot acquisition. Once the mobile station achieves pilot PN sequence synchronization by acquiring the forward pilot channel, the synchronization for the sync channel is immediately known. This is because the sync channel frame and interleaver timing on the sync channel are aligned with the pilot PN sequence.

The SYNCH is also a broadcast channel, which unlike the pilot channel, carries encoded messages. The messages received on the synch channel are the first set of information needed by the mobile to synchronize with and identify the base station and the supporting network.

The only message sent on the SYNCH is the *sync channel message*. This message includes the following parameters:

- Identification parameters
- System identification (SID)
- Network identification (NID)
- Pilot's PN sequence offset index (PILOT_PN)
- Timing parameters
- Long code state (LC_STATE): long code at the time specified in the system time parameter
- System time (SYS_TIME) and offset of local time from the system time
- Leap seconds (LP_SEC): number of leap seconds that have occurred since the start of the system
- Daylight saving time indicator
- Paging channel data rate (PRAT): 4.8 or 9.6 kbps

After the successful reception of the *sync channel message*:

- The mobile can identify the system and network to which the base station belongs and, on the basis of its preset preferences, determine whether system is a preferred network or whether it should look for an alternative network.
- The mobile can identify the base station and synchronize its timing at the frame and slot levels with system time.
- With the timing parameters, the mobile can start running the long PN sequence at the state given by the LC_STATE. As a result, the mobile also becomes synchronized with the long PN sequence.
- The mobile is also informed of the data rate of the paging channel, to be monitored right after synchronization.

The sync channel transmissions are structured in groups of 80-ms *sync channel superframes* each containing 96 bits, resulting in a fixed data rate of 1200 bps. Each superframe contains three *sync channel frames*, shown in Figure 3.11, and the sync channel frames are aligned with the short PN sequence period of 26.67 ms. Thus, once the mobile achieves synchronization with the pilot channel, the alignment or the sync channel frame follows immediately and the mobile can start reading messages on the synch channel.

The sync channel message may be longer than a frame or superframe and therefore, it is structured within a *sync channel message capsule*. A sync channel message capsule consists of the *sync channel message* and some *padding*. The sync channel message has an 8-bit message length header, a 2- to 1146-bit message body, and a 30-bit cyclic redundancy check (CRC).

If the sync channel message occupies more than a frame, some padding bits are used to fill up the bit positions up to the beginning of the next sync channel superframe, where the next sync channel message starts. The sync channel always starts at the beginning of a superframe.

Each sync channel frame begins with the *start-of-message* (SOM) bit. Whereas an SOM of 1 indicates the start of the sync channel message, an SOM of 0 indicates the continuation of message started in previous frames.

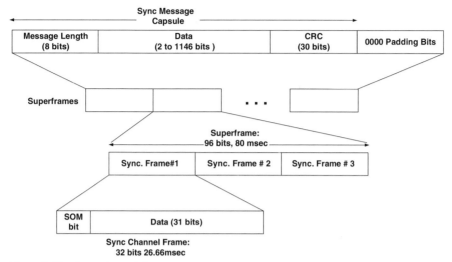

Figure 3.11   Sync channel frame structure.

3.4.3   Paging Channel

The paging channel (PCH) in IS95A has a dual role. It serves as a broadcast control channel when it carries the overhead messages and system parameters, and it serves as a forward common channel when used for mobile-directed paging messaging.

The mobile starts monitoring the paging channel right after synchronizing with the base station, after the synch channel message, to obtain system and access parameters and other overhead information. The mobile continues monitoring the paging channel for paging and/or updates overhead messages as long as it is the idle state waiting for a mobile-directed page message or call origination by the mobile.

Up to seven paging channels can be defined per sector; however, each mobile only monitors one paging channel. When multiple paging channels are defined as indicated as part of systems parameters, a hashing algorithm based on a mobile's ID number (MIN) is used to determine which paging channel should be monitored by the MS. This allows the system to distribute the signaling load among all paging channels.

The paging channel rate (PRAT) may be 4.8 kbps or 9.6 kbps, and it is indicated in the synch channel message. The paging channel is slotted, and the 80-ms paging channel slot is divided into four frames (see Fig. 3.12). Each paging channel frame is further divided into two paging channel half-frames. A message on the paging channel may occupy more than one paging channel half-frames, and it may end in the middle of a paging channel half-frame. The entire paging channel message is sent in 1–2048 slots. Thus a group of 2048 slots is called a maximum slot cycle.

Similar to the synch channel message, a message on the paging channel is structured as paging message capsules. A message can be transmitted as a synchronized capsule. Whereas an unsynchronized capsule can end anywhere within a half-

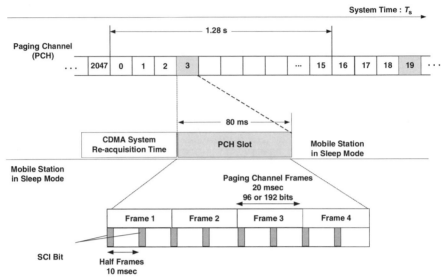

Figure 3.12   Paging channel frame structure.

frame, a synchronized capsule always ends on a half-frame boundary and therefore may need some padding. A message capsule includes a message body followed by some padding, if it is a synchronous capsule. A paging channel message includes an 8-bit message length header, a message body of 2–1146 bits, and a 30-bit CRC.

A PCH message is segmented and transmitted in a number of half-frames, each starting with a single-bit synchronized capsule indicator (SCI) field and 47 or 95 bits of the paging channel frame body for 4.8 or 9.6 kbps respectively. An SCI = 1 indicates the start of a new message capsule in the current half-frame and an SCI = 0 would imply any other case, for example, the case in which the current frame is not the start of a message and may include a message and/or padding bits. The base station always starts a slot with an unsynchronized message capsule with an SCI = 1 at the beginning of the slot.

A mobile may monitor the paging channel in one of two modes:

- Nonslotted Mode: In this mode the mobile is required to monitor all paging slots.

- Slotted Mode: In this mode a mobile listens for pages only at during certain paging slots and as a result the mobile saves its battery power and increases its standby time.

A mobile station operating in the slotted mode generally monitors the paging channel for one or two slots per slot cycle. The mobile can specify its preferred slot by using the SLOT_CYCLE_INDEX field in the registration, origination, or page response message, or as part of traffic channel signaling to the base station.

The minimum-length slot cycle is 1.28 s, consisting of 16 slots of 80 ms each. A slot cycle T, measured in units of 1.28 s, is given by $T = 2^{\text{SLOT\_CYCLE\_INDEX}}$. For example, a SLOT_CYCLE_INDEX = 1 implies a paging lot cycle of $2^1 \times 16 = 32$ slots and makes the mobile listen to every 32nd paging slot. For each mobile, the specific slot out of the $16 \times T$ slots to be monitored is determined by a parameter PGSLOT, which is a fixed but randomly calculated number.

The paging channel carries both broadcast and mobile-directed messages. Examples of broadcast messages are:

- System Parameter Message: includes various parameters, for example, base station ID and number of paging channels
- Access Parameters Message: defines parameters needed by a mobile station to structure its uplink transmission on an access channel
- Neighbor List Message: provides information about neighboring base stations' pilot PN offset index
- CDMA Channel List Message: provides the list of CDMA carriers.

Examples of mobile-directed messages are:

- Page Message: provides signaling messages addressed to specific mobile(s) notifying them of an incoming call
- Channel Assignment Message: directs the mobile station to tune to a specific traffic channel, that is, radio frequency and Walsh code channel.
- Authentication Challenge Message: seeks the mobile's response to an authentication test. The mobile uses specific authentication keys, the last shared secret data (SSD) provided by the system, along with the mobile's ID and the handset's electronic serial number, to calculate an authentication challenge response to be sent to the base station in the uplink. The authentication test must be passed before a traffic channel can be assigned to the user.
- SSD Update Message: an order by the base station to the mobile station to update the SSD to be used as part of the authentication process
- Data Burst Message: a Layer 3 user data message sent by the base station to the mobile station
- Order and Notification Messages: a number of different messages fall in this group including alerts, confirmations and acknowledgment, and maintenance-related signaling while the mobile is idle, that is, is not on a traffic channel. This group also includes notification messages providing information to be displayed by the mobile such as caller's ID or waiting messages.

3.4.4 Forward Traffic Channel

The forward traffic channel has the same frame structure and data rates as the reverse traffic channel. It can carry voice, data, and associated signaling messages during a call. The main difference is multiplexing of a reverse power control subchannel with the forward traffic channel bits. With this power control subchannel the base station continuously sends binary power up/down commands or bits to the mobile station in every 1.25 ms.

The forward traffic channel supports Rate Set I and optionally Rate Set II. The traffic channel also supports associated signaling in the form of blank and burst or dim and burst. A signaling message on the forward traffic channel includes an 8-bit header, a 16- to 1160-bit message body, and a 16-bit CRC code. Some padding bits may also be used to end the message on a frame boundary. The entire message may be multiplexed with traffic and transmitted over one or several frames based on blank and burst or dim and burst frameworks.

The following are some of the main messages sent as part of associated signaling on the forward traffic channel.

- Handoff Direction Message (HDM): provides the mobile with information to begin the handoff process
- Extended Handoff Direction Message (EHDM): same as HDM with additional features and parameters
- Analog Handoff Direction Message: directs the mobile to switch to the analog mode and begin the handoff process
- In-Traffic System Parameters Message: updates some of the parameters set by the system parameters message in the paging channel
- Neighbor List Update Message: updates the neighbor base station parameters set by the neighbor list message on the paging channel
- Power Control Message: tells the mobile how long the period is, or what threshold is to be used, in measuring frame error statistics that will be sent in the mobile's power measurement report message
- Order Message: mostly similar to the order message on the paging channel
- Data Burst Message: sent by the base station to the mobile
- Retrieve Parameters Message: requests the mobile to report on any of the retrievable and settleable parameters
- Set Parameter Message: informs the mobile to adjust any of the retrievable and settleable parameters
- SSD Update Message: request from the base station for the mobile to update the shared secret data
- Flash with Information Message: contains information that allows the network to supply display information to be displayed by the mobile, to identify the responding party's number (the connected number), to convey information to the mobile by means of tones or other alerting signals, and to indicate the number of messages waiting
- Mobile Registration Message: informs the mobile that it is registered and supplies the necessary system parameters
- Authentication Challenge Message: when the base station suspects the validity of the mobile, it can challenge the mobile to prove its identity.
- Alert with Information Message: allows the base station to validate the mobile identity

Figure 3.13   Synch channel coding.

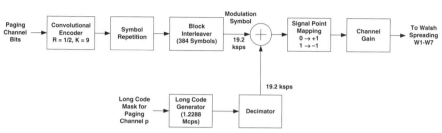

Figure 3.14   Paging channel coding and scrambling.

3.4.5   Forward Physical Channel Processing

Figure 3.13 shows the channel encoding process for the synch channel. The synch channel bits at 1.2 kbps, structured as 32 bits per 26.67-ms frame, are encoded using a 1/2-rate convolutional encoder followed by symbol repetition and block interleaving. The resulting modulation symbols are then converted from binary (1/0) into bipolar (−1/+1) form and the SYNCH channel gain is digitally adjusted before any spreading is used. The Walsh and quadrature spreading and modulation are described below in this section.

For the paging channel, the baseband information bits are first encoded for error protection, then repeated once, only if the 4.8 kbps is used, and then interleaved (see Fig. 3-14).

After interleaving, the data are first scrambled by a decimated long PN sequence. A long PN code is generated by using a mask specific to each unique paging channel number (i.e., 1 through 7), and then it is decimated with 64-to-1 ratio to form a 19.2-ksps (= 1.2288 Msps/64) stream that then directly scrambles the paging message data symbols. After scrambling, the signal is spread by specific Walsh codes (W1–W7) assigned to the paging channel and further spread by the short PN sequence assigned to the serving sector.

Figure 3.15 and Figure 3.16 show the physical layer processing of information bits on the forward traffic channels for Rate Set I and Rate Set II. The process is similar to the paging channel processing with some differences including power control multiplexing and variable repetition.

For RS-1 the channel coding includes a CRC as frame quality indicator followed by a half-rate convolutional encoding and symbol repetition. Figure 3.15 shows the process for all four vocoder rates. When repetition is used at lower vocoder rates, the receiver can combine power over repeated symbols to improve the overall coding gain. For example at 1/4 rate, each symbol is repeated four times,

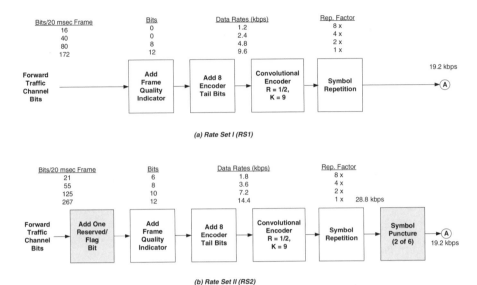

(a) Rate Set I (RS1)

(b) Rate Set II (RS2)

Figure 3.15   Traffic channel coding and rate matching.

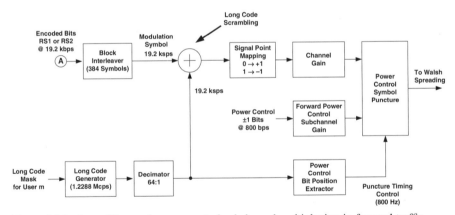

Figure 3.16   Scrambling and power control subchannel multiplexing in forward traffic channel.

and because the receiver integrates power over the repeated symbols, each symbol can be transmitted at 1/4th of the power needed for the full-rate transmission.

In general, the modulation symbols that are transmitted at lower data rates are transmitted with lower energy, such that the product energy per modulation symbol (E_s) times data rate R, $E_S \times R$, is the same for all rates. Therefore, the symbol repetition is used together with lower per-symbol power transmission, to reduce the overall interference caused by "nonactive speakers." This reduced interference based on voice activity would effectively increase the voice capacity.

For Rate Set II the coding process is similar, except for the addition of a single reserved bit before any coding and some symbol puncturing after symbol repetition. The puncturing takes out two of six symbols such that the final symbol rates is kept at 19.2 kbps, matching the symbol rate for the RS-1 number. With this design the same data processing, including the spreading and modulation, can be used for both Rate Sets.

After the channel encoder, the encoded bits at 19.2 kbps are block interleaved over every 384 symbols or 20 ms. The interleaving randomizes burst errors in time and improves the channel coding performance over a fading channel. The interleaved data are scrambled by a decimated long PN sequence (see Fig. 3.16).

The output of the long PN sequence generator that runs at 1.2288 Mcps is decimated by a ratio of 64:1, resulting in a rate of 19.2 kcps, which matches the symbol rate at the output of the interleaver. Thus data scrambling is achieved by multiplying the interleaved data by the decimated long PN sequence at 19.2 ksps. Because the mask used for generating the long PN sequence is user specific, only the desired user can de-scramble the data. Therefore, this scrambling provides some level of message security at the physical layer.

The binary symbols after scrambling are converted to bipolar $(-/+1)$ form, and a digital gain specified by the forward link power control process is applied to the symbols. The resulting symbols are then multiplexed with the power control bits (PCBs).

The PCBs are punctured into the traffic symbol stream at the rate of 800 bps or every 1.25 ms; thus each PCB replaces two data symbols. A PCB = 0 indicates that the mobile has to increase its mean output power level by 1 dB, and a PCB = 1 indicates a 1-dB decrease in the mean output power.

Each 20-ms frame consists of 16 power control groups of length 1.25 ms or 24 symbols. Although a PCB can be punctured into any one of the first 16 symbol positions of a power PCG, the exact location of the PCB in the PCG is randomized. The PCB position randomization is based on the decimal value of the four most significant bits of the long PN decimator's output.

For each user, the resulting multiplexed stream after PCB addition is orthogonally spread by an assigned Walsh code for the traffic channel. Because the forward link uses Walsh codes of length 64 for channelization spreading, each symbol is spread by a factor of 64, and with a 19.2-ksps input, the resulting output runs at the rate of $19.2 \times 64 = 1.2288$ Mcps.

After the Walsh code spreading, all data streams for pilot, synch, paging, and traffic channels are summed up and further spread by the assigned pair of short PN sequences of the transmitting sector (see Fig. 3.17). Each of the I and Q branches uses a different short PN code. These short PN codes are two maximum-length linear feedback shift register sequences $i(n)$ and $q(n)$ of length $2^{15} - 1$, generated according to the following linear recursions:

$$i(n) = i(n-15) \oplus i(n-10) \oplus i(n-8) \oplus i(n-7) \oplus i(n-6) \oplus i(n-2)$$
$$q(n) = q(n-15) \oplus q(n-12) \oplus q(n-11) \oplus q(n-10) \oplus q(n-9)$$
$$\oplus q(n-5) \oplus q(n-4) \oplus q(n-3) \tag{3.2}$$

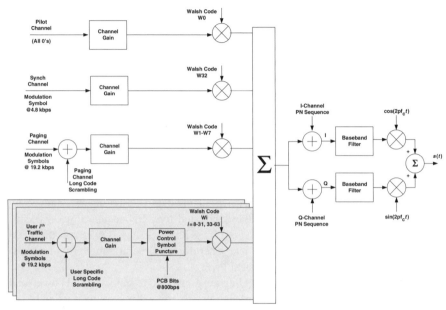

Figure 3.17 Walsh channelization and PN quadrature spreading and modulation of forward link channels.

where i(n) and q(n) are binary valued ("0" and "1") and the additions are modulo-2. The short PN sequences provide another layer of code isolation for the combined spread spectrum signal to allow other-cell interference rejection.

The composite spread spectrum signal including all control and traffic channel is modulated based on QPSK and transmitted over the air.

3.5 RANDOM-ACCESS CHANNEL OPERATION

The reverse access channel in IS95A, or R-ACH for short, is a contention-based random-access channel for uplink transmissions during the access state when no dedicated resources are available to the mobile. All transmissions on the R-ACH are slotted, based on R-ACH slot, and organized within an access attempt structure, which is described in this section.

Transmission of any message on a R-ACH is conditioned on the primary pilot E_c/I_o, chip energy over interference spectral density, of the base station exceeding an access threshold set by the system. If the threshold is not exceeded, the transmission of the probe is delayed until this condition is satisfied.

During transmission on the R-ACH, the mobile station continuously monitors the pilot E_c/I_o. If the pilot E_c/I_o falls below a drop threshold for a specified period, the mobile station ceases transmitting and restarts the "system determination" process.

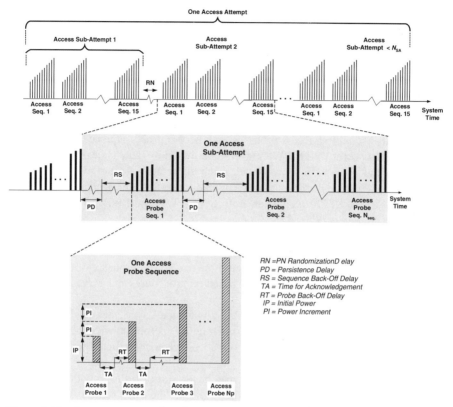

Figure 3.18 Access attempt and probe sequence structure on the R-ACH.

While waiting for the Layer 2 acknowledgement, the mobile station also monitors the pilot E_c/I_o and follows procedures defined for the access state handoffs, which are discussed later in this chapter.

Access Attempts: The entire process of sending one Layer 2 encapsulated PDU and receiving (or failing to receive) an acknowledgment for that message constitutes an access attempt. The Layer 2 encapsulated PDU, sent on the R-ACH, may be the result of call origination, registration, or responses to pages and other order messages received from the base station.

One access attempt consists of one or more access subattempts, and each transmission in the access subattempt is called an access probe. Within an access subattempt, access probes are grouped into access probe sequences (see Fig. 3.18).

Each access probe consists of an R-ACH preamble and an R-ACH message capsule, containing a Layer 2 encapsulated PDU. The mobile stops sending access probes once it receives the L2 acknowledgement from the base station.

The maximum number of access subattempts within an access attempt (N_{SA}), the number of access probe sequences within an access subattempt ($N_{Seq.}$), and the number of access probes within a sequence (N_{Probes}) are defined as part of the access parameters.

There is one or more R-ACH associated with each F-PCH in the forward link. For each access probe transmission, the mobile pseudo-randomly chooses one of the R-ACHs associated with the mobile's current F-PCH. Therefore, if multiple R-ACHs are associated with a mobile current F-PCH, the mobile access probes within an access probe sequence may be transmitted on any of those R-ACHs.

The access probe transmission power on the R-ACH is subject to an open loop power control. The first access probe of each access probe sequence is transmitted at a low initial power (INIT_PWR) level, determined by the physical layer, relative to the nominal open loop power level. Each subsequent access probe within an access probe sequence is transmitted at a higher power level determined by the MAC based on a (PWR_STEP) parameter. Both INIT_PWR and PWR_STEP are part of the access parameters defined by the base station.

The timing of all transmissions on the R-ACH is expressed in terms of R-ACH slots. Each access probe transmission begins at the start of an R-ACH slot. To reduce the chance of collisions, the transmission timing of access probes, probe sequences, and access subattempts are randomized based on procedures and parameters defined in the MAC layer.

After each access probe the mobile waits for the Layer 2 acknowledgment for TA slots followed by the random back-off time RT, before it sends the next probe. This is part of the ALOHA-based random access protocol adopted for the R-ACH channel.

Note that any failed access probe may be a result of collision with other user's transmission and/or insufficient received signal at the base station. Whereas the random wait times between probes avoid repeated collisions in subsequent transmissions, the power increments address the insufficient transmission power. Together, the random wait time and power increases improve the chance of success in the next probes.

There is clearly a trade-off in choosing the best value for initial power and power increments. A large value for these parameters increases the chance of success after a smaller number of probes at the expense of too much increase in the interference caused by the uplink access transmissions. On the other hand, setting IP and PI to very small numbers increases the average number of probes needed for a successful access, which increases the latency and also the interference caused by too many weak access probes.

The timing of the start of each access probe sequence is also randomized based on a pseudo-randomly generated back-off delay RS and an additional persistence delay (PD). After the back-off delay of RS slots, a PD is imposed as a result of a pseudo-random test by the MAC layer. If the test is passed, the first access probe of the sequence begins in that slot; otherwise, the access probe sequence is deferred until at least the next slot. The persistence test is not used for response messages.

At the access attempt level, the precise timing of each access subattempt transmission on R-ACH is determined by a procedure called PN randomization. For each access subattempt, the MAC computes a delay, RN, from 0 to 2 PROBE_PN_RAN$_s$ − 1 PN chips, with a hash function.

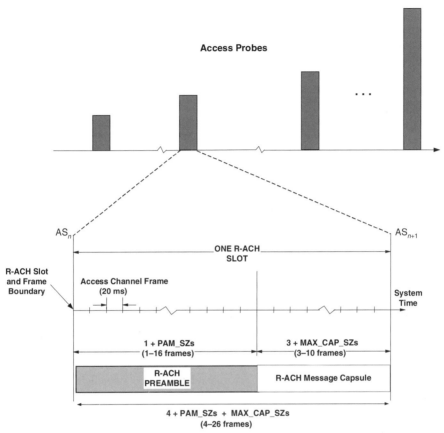

Figure 3.19   Access channel probe structure.

In summary, on determination of access failure, the mobile station reattempts access by resending the access probes at increased power levels, each after a random back-off time, until it reaches the maximum permissible number of probes. The mobile may then initialize its probe counter and start another sequence of access probes up to a maximum number of sequences, and if it still fails to access, it enters the system determination substate.

R-ACH Slot Structure: An R-ACH transmission consists of a preamble and a message capsule (see Fig. 3.19). The R-ACH message capsule consists of a Layer 2 encapsulated PDU plus some padding bits added so that the length of the R-ACH message capsule is an integer number of R-ACH frames.

The maximum size of the message capsule and the preamble size are from 1 to 16 frames and from 3 to 10 frames, respectively. The preamble size of $1 + PAM\_SZ_s$ frames and the maximum message capsule size of $3 + MAX\_CAP\_SZ_s$ frames are defined by the base station. An R-ACH slot, therefore has a length of $L = 4 + MAX\_CAP\_SZ_s + PAM\_SZ_s$ frames. The actual size of a capsule depends on the L2 message size.

All R-ACH slots begin and end on an R-ACH's 20-ms frame boundaries. In addition, all R-ACHs associated with a particular F-PCH have the same slot size, and all of the slots begin at the same time.

3.6   POWER CONTROL SCHEMES IN IS95A

The IS95A standard defines procedures for power control in both the forward and the reverse link [2].

Whereas the forward link power control aims at reducing interference and power management for higher forward link capacity, the reverse power control is critical for interference management and system capacity as well as longer mobile battery life.

The reverse power control mechanisms in IS95A include the power control during access and the traffic channel power control. The reverse traffic channel power control is a combination of an open-loop and a closed-loop mechanism.

The open loop allows the mobile to autonomously change its transmit power based on the downlink power measurement to quickly compensate for any major path loss changes due to shadowing and distance without the need for any feedback from the base station.

The closed loop, on the other hand, attempts to provide a fast and fine correction mechanism to compensate for the effect of fading based on power control commands from the base station.

The two mechanisms together provide at least 80 dB of dynamic range for the mobile power. The following describes the reverse access and traffic channel as well as forward link power control in more detail.

3.6.1   Access Power Control

When the mobile first attempts to access the system, the base station has no information about the location of the mobile and thus the power at which the mobile should access the system. Therefore, the mobile station takes an open loop power control approach during its access state.

The mobile attempts to access the system by transmitting a series of access probes. The first access probe is transmitted at a relatively low power and is followed by a series of successive probe transmissions of progressively higher power until an acknowledgment is received.

The mobile's initial power for the first access probe is calculated at:

$$\text{Mean Output Power (dBm)} = -\text{Mean Received Power (dBm)} + \text{P\_Offset(dB)}$$
$$+ \text{NOM\_PWRs} - 16 \times \text{NOM\_PWR\_EXTs}_{(dB)}$$
$$+ \text{INIT\_PWRs}_{(dB)} \qquad (3.3)$$

Where:

- P_Offset is a default constant offset power specified by the standard using generic assumptions of typical loading and interference level as well as the imbalance between mobile and base station ERPs and receiver sensitivities. The default values of P_Offset for cellular (800 MHz) and PCS (1900 MHz) bands are −73 dB and −76 dB, respectively.

- NOM_PWRs − 16 × NOM_PWR_EXTs is another parameter that allows the open-loop estimation process to be adjusted for different operating environments. NOM_PWRs is the nominal power ranging between −8 and 7 dB, which if combined with NOM_PWR_EXTs, for the PCS band only, can be between −24 and 7 dB.

- INIT_PWRs is the adjustment that is made to the first access channel probe so that it should be received at somewhat less than the required signal power. This conservatism partially compensates for some of path loss de-correlation between the forward and reverse CDMA channels. The range of the INIT_PWRs parameter is −16 to +15 dB, with a nominal value of 0 dB

After each probe transmission, the mobile waits for acknowledgment from the base station and waits for a random time interval before sending the next access probe at a slightly higher power defined by the system parameter PWR_STEP. The range of the PWR_STEPs parameter is 0–7 dB.

The values of P_Offset, NOM_PWR, NOM_PWR_EXT, INIT_PWRs as well as the parameter PWR_STEP are broadcast by the base station (in the *access parameters message*) and received by the mobile before the first access probe transmission. During an access probe transmission and for any subsequent probe, the mobile station needs to update the mean output power in response to changes in the mean input power or other parameter updates.

3.6.2 Open-Loop Power Control on the Traffic Channel

The open-loop power control on the reverse traffic channel is the continuation of the mobile open-loop power control during access attempts. After a call is established, the mobile may move around the cell while on the traffic channel. As a result, the path loss and the received power at the mobile change, requiring the open-loop power control adjustments to the mobile transmit power according to the following equation:

$$
\begin{aligned}
\text{Mean Output Power (dBm)} = &-\text{Mean Received Power (dBm)} + \text{P\_Offset(dB)} \\
&+ \text{NOM\_PWRs} - 16 \times \text{NOM\_PWR\_EXTs}_{(dB)} \\
&+ \text{INIT\_PWRs}_{(dB)} \\
&+ (\text{Sum of all access probe corrections}) \quad (3.4)
\end{aligned}
$$

This is the same equation used for the access state and includes all the power adjustment made during the access probing process.

Assuming a reciprocity between uplink and downlink path losses, the open-loop power control points to a higher reverse transmit power when it measures a weaker downlink pilot power from a base station. The open-loop power control is aimed at compensating for slow-varying and lognormal shadowing effects where there is a correlation between the forward-link and reverse-link fades. For multipath and fast Rayleigh fading, however, this reciprocity assumption does not hold. The fast fading effects on the forward and reverse links are frequency dependent and therefore uncorrelated to each other. Also, the fast rate of power change as result of multipath fading cannot be compensated effectively by the open-loop slow response time that is due to the longer-term averaging process needed for forward-link measurement.

The closed-loop power control attempts to address these shortcomings and to further fine-tune the mobile station's output power on the traffic channel.

3.6.3   Closed-Loop Power Control on the Reverse Traffic Channel

The closed-loop power control is used to compensate for fast radio link changes due to multipath fading. The closed-loop process involves uplink measurement and power control decisions made by the base station followed by mobile's power up/down changes according to base station power control commands.

Each traffic channel frame is 20 ms long, consisting of 16 PCGs. During a PCG the MS's transmit power is constant. The binary up/down power control commands are sent as power control bits punctured into forward traffic channel in every power control group or every 1.25 ms. The power steps for in each PCG are +/−1 dB. If the received bit is "0," the mobile increases its power by 1 dB; otherwise, it decreases the power by 1 dB.

Although the FER is the appropriate measure of link quality, it can only be measured over a relatively long period of time and, by definition, can only be updated once every frame. On the other hand, fast power control to respond to fading would require faster power control rates that can rely on shorter-term (E_b/N_o) measurements.

In general, in a mobile radio channel, the relationship between the FER and E_b/N_o depends on the environment and the mobile's speed and one cannot find a one-to-one mapping between FER and E_b/N_o. Therefore, the right threshold for E_b/N_o to maintain an acceptable FER needs to be defined dynamically.

To address these issues, the process of closed-loop power control is designed as a combination of two loops, as shown in Figure 3.20:

- The outer power control loop estimates and updates a set-point value for E_b/N_t to achieve the target frame error rate (FER) on reverse traffic channel, in every frame or every 20 msec.

- The inner power control loop, in every 1.25 msec PCG, compares the E_b/N_t measured on the received reverse traffic channel with the outer-loop E_b/N_t set point to determine the value of the power control bit to be sent to the mobile.

The details of outer-loop power control depend on the implementations and may vary by equipment vendor. However, in principle, the objective of the outer loop is to balance the desired quality or FER, typically between 0.2% and 3%, and the cell capacity.

For instance, the base may slightly reduce the E_b/N_o set point, for example. by δ_D dB, in every frame until it detects a higher-than-accepted FER (see Fig. 3.21).

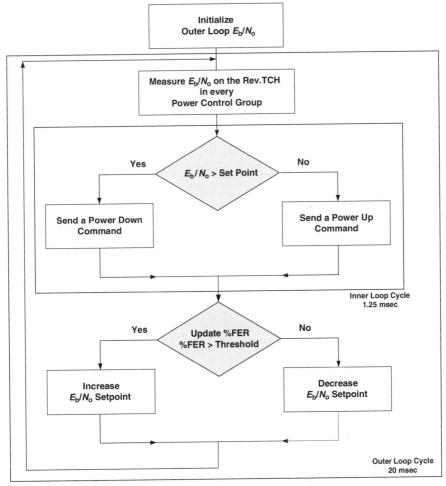

Figure 3.20 Reverse traffic channel closed-loop power control process.

Once the high FER is detected, the base station may immediately increase the set point by a large amount, for example, by δ_U dB. Typically power-up steps δ_U are much larger than power-down steps δ_D. The base station typically maintains a minimum and a maximum E_b/N_t allowable set point for each traffic channel.

After the channel coding and before spreading, the PCBs at 800 bps are multiplexed or punctured onto the baseband traffic channel stream of 19.2 Kbps. Because the PCBs are multiplexed onto the forward traffic channel *after* the convolutional encoder, they are not error protected. This design choice was made to avoid delays that are inherent in the frame decoding process, which could slow down the speed of power control.

Each PCG contains 24 bits, and the location of the PCB within a PCG is determined in a pseudo-random fashion. There are three additional points to mention

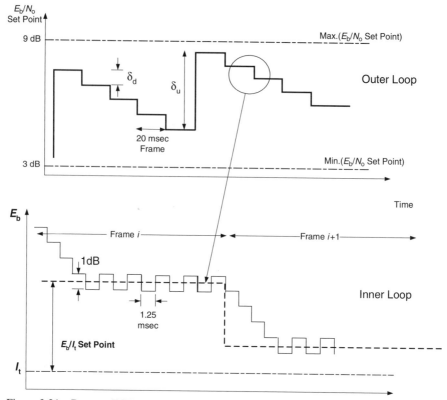

Figure 3.21 Reverse TCH power change as a result of inner and outer loop.

regarding closed-loop power control. The stream of PCBs at 800 bps creates a power control subchannel, which is associated with the forward traffic channel.

The mobile's behavior in response to power control commands changes once it is in soft handoff. The specific process of power control in soft handoff is explained in the section on soft handoff.

3.6.4 Forward Link Power Control

Although there is no battery life problem for the base station, there still needs to be power control in the forward link. It is important for the base station to manage its power resources to make sure enough power is available to support all traffic channels for the users across the cell area. The power is one of the key limitations that define the IS95A forward link capacity. Therefore, forward power control is needed to provide a forward link capacity at levels equal to or above the reverse link capacity.

Forward power control also avoids excessive power transmission at the base stations, which could contribute to an increased other-cell interference and thereby

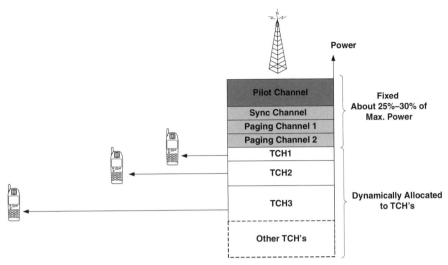

Figure 3.22   Power allocation in the forward link.

reduces capacity. In general, however, for voice application of IS95A the power control requirement for the forward link is not as stringent as that for the reverse link.

The total transmit power of the base station is divided among control and traffic channels (see Fig. 3.22). The control and overhead channels including pilot, synch, and paging channels take about 25–30% of the maximum power. This allocation is static and does not change with user mobility and number. The remaining power, about 70–75% of total power, is a shared resource for all forward traffic channels and needs to be dynamically allocated to users according to their radio link conditions. On average, users close to the base station require less power than those far from the base.

The forward link power control in IS95A is simpler and slower than the reverse link power control. The basic idea is very similar to that presented for the outer loop portion of reverse closed-loop power control.

The base station directly or indirectly estimates the forward link FERs in every N frame and responds accordingly. The base station reduces its power, that is, the traffic channel's gain, by Δ_{Down} in every N frame and maintains the same power during the next N frame unless there is an indication that downlink FER has exceeded the acceptable threshold (see Fig. 3.23). In the case in which the FER exceeds the threshold the base station immediately increases the traffic channel power by Δ_{Up}.

The details of power control procedure for Rate Set I and Rate Set II systems are somewhat different (see Fig. 3.23).

- In Rate Set I (RS-1), the mobile continuously measures the quality of forward link frames and keeps track of the frame erasure rates over a period N defined

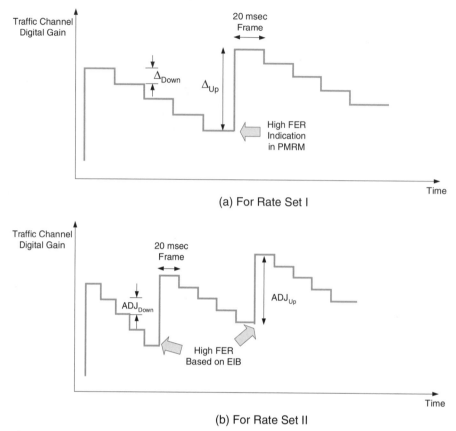

Figure 3.23   Forward link power control.

by parameter pwr_re_frame. The mobile station periodically, or whenever the FER exceeds a certain threshold, sends a power measurement report message (PMRM) to the base station. The PMRM contains the number of errors, the total number of frames measured, and the FER. The mobile then waits for a period specified by pwr_rep_delay before it sends the next report. The PMRM provides the information needed by the base station to determine whether the acceptable level of FER is exceeded, in which case it follows the process described above and shown in Figure 3.23a. In this case, the traffic channel digital gain can be from 34 to 108 with an average value of around 60.

• In Rate Set II (RS-2), the mobile sends an erasure indicator bit (EIB) in every frame, flagging bad frames to the base station by setting EIB = 1. The EIBs are then used by the base station to calculate FER and adjust the traffic channel gain accordingly. In this case, the FER measurement and response processes are much faster than RS-1 because updates are available in every 20-ms frame, that is, $N = 1$. Therefore, the step-up and step-down values can be smaller, resulting in faster and finer tuning of power in RS-2. In this case, the traffic channel digital gain can be from 40 to 108 with an average value of around 80.

In general, the details of forward link power control as it relates to base stations behavior are not specified in the standard and they are subject to implementations.

The forward link power control is only applied to the individual forward traffic channels. The pilot, synch, and paging channels are transmitted at fixed but selectable power levels.

3.7   TRAFFIC CHANNEL HANDOFF IN IS95A

This section describes a number of handoff processes defined in the IS95A standard. The handoff scenarios on traffic channels can be broadly classified into soft and hard hand-offs.

In soft handoff the mobile moves from a source cell to a target cell that is in the same operator's network, in the same carrier frequency, and having the same frame offset as the source cell. In this case, the mobile makes a new connection with the target cell before breaking the link with the source cell, to allow for a make-before-break (MBB) transition. As a result, there is a window of time during which the mobile communicates with multiple base stations. In IS95A, up to three base stations can be involved in soft handoff. The details of the soft handoff process under different scenarios as well as soft handoff parameters are described below in this section.

In the case of hard handoff, the connection with the source cell is broken before the link with the target cell is established. This is a break-before-make (BBM) scheme, which is used whenever the mobile performs an interfrequency handoff, or when it hands off or "hands down" to an analog channel. The interfrequency, CDMA-to-CDMA or D-to-D handoff occurs when the mobile moves, in a multi-carrier CDMA network, from one cell to another cell in which the current carrier is either full or not available. In this case, the mobile cannot perform the soft handoff because it cannot simultaneously transmit or receive on two frequency carriers. CDMA-to-analog or D-to-A handoff occurs when the mobile is traveling into an area where there is analog service but no CDMA service. CDMA-to-analog handoff is sometimes called. Depending on the implementation a hard handoff may also occur as the mobile crosses the MSC service areas or moves across network operator's boundaries.

3.7.1   Soft Handoff Scenarios

Depending on the number and locations of sectors involved in soft handoff, the reverse link traffic channel frames may be processed differently. Figure 3.24 shows the following scenarios:

Two- or Three-Way Soft Handoff: In these cases, each of the two or three sectors in the active set and involved in the soft handoff is located at different cell

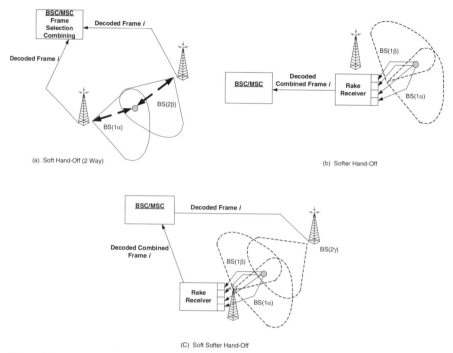

Figure 3.24   Soft handoff scenarios.

sites. Each base station's receiver decodes each frame independently and sends it to the MSC, where the best frames are selected and passed to the rest of the network.

Softer Handoff: In this scenario, the two base stations involved in the soft handoff are two sectors on the same cell site and therefore the two receivers are colocated. In this case, the signal received by the two sector antennas can be soft combined and the combined signal is decoded as one frame, which is then sent to the MSC.

The Soft-Softer Handoff: This is a three-way handoff in which two sectors in the active set are on one site and the third sector is on a separate site. The process in this case is a combination of two-way and softer handoff. The signals received by the co-site sectors are locally combined, with a rake receiver and based on maximum ratio combining, and one frame is sent to the MSC. The MSC performs the selection combining of this frame and the frame sent by the third sector.

In the forward link, the frame processing is the same for all scenarios.

3.7.2   Soft Handoff Process and Parameters

The soft handoff in IS95A is base station controlled and mobile assisted. The base station specifies the parameters to be used for soft handoff process as well as the list of pilot offsets used by the neighboring station.

As part of the mobile-assisted Handoff (MAHO) process, the mobile provides the serving base station with pilot E_c/I_o measurements from the strongest neighboring cells.

To properly detect and select the best candidate base station for handoff, a mobile station in the dedicated traffic channel mode groups the received pilot channels into four disjoint sets and updates these sets frequently. These sets, namely, the active, candidate, neighbor, and remaining sets, are described as follows:

- *Active Set:* This includes the pilot offsets of those base stations that are actively in communication with the mobile. Therefore, the active set points to pilots of sectors from which traffic channels are being demodulated.

- *Candidate Set:* This is the set of pilots that are strong enough to be considered for soft handoff but are not being demodulated

- *Neighbor Set:* This is the set of pilot offsets for all remaining neighboring sectors that are not included in the active or candidate sets. Because these pilots are used by neighbor stations, they may become candidates and therefore need to be monitored in every measurement. This set reflects the pilots originally specified by the serving base station in the neighbor list message.

- *Remaining Set:* This is the set of all possible pilot offsets in the current system and on the current CDMA Frequency Assignment, excluding the pilots in the neighbor, candidate, and active sets. These pilot offsets are integer multiples of PILOT_INCs, which is a parameter defined in the *neighbor list* and *neighbor list update messages.*

The following describe the rules for pilot transitions across sets:

Adding to Neighbor Set: Any pilot specified in the neighbor list message, from the serving base station, and not in the active or candidate set should be added to the neighbor set.

The mobile keeps an aging counter for each pilot in the neighbor set. This counter is initialized to zero when the pilot is moved from the active or candidate set to the neighbor set, and it is incremented whenever a neighbor list update message is received.

The pilot is moved from neighbor to the remaining set if the counter exceeds a threshold specified by the parameter NGHBR_MAX_AGE. The neighbor set can contain at least 20 pilots.

Transition Between Neighbor and Candidate Sets: A pilot with E_c/I_o above a *pilot-add threshold* T_ADD is considered a candidate. A pilot is removed from this set and placed in the neighbor set if the strength of that pilot drops below the *pilot-drop threshold* T_DROP for longer than the duration specified by the *handoff drop timer expiration* T_TDROP. The candidate set can contain at least six pilots.

Transition Between Candidate and Neighbor Sets: The decision about when and which base stations can be added to or removed from an active set is directly made by the base station and indicated to the mobile in a message. A pilot can only be added to the active set if the base station's handoff direction message (HDM) contains that particular pilot to be added to the active set. The pilot is removed from the active set and placed in the candidate set if the received HDM does not include

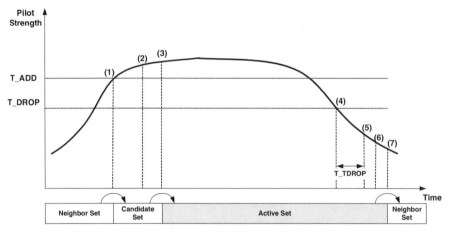

Figure 3.25   Soft handoff thresholds and their effect on pilot transition between sets.

that particular pilot. In this case, if T_TDROP for that pilot has not expired, the pilot is removed from the active set and placed in the candidate set; otherwise, it will move to the neighbor set.

An active set with only one pilot means the mobile is not in soft handoff, and an active set with three pilots means the mobile is in a three-way soft handoff. In a typical CDMA system the mobile with a three-finger rake receiver can demodulate signals from up to three base stations, but the active set can include up to six pilots. At each point the top three pilots from the list are chosen for demodulation.

Figure 3.25 and the following steps show the process of pilot transitions from neighbor set to candidate and active sets:

1. Pilot strength exceeds T_ADD. Mobile station sends pilot strength measurement message, as request to add to BS, and transfers pilot to candidate set.

2. Base station sends HDM or extended handoff direction message (EHDM).

3. Mobile station transfers pilot to active set and sends handoff completion message.

4. Pilot strength drops below T_DROP. Mobile station starts handoff drop timer.

5. Handoff drop timer expires. Mobile station sends pilot strength measurement message as request to remove to BS.

6. Base station sends HDM or EHDM.

7. Mobile station moves pilot from active set to neighbor set and sends handoff completion message.

Search Windows: In a typical fading channel different multipath components of pilot signals arrive at the mobile receiver within a few chips' delay following the direct-path component. The delay profile, namely, the relative magnitude and time offset, of multipath components also changes as the mobile moves. To capture these

TABLE 3.3   Supported Search Window Sizes

Search Window Setting	Window Size (chips)	Search Window Setting	Window Size (chips)
0	4	8	60
1	6	9	80
2	8	10	100
3	10	11	130
4	14	12	160
5	20	13	226
6	28	14	320
7	40	15	452

multipath signals the mobile station tracks the received pilots within a search window specified by the base station.

The size of the search window, typically expressed in number of chips, needs to be optimized based on cell sizes and expected delay spread in the propagation environment. This window size has to be wide enough to ensure that all usable multipath components of the signal are captured, but it needs to be as small as possible to allow the fastest pilot search.

The EIA-TIA95 system can define different search windows for tracking pilots in the active/candidate, neighbor, or remaining sets. Specifically, the parameters SRCH_WIN_A, SRCH_WIN_N, SRCH_WIN_R define the search window size used to search for pilots in the active/candidate sets, pilots in the neighbor sets, and pilots in the remaining sets, respectively. These three parameters are sent to the mobile in the *system parameters message* and handoff direction message, and they can be set to any value in Table 3.3. Also note that in EIA-TIA95 the chip rate is 1.228 Mcps; therefore, the chip duration is $1/1.228\,\text{Mcps} = 0.814\,\mu\text{s}$.

The search window for the active and candidate sets is typically referenced to and centered on the earliest-arrived pilot multipath component (see Fig. 3.26). Therefore, the size of SRCH_WIN_A is typically set to be equal to or slightly larger than twice the expected delay spread measured in number of chips.

For example, consider a propagation environment in which the effective delay spread τ_d is about $8\,\mu\text{s}$. Therefore, in this case the search window can be set to $2 \times 8/0.814\,\mu\text{s}$ or about 20 chips, which, based on Table 3.3. corresponds to parameter SRCH_WIN_A = 5.

The search window for the neighbor set, SRCH_WIN_N, has to be wide enough to capture not only the usable multipath signals from the serving cell but also the usable signals from all neighboring cells. Once the mobile identifies a known pilot offset to be searched, the mobile's searcher subtracts the PN offset from the received pilot phase before searching. As a result, for each pilot in the neighbor and remaining sets, the mobile centers its search window around the target pilot's PN offset so that only the propagation delay needs to be captured by within the SRCH_WIN_N.

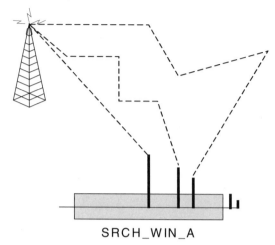

SRCH_WIN_A

Figure 3.26   Search window is set to capture usable multipath components.

The appropriate size of SRCH_WIN_N needs to be estimated on the basis of not only the delay spread but also the path delay difference between pilots arriving from the serving base station and its neighbors. Therefore, the size of SRCH_WIN_N is typically larger than that of SRCH_WIN_A.

SRCH_WIN_R for the remaining set is typically set to be equal to or larger than SRCH_WIN_N.

Considering our previous example for delay spread and assuming maximum distance d between cells to be about 10 km, we can estimate the SRCH_WIN_N. Note that the distance corresponding to 1 chip time in the CDMA system, calculated as 0.814 ms chip time multiplied by the speed of light (3×10^8 m/s), is about 244.14 m/chip. Therefore, the neighbor set search window can be as large as $2 \times (d/244.14\,\text{m})$ or about 41 chips, which, based on Table 3.3, corresponds to parameter SRCH_WIN_N set to 7.

3.8   REFERENCES

1. Mobile Station-Base Station Compatibility Standard for Dual-Mode Wideband Spread Spectrum Cellular System, TIA/EIA IS95-A, May 1995.
2. *IS-95 CDMA and cdma 2000: Cellular/PCS Systems Implementation*, Vijay K. Garg, Prentice Hall PTR, 1st edition, 1999.

CHAPTER *4*

NEW CONCEPTS AND TECHNOLOGIES IN CDMA2000

4.1 INTRODUCTION

The design of cdma2000 air interface and its evolution to high-rate packet data (HRPD) involves a number of new concepts and techniques, which together contribute to overall system performance efficiency. These techniques take advantage of higher affordability of complex integrated circuit designs and advances in coding and signal processing algorithms to create very flexible and adaptive physical layer structure.

The enhancements in the physical layer are combined with, and utilized by, sophisticated QoS-aware media and link access control protocols. The resulting air interface protocol stacks provide much higher spectrum and coverage efficiency than conventional circuit-switched voice-based systems, and they support a much wider range of multimedia and concurrent applications with mixed QoS requirements.

This chapter introduces some of the key component technology concepts utilized in the design of the cdma2000 family of standards. Our emphasis will be on the new physical and MAC layer features and techniques, many of which have also been used similarly in other 2.5/3G radio access technologies. We will discuss the following techniques:

- Link adaptation based on adaptive modulation, coding, and spreading
- Physical layer fast hybrid ARQ
- Enhanced channel coding and turbo codes
- Adaptive scheduling and multiuser diversity
- Space and antenna diversity techniques
- Enhanced vocoder with network-based rate control capability
- Fast forward link power control and coherent uplink demodulation

Each technique is presented generically here to provide an overall understanding of its concepts, and specific implementations of each technique within IS2000 and HRPD designs are discussed later in appropriate chapters.

CDMA2000® Evolution: System Concepts and Design Principles, by Kamran Etemad
ISBN: 0-471-46125-3   Copyright © 2004 John Wiley & Sons, Inc.

4.2 LINK ADAPTATION

The concept of link adaptation in general may encompass many techniques used in modern wireless communication systems. The basic idea is to dynamically predict or estimate the radio channel condition that is expected during a data transmission and use the most efficient combination of power, modulation, coding, and spreading rates supported by the system. In this context, maximum efficiency implies highest average throughput and lowest average latency or achieving maximum system capacity subject to acceptable quality of service (QoS) for all users.

In conventional voice-based 2G networks, the physical voice channels have fixed coding, spreading, modulation, and therefore data rates. Therefore, the process of link adaptation in 2G systems is limited to adaptive power control. In most 2.5G and 3G wireless technologies, however, in addition to the fixed-rate voice channels, the system also defines physical data channels with variable rates with or without power control, which are more suited for high-speed packet data services.

A variable-rate data channel is a physical channel that supports multiple combinations of coding, modulation, and spreading rate options. The data rate can be chosen based on some information about the radio link condition. The link adaptation typically requires some feedback from the receiver to help estimate the current channel condition and to select the best rate option possible.

The effectiveness of any link adaptation scheme depends on the accuracy of channel quality measurements and the freshness of the corresponding feedback provided to the transmitting node.

For forward link transmissions, the base station typically selects the best data rate based on the mobile station's measurement reports in the uplink. For uplink transmissions, however, the rate may be determined autonomously by the mobile station subject to the rule and limitation set by the base station or the mobile may wait for the base station to allocate the rate for each uplink transmission.

4.2.1 Adaptive Modulation and Coding

Generally the rate of a channel-coding scheme is defined as the ratio of the input bit rates to the encoder's output symbol rate, which is a number less than 1. Whereas a lower-rate channel coding provides higher data protection at the expense of more redundant bits and lower spectral efficiency, a higher-rate or lighter channel coding adds less redundancy and provides lower channel robustness.

Similarly, the order of a modulation scheme has a direct impact on the channel robustness and the spectrum efficiency. In a high-order modulation scheme each symbol represents more bits, resulting in more spectrum efficiency and requiring tighter symbol constellations. The tight constellation, however, limits the demodulation robustness to any channel impairment such as noise, fading, and interference.

Figure 4.1 shows the symbol constellations for BPSK, QPSK, 8PSK, and 16QAM, which are some of the main digital modulation schemes used in the design of modern mobile or fixed wireless access networks, including the cdma2000 family.

Comparing 16QAM and QPSK, for example, one can realize that with 16QAM each modulation symbol carries four channel bits and, therefore, it is two times more

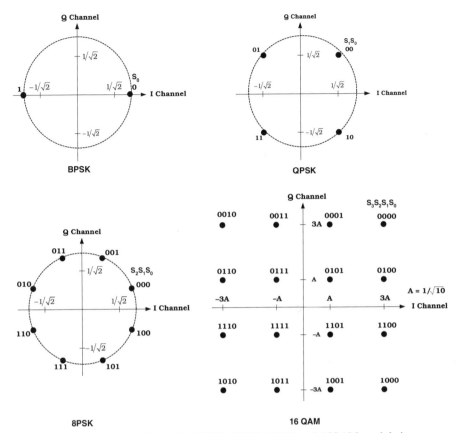

Figure 4.1   Symbol constellation for BPSK, QPSK, 8PSK, and 16QAM modulation schemes.

efficient than QPSK, which maps two channel bits to each modulation symbol. From a robustness perspective, however, the 16QAM modulation requires a clearer channel or higher C/I than a QPSK modulation for the same BER. This is because of the smaller distance between 16QAM symbols and thus the higher probability of their misdetection compared with QPSK. Therefore, 16QAM is less power- or coverage efficient but more spectrally efficient than QPSK.

Traditionally, the design of the physical layer for a wireless communications system has involved a trade-off between spectral efficiency for higher capacity and channel robustness for better coverage.

Specifically, in 2G mobile wireless technologies all physical traffic channels have the same channel coding, spreading, and modulation rate. This design inevitably has to be tailored to meet the channel requirements for the worst-case user in a cell. In 2G systems, the transmit power of the base station or mobile station is the only variable that can be dynamically controlled and changed based on channel

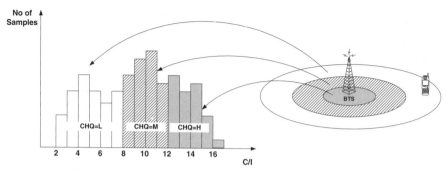

Figure 4.2   Channel gain or C/I variation in a cell.

conditions. In these systems, the link adaptation therefore is limited to adaptive power control.

In a typical cellular network, however, the C/I or the radio channel gains vary significantly because of path loss, fading, and interference across the cell area. Figure 4.2 schematically illustrates various channel conditions in a cell, which are categorized as high-, medium-, and low quality based on the receiver's measured C/I.

The users with lowest channel quality or smallest C/I are those far from the serving base station and/or those experiencing severe shadowing, multipath fading, or interference conditions. Providing sufficient QoS to these users would require a robust physical channel that is highly protected with strong channel coding and low-rate modulation. However, there are many other users with much better channel conditions who can benefit from lighter coding and higher modulation rates to receive higher throughput with fewer radio resources.

From this perspective, the 2G systems' approach of providing an equal channel protection to all traffic channels regardless of the users' radio environment is not efficient, as it penalizes or limits most users because of a small percentage of users with poor channel conditions. This inefficiency becomes more pronounced if one also aims at providing high-speed packet data services, which primarily have variable rates. A physical layer design with fixed coding and modulation structure, therefore, not only lowers the average user throughput but also lowers the system capacity.

In light of these observations, most mobile wireless technologies beyond 2G systems are designed with multiple choices of modulation, spreading, and channel coding schemes. With this flexible framework, the radio resource control entity can dynamically select the best coding and modulation scheme to be used for each transmission based on the most recent information about channel quality and system loading.

For each choice of coding and modulation scheme (CMS) the throughput is an increasing function of C/I. At low C/Is, the throughput is low because of the high frame erasure rate (FER) and the large number of retransmissions required. As C/I increases the FER is reduced and throughput improves because of the lower number of retransmissions. At some point further increase in C/I levels does not result in

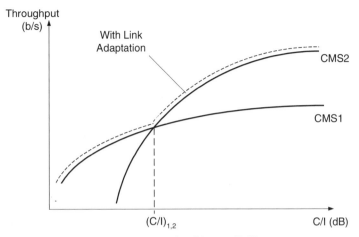

Figure 4.3   Link adaptation example with two CMSs.

significant increase in throughput as all transmitted frames are successfully received. At that point using the same CMS becomes inefficient and switching to a more "aggressive" CMS, that is, with lighter channel coding and/or higher modulation, can increase spectrum efficiency.

Figure 4.3 provides another graphic representation of the link adaptation concept for a system with only two options of coding and modulation, namely CMS1 and CMS2. Here, CMS1 represents a lower channel coding rate and/or lower-order modulation compared with CMS2. When link adaptation is used, the system switches between CMS1 and CMS2 according to the C/I values. When C/I is higher than $(C/I)_{1,2}$ the system uses the CMS2 scheme; otherwise, it uses CMS1. In other words, the system always uses the most "aggressive" modulation and coding that for a given C/I condition meets the FER or BER criterion. As a result, the achievable throughput as a function of C/I follows the dashed curve in Figure 4.3, which is the outer envelope to the C/I curves for all supported CMSs.

With this flexibility less compromise needs to be made between coverage and spectrum efficiency. At low-C/I conditions, for example, at the cell edge, where coverage is the main concern, a lower-rate but more robust CMS is used, but as soon as C/I condition allows the system switches to higher-rate but less protecting CMSs.

4.2.2   Variable Channelization Spreading

A CDMA-based air interface supporting variable-rate physical channels needs to have means of flexible spreading factors. Because the chip rate of the spread spectrum signal is fixed and limited by the carrier bandwidth, for example, to 1.228 Mcps for a 1.25-MHz cdma2000 carrier, low-data-rate physical channels will have more room for spreading than high-rate channels. The code channelization therefore needs to use spreading codes that not only are mutually orthogonal but also have different lengths. These codes can be defined and constructed according to a tree structure, using the same principles used to construct Walsh codes from Hadamard matrices described in Chapter 3 for cdmaONE systems.

Remember that Hadamard matrices can be generated by means of the following recursive equation:

$$H_1 = [0] \rightarrow H_2 = \begin{bmatrix} 0 & 0 \\ 0 & 1 \end{bmatrix} \rightarrow H_4 = \begin{bmatrix} 0 & 0 & 0 & 0 \\ 0 & 1 & 0 & 1 \\ 0 & 0 & 1 & 1 \\ 0 & 1 & 1 & 0 \end{bmatrix} \rightarrow H_{2N} = \begin{bmatrix} H_N & H_N \\ H_N & \overline{H_N} \end{bmatrix} \quad (4.1)$$

where N is a power of 2 and $\overline{H_N}$ denotes the binary complement of H_N.

In cdma2000 systems the target Walsh codes will have different lengths; therefore, two indices are needed to identify each code of each length. The Walsh function $W_{N,n}$ or W_n^N represents a Walsh function of length N that is serially constructed from the nth row of an $N \times N$ Hadamard matrix with the 0th row being Walsh function 0, the first row being Walsh function 1, etc.

Alternatively, the codes can be constructed directly on a binary tree according to the expansion rule shown in Figure 4.4. Each Walsh code $W_{2,1}^n$ of length 2^n at the nth level of the tree, which is considered as the parent code, can be split into two sibling codes of twice the length. Although the two siblings are orthogonal to each other, they are highly correlated with their parent code by construction. Figure 4.4 shows a three-layer expansion of the Walsh code tree as an example.

Obviously, not all the codes on the tree are orthogonal to each other. In fact, the tree representation is particularly useful for recognizing which set of codes can be selected from the tree to maintain mutual orthogonality of selected codes. Once a code is assigned to a physical channel in the system, none of the codes derived from that code, that is, those directly or indirectly spun off from that code, shall be used.

For example, one can use $\{W_{4,0}, W_{8,2},$ and $W_{2,1}\}$ as a valid set of orthogonal codes to be assigned at the same time. However, $\{W_{4,0}, W_{8,4},$ and $W_{2,1}\}$ is not an acceptable set because $W_{4,0}, W_{8,4}$ are not mutually orthogonal.

In the case of WCDMA, the same concept of variable-length Walsh code-based channelization is used, where it is referred to as orthogonal variable spreading factors (OVSF).

Comparing this variable spreading for channelization with the 64 Walsh codes used in the forward link of the EIA-TIA95 system, one can see the generalization of the same concept that brings both flexibility and complexity to the resource allocation process.

Also note that cdma2000 utilizes variable-length orthogonal spreading for channelization not only in the forward link but also in the reverse link. Unlike 2G CDMA mobiles, which transmit only one uplink channel, that is, traffic or access channel, at a given time, a cdma2000 mobile may transmit multiple uplink physical channels simultaneously. Therefore, to define and isolate the uplink physical channels, some of which support variable rates, the system uses variable-length Walsh spreading codes, sometimes referred to as Walsh covering, similar to the downlink.

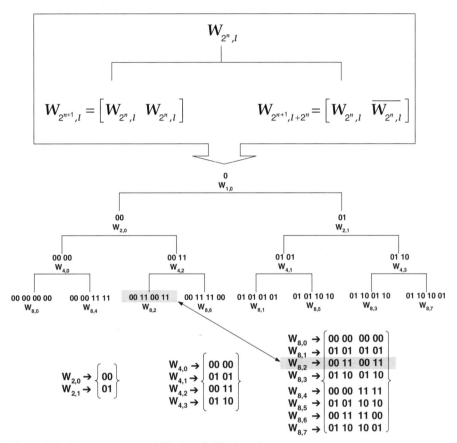

Figure 4.4   Tree structure variable-length Walsh codes.

4.2.3   Physical Layer Hybrid ARQ and Incremental Redundancy

The link adaptation can be effective only when channel quality measurements by the receiving node and the corresponding feedback to the transmitting node can be provided accurately and quickly.

In the basic automatic repeat request (ARQ) protocol, the receiver detects frame errors and requests retransmissions of the unsuccessfully received packets. Each retransmission increases the probability of success because of the inherent time diversity. In fact, ARQ is a powerful time diversity scheme that allows the system to operate at relatively high frame erasure rate (FER) because the occasional missing packets can be delivered as part of retransmission protocols.

There are three conventional ARQ schemes that can be associated to the packet buffering capacities at the transmitter and receiver. At the transmitter, a buffer can be used to store unacknowledged packets for possible retransmissions. At the receiver, a buffer can be used to correctly store and order those received packets, which cannot be released in the correct order until packets ahead in the sequence are successfully received.

- The *Stop and Wait* scheme corresponds to a transmitter capable of memorizing only one packet at a given time.

- In the *Go Back N* scheme, the transmitter is able to store packets. It can send several packets, but the receiver is not able to store packets. Therefore, if the leading packet is erroneous, the next packets will be dropped and need to be retransmitted, because there is no possibility of reordering them.

- The *selective repeat ARQ* scheme uses two buffers, the first one at the transmitter side, for possible retransmissions, and the second one at the receiver, for resequencing successfully received packets.

The selective repeat ARQ offers the best performance because it retransmits only those packets that are found to be corrupted and it efficiently utilizes the channel by avoiding unnecessary wait times.

The average number of retransmissions in an ARQ protocol depends on the FER, and it can be estimated as $1/(1 - FER)$. The higher the number of retransmissions, the lower the effective throughput and the higher the latency perceived by the application.

To provide increased throughput and lower latency in packet transmission, hybrid ARQ (HARQ) schemes are designed to combine ARQ protocols with forward error correction codes (FEC). In this approach, the average number of retransmissions is reduced by the FEC but each transmission carries some redundant symbols. In the HARQ scheme, although the FEC provides correction of the most likely errors and thus avoids the conventional ARQ scheme stalling, the ARQ process allows the link to operate at higher FER as occasional errors are corrected by retransmissions.

HARQ schemes may be classified as Type I, Type II, or Type III depending on the level of complexity employed in their implementation.

- *Type I Hybrid ARQ:* On a decoding error, this ARQ scheme discards erroneous packets and sends a retransmission request to the transmitter. The entire packet is retransmitted on receipt of the negative acknowledgement (NAK). The retransmitted packets may be combined with previous copies based on the weighted SNRs of individual bits or soft bit energy values before decoding, in which case the technique is termed chase combining.

- *Type II Hybrid ARQ:* In this ARQ scheme, retransmitted packets consist only of additional parity bits. The receiver keeps the first transmitted packet, which includes all information bits, and combines it with additional parity bits from retransmissions to form lower-rate and stronger codes. It then attempts to decode the combined codeword after each new set of parity bits are received. This approach is also called incremental redundancy (IR) and will be discussed further in this section.

- *Type III Hybrid ARQ:* In Type III ARQ schemes, individually transmitted packets are self-decodable and each packet differs in coded bits from the pre-

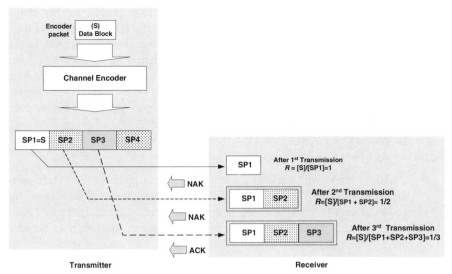

Figure 4.5   Hybrid ARQ Type II at the transmitter and the receiver.

vious transmission. In this scheme, packets are only combined after decoding has been attempted on the individual packet.

The HARQ protocols used in packet data channels in IS2000 and 1xEV-DO systems are based on selective repeat ARQ. The 1xEV-DO and 1xEV-DV systems also use different variations of HARQ Type II or incremental redundancy to further increase the transmission efficiency. The remainder of this section is therefore focused on describing the concepts of incremental redundancy along with its implementation as it relates to our technologies of interest.

Figure 4.5 shows the basic process of incremental redundancy (IR) with an example. The encoder at the transmitter side generates coded bits and groups them in four subpackets. The first subpacket (SP1) always includes the information bits with or without a few parity bits as well as a cyclic redundancy code that is used as the frame quality indicator for the encoder packet S. In principle, this first subpacket may be protected very lightly with a high-rate channel code. The other subpackets SP2, SP3, and SP4 include additional parity bits to provide additional redundancy and protection for information bits in S if needed. The receiver decodes SP1 and checks the CRC to see whether the packet is acceptable.

If the received subpacket is decoded successfully an ACK is sent to the transmitter indicating that the remaining subpackets of the current encoder packet can be discarded and the next encoder packet can be sent.

In the case where SP1 does not pass the CRC test, a NAK is sent, asking transmitter to send the second subpacket SP2, which provides additional redundancy, to decrease the code rate seen by the receiver. The decoder then appends the newly received SP2 to the previously received SP1 and attempts to decode the combined codeword.

Assuming that subpackets have the same number of bits, the combined code-word has twice the number of bits compared with the original data block and thus represents a 1/2-rate channel code. In most cases, this attempt is successful, but if it is not, the receiver sends another NAK indicating that more redundant bits are needed and causing SP3 to be transmitted. The decoder attempts to decode the combined [SP1-SP2-SP3] codeword that now represents a 1/3-rate channel code. Figure 4-5 shows a case in which three out of four subpackets were required for the receiver to decode the encoded data block.

The process of HARQ may be further illustrated by the three scenarios in Figure 4.6 as follows:

- Scenario 1 starts with encoder packet (i), which is received after three sub-packets SP1(i), SP2(i), and SP3(i) similar to the example described above. The encoder packet ($i + 1$), however, is successfully received after the first sub-packet is decoded and as a result, the transmitter moves on to the next encoder packet ($i + 2$) on the next transmission.

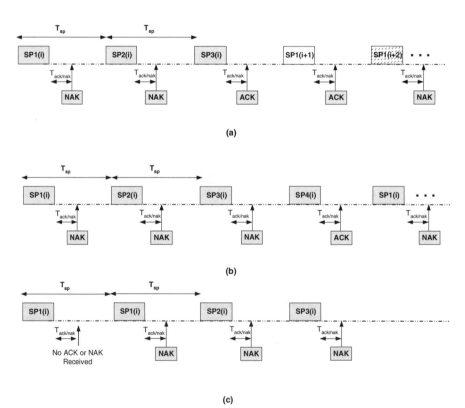

Figure 4.6 Incremental redundancy examples: (a) Scenario 1, (b) Scenario 2, and (c) Scenario 3.

- Scenario 2 shows a case in which all subpackets associated with the ith encoder packet are transmitted and the data block has not yet been received successfully. In this case the process repeats, starting with resending SP1(i).

- Scenario 3 illustrates that if the ACK or NAK is not received within the expected time frame $T_{\text{ack/nak}}$, the previously sent subpacket may be sent again.

The IR schemes take advantage of the fact that as the mobile radio channel changes in many instances of time, the radio link may not require strong channel coding. Therefore, as the protocol starts with sending a data block with no or very small redundancy and adding more redundancy only if needed, the effective average coding rate becomes higher. For example, if during a packet data session 50% of the time the first SP1 is successfully received, 30% of the time SP2 is needed, and 20% of the time SP2 and SP3 are needed, then the overall coding rate will be:

$$\overline{R} = 50\% \times 1 + 30\% \times 1/2 + 20\% \times 1/3 = 0.72 \qquad (4.2)$$

So although the effective code rate is close to 3/4 the system can still protect the data blocks in a weak channel condition with a strong 1/3-rate code.

Please note that the basic IR protocol, shown in Figure 4.6, is synchronous as the time intervals between retransmissions T_{sp} are the same. Once the first subpacket is transmitted, it locks the specific time slots over which future subpacket transmissions should occur.

An asynchronous variation of IR has also been proposed and adopted for high-speed packet channels. The asynchronous IR (AIR) takes advantage of radio channel changes by allowing the retransmission times to be flexible and to be chosen by the transmitter based on the radio link conditions. This form of AIR would typically require some feedback from the receiving terminal indicating measured conditions. When AIR is used in the forward link, one could think of it as a combination of multiuser diversity and incremental redundancy concepts.

This AIR scheme can be enhanced further by allowing link adaptation to be applied independently to each subpacket, that is, different subpackets of the same encoder packet can be transmitted at different spreading and modulation rates. The combination of adaptive and asynchronous IR (AAIR or A^2IR) can be a powerful scheme to maximize throughput and delay performance of a packet-based air interface.

The basic concept of AAIR is shown with an example in Figure 4.7. In this example, the first subpacket of encoder packet (i) is initially transmitted with QPSK modulation but based on feedback from the receiver the transmitter determines that the channel condition is good enough to allow for a higher-order modulation. So the

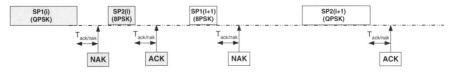

Figure 4.7   Adaptive and asynchronous incremental redundancy (AAIR example).

next subpacket SP2(i) is sent with 8PSK modulation. After packet (i) is received the next encoder packet transmission starts as SP1($i + 1$) with 8PSK modulation, but after the NAK is received the channel condition is determined not to be favorable so the transmitter waits for a relatively longer time before it sends the next subpacket with QPSK modulation.

The hybrid ARQ and IR techniques are primarily used for forward link scheduling on packet data channels; however, they are also being considered for reverse link transmissions.

4.3 MULTIUSER DIVERSITY

The concept of multiuser diversity is based on the following observations:

- In a wireless system serving a relatively large number of randomly distributed users with channel fading independently, there is likely to be a user with a very good channel at any time.

- The long-term system throughput can be maximized by serving the user(s) with the best channel condition(s) at the time of the transmission.

The challenge, however, is to exploit multiuser diversity while maintaining fairness and QoS for all users despite their radio link condition.

Multiuser diversity can be introduced by using the generic model shown in Figure 4.8. In this model the traffic data can be buffered at the transmitter side and sent to the target user based on a scheduling algorithm. There are multiple users with

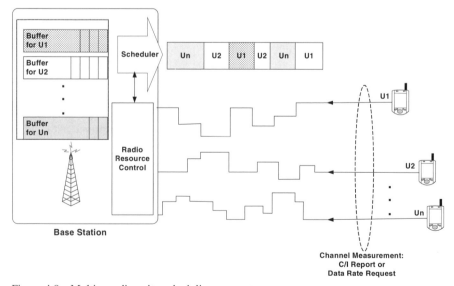

Figure 4.8   Multiuser diversity scheduling concept.

different and time-varying channel conditions. Each user periodically or continuously measures the channel gain or C/I and reports that as a channel feedback to the transmitter. In some systems such as HRPD the user terminals may estimate their achievable data rate and then send that as their requested data rate instead of sending their measured C/I.

The packet scheduler at the transmitter takes the channel feedback from all users into account in making decisions as to which user needs to be served when.

A number of different scheduling strategies can be used, each with a different implication for system throughput and fairness to users. Some of the basic scheduling schemes are described in the following:

- *Round-Robin (RR) Algorithm:* gives the same number of time slots to all the users in a round-robin or first come first serve basis, regardless of their channel conditions. This strategy achieves fairness by design but at the expense of lower throughput.

- *Best Rate (BR) or Best Effort Algorithm:* always transmits to users with the best channel condition requesting for highest data rate. The main idea of the best rate algorithm is to take advantage of multiuser diversity to maximize the sector throughput. This method, however, does not provide any type of fairness protection for users with relatively worse channel conditions, as they may wait for a very long time in the queue before they are served.

- *Proportional Fairness (PF) Algorithm:* attempts to balance a trade-off between the round-robin and best rate methods by using a mixed criterion. For each user, this criterion can be defined based on the ratio of requested data rate $R_{req}(i)$ to the user's average received data rate $R_{avg}(i)$ at the ith instance:

$$r(i) = \frac{R_{reg}(i)}{R_{avg}(i)}$$

$$R_{avg}(i) = (1 - \alpha) \times R_{avg}(i - 1) + \alpha \times R_{rec}(i) \tag{4.3}$$

where $R_{rec}(i)$ is the user's received data rate and it is set to zero for all users not served in the current cycle (i). Thus, as the scheduler keeps track of the moving average of each user's received data rate, it applies that as a fairness normalization factor to the requested data rate and uses this normalized requested rate $r(i)$ as the criterion for its scheduling. Before each transmission, the scheduler updates the value of scores $r(i)$ for all users with nonempty data buffers and transmits to the user with the highest score.

The proportional fairness score is a combination of good channel condition and longer queuing time for each user. The longer each user terminal waits in the scheduler's queue, the lower its average data rate $R_{avg}(i)$ would become, which in turn increases the score $r(i)$, and thus the user's chance of being served in the next transmission cycle.

Fairness and maximum packet delay can be controlled with the averaging time constant of $1/\alpha$. A small value of α puts less weight on fairness and results in higher throughput, whereas a large value of α gives higher weight to fairness and results in lower throughput efficiency.

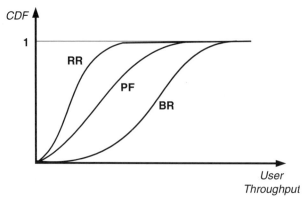

Figure 4.9 Rate distribution comparison of round-robin, proportional fair, and best rate schedulers.

In the short term, the scheduler will attempt to serve client terminals when they are experiencing better than "their average" channel conditions. However, over the long term, client terminals will get a roughly equal number of available time slots.

The average user throughput with proportional fair (PF) scheduler in most cases is lower than the best rate (BR) and higher than round-robin (RR). Whereas Figure 4.9 shows this comparison schematically for all three schedulers, the example in Figure 4.10 shows the different rate allocations for PF and RR derived from a 1xEV-DO systems simulation.

Note that these basic scheduling algorithms are described here as examples only, and the actual implementations of scheduler vary by implementations. There are even various ways of applying the proportional fairness criterion, by defining alternative rate-averaging equations and/or other means of imposing fairness on the scheduler's score calculation.

The multiuser diversity gain can be defined as the improvement in aggregate throughput or spectrum efficiency as a result of incorporating channel information in the scheduling. Figure 4.11 shows that the concept of multiuser diversity gain applies to fading channels, and it may result in performance levels better than those obtained in additive white Gaussian noise (AWGN) channels. The proportional fairness approach maximizes this gain subject to uniform user experience and fairness.

In general, the diversity gain increases with the number of users, but as the number of users becomes large the increase in diversity gain becomes less significant. Note that for a very large number of users the effective throughput and thus spectrum efficiency will decrease as result of resources needed by the overhead channels.

Although the concept of multiuser diversity is presented in this section in the context of forward link scheduling, similar ideas with different implementations may also be used to schedule uplink data transmissions.

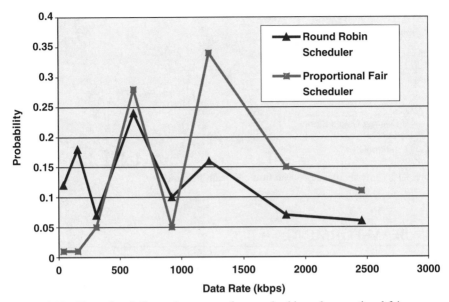

Figure 4.10   Example of allocated user rates for round-robin and proportional fair
scheduling.

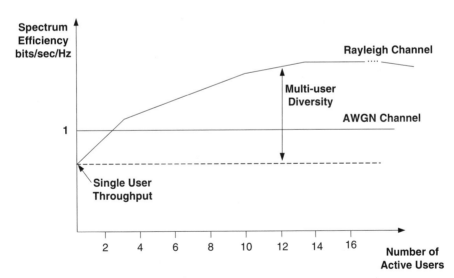

Figure 4.11   Multiuser diversity gain.

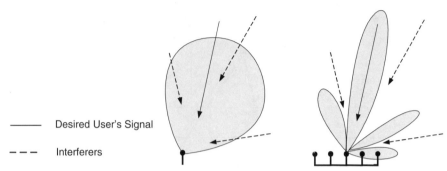

—————— Desired User's Signal

– – –   Interferers

Figure 4.12   Interference reduction through beam forming.

4.4   BEAM-FORMING

Any wireless access network technology would benefit from interference reduction techniques to improve its capacity and coverage performance. Beam forming is one of these techniques that relies on antenna arrays and spatial diversity (see Fig. 4.12).

An antenna array typically consists of several antenna elements each with its own radiation pattern. The combination of antenna elements can provide a composite radiation pattern. The shape of this composite pattern depends on the complex gains applied to the input signal as it feeds to each element of the antenna array.

The beam-forming algorithm estimates the angle of arrival for the desired signals as well as interferers and calculates the best set of complex gain factors that forms the desired antenna pattern. The desired pattern should maximize SINR by maximizing the antenna gain in the direction of desired users and minimizing or "nullifying" the gain in any other direction, particularly in directions from which signals from a main interferer may be received.

Many variations of beam-forming algorithms and smart antenna structures are discussed in the literature. In this section, however, our focus is on the type of beam forming used in cdma2000 systems, which can broadly be categorized as a switched-beam antenna scheme.

The switched-beam method is a simple beam-forming approach in which the beams and nulls of the antenna pattern can only be chosen from a limited number of positions and cannot be set to arbitrary directions. For example, in a switched-beam antenna system with four possible beam patterns the base station can identify one of four possible directional beams that best matches the direction of the mobile and use only that beam for any data transmission to or reception from the mobile.

In cdma2000 beam forming is an optional feature to increase the C/I experienced by each active user and therefore improve system capacity and achievable user throughput. If beam forming is used in cdma2000, the system needs to define a special type of pilot channel in addition to the common pilot channel used for system acquisition and handoff.

Remember that the principal operation of the coherent demodulation by the rake receiver is based on estimating the relative strength and timing of multipath

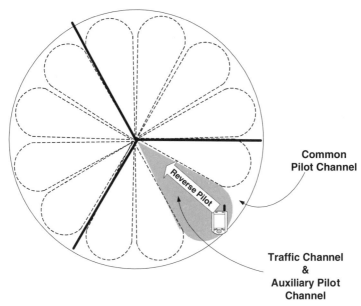

Common
Pilot Channel

Reverse Pilot

Traffic Channel
&
Auxiliary Pilot
Channel

Figure 4.13   Auxiliary pilot and switched beam antenna.

components of the pilot channel and applying those estimates to the demodulation of traffic channel. However, the common pilot channel transmitted continuously and uniformly across the cell will have a different multipath profile than a traffic channel that is beam formed toward a particular user or a portion of the cell. This difference of multipath profiles will not allow proper operation of the rake receiver and will cause performance degradation. Therefore, to benefit from beam forming, the CDMA system needs to define dedicated or semidedicated pilot channels that can be transmitted on the same antenna beam used for the traffic channel.

Some CDMA systems such as WCDMA have defined such dedicated forward link pilots for each user, and it is this pilot that is used by the mobile for coherent demodulation of traffic channels.

Of course, each dedicated pilot takes some of the power and code resources of the forward link, and with the cdma2000 channelization structure such an approach will not be feasible and efficient. Instead, the cdma2000 makes a compromise and allows optional use of limited number of auxiliary pilot channels, which are beam formed along with the traffic channels (see Fig. 4.13). Each switched beam can be considered as the "subsector" within a sector, and users contained within each subsector use the corresponding auxiliary pilot. Each sector can define up to four auxiliary pilots, one for each of four possible switched beams, or "subsectors."

The mobile's rake receiver will use the auxiliary pilots for its coherent demodulation of traffic channel, while it uses the main common pilot for system acquisition and handoff measurements.

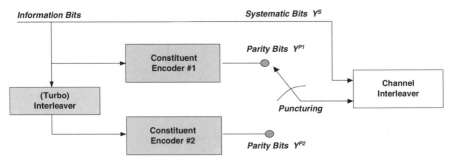

Figure 4.14   Turbo encoder.

4.5   TURBO CODES

Turbo codes, first introduced in 1993, are essentially parallel-concatenated convolutional codes with an internal interleaving mechanism combined with an iterative, soft decoding algorithm. These codes have shown near-Shannon capacity limit performance over additive white Gaussian noise (AWGN) channels, significantly outperforming conventional convolutional codes of similar decoding complexity [1,2].

The general turbo code encoder is shown in Figure 4.14. The turbo encoder employs two systematic recursive convolutional encoders connected in parallel, with an interleaver (the "turbo interleaver") preceding the second encoder. The term "systematic" implies that the encoder output always includes the original information bits plus some additional parity bits. The two recursive convolutional codes are called the constituent codes of the turbo code. The information bits are coded by both constituent encoders to generate one set of systematic bits and two sets of parity bits, one set out of each constituent encoder.

Whereas the first encoder operates on the input bits in their original order, the second encoder operates on the input bits as permuted by the turbo interleaver. As a result of interleaving the input to the second encoder appears to be a different sequence.

Based on this structure, the turbo encoder uses two identical encoders with almost independent inputs.

The interleaver design is an important part of a turbo encoder design. The turbo interleaver must be designed such that its output, which is input to the second encoder, appears as a random sequence independent from the original information sequence. As a result, if the output of one encoder produces a small-weight codeword, the second encoder that observes a "different" input sequence is likely to produce a codeword with a large weight. The effective number of codewords with small hamming weight is reduced, allowing the bit error rate (BER) curve to drop at a faster rate as the signal-to-noise ratio (SNR) increases. Thus the turbo coding utilizes a form of internally generated time diversity to statistically increase the minimum codeword distance and thus improve the combined coding gain.

A good choice of an interleaver is a random interleaver; however, other types of interleavers exist for small buffer sizes and offer comparable performance. In any

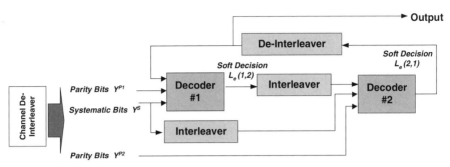

Figure 4.15   Iterative decoding for turbo code.

case, the interleaving pattern must be known at the receiver for a proper decoding process. Note that the turbo interleaver buffer size must also be based on the minimum encoder packet size and the maximum acceptable delay in the coding/ decoding process.

Based on the described turbo code structure, the information bits are always transmitted across the channel. However, depending on the desired code rate, the parity bits from the two constituent encoders may be punctured before transmission. For example, if the constituent encoders are of rate 1/2:

- For a rate 1/3 turbo code, all parity bits are transmitted along with information bits, that is, for every systematic bit X there are three outputs $Y^S = X$, Y^{P1}, and Y^{P2} (see Fig. 4.14).

- For a rate 1/2 turbo code, the parity bits from the constituent codes can be punctured alternately, that is, for every systematic bit X there are two outputs, $Y^S = X$ plus either Y^{P1} or Y^{P2}.

- For a rate 2/3 turbo code, the parity bits from the constituent codes can be punctured alternately and with one out of every four parity bits removed pseudo-randomly, For example, for every 2 information bits there are $4 - 1 = 3$ parity bits and the omitted parity bit is selected pseudo-randomly from Y^{P1} or Y^{P2} sets.

The puncturing pattern for each code rate needs to be known by the decoder for proper interpretation of received symbols.

For transmission over a fading channel, the coded bits from the turbo encoder are further interleaved by a channel interleaver, not to be confused with the internal turbo interleaver, before transmission.

Figure 4.15 gives a general block diagram for a turbo code decoder, which works based on soft-input/soft-output and iterative decoding algorithms. Although the details of decoding algorithms are subject to implementations and are not specified in a standard, the general idea is as follows:

- The received soft-decision (likelihood) information for the systematic and parity bits from the first constituent code, that is, Y^S and Y^{P1} estimates, is sent to the first decoder.

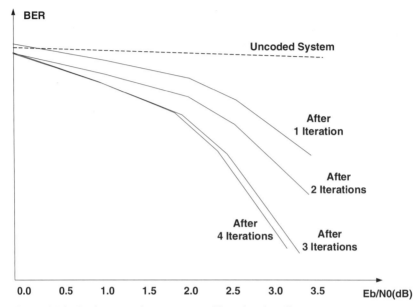

Figure 4.16   Performance improvement of iterative decoding.

- Decoder 1 generates updated extrinsic soft-decision likelihood values $L_e(1,2)$ for the information bits Y^S, which after reordering in accordance with the turbo interleaver are passed to Decoder 2 as a priori information.

- Decoder 2 accepts the received this interleaved likelihood information for the systematic bits Y^S along with the likelihood information for Y^{P2} corresponding to the parity bits from the second constituent encoder.

- Decoder 2 then generates its own extrinsic soft-decision likelihood values $L_e(2,1)$ as updated likelihood information for the systematic bits, which after de-interleaving is fed back to the first decoder.

- The process can be repeated as many times as desired, up to the point when additional iterations show insignificant changes to likelihood numbers.

For example, Figure 4.16 shows schematically the performance refinements of an iterative turbo decoder for 1, 2, 3, and 4 iterations, where the decoding process can stop after the third iteration without any performance impact. Hard decisions on the systematic information bits are then made after the last decoder iteration is completed. Typically, only few iterations are needed. The design of a turbo encoder includes the design of constituent encoders, turbo interleaver, and puncturing patterns for each desired coding rate. This design must take into account the appropriate channel models, packet sizes, and delay and complexity constraints and thus needs to be optimized for each specific air interface.

The turbo code advantage over conventional codes depends on the channel model and data rates. Figure 4.17 "schematically" shows the general comparison for different data rates or block sizes. The results are based on an AWGN channel and

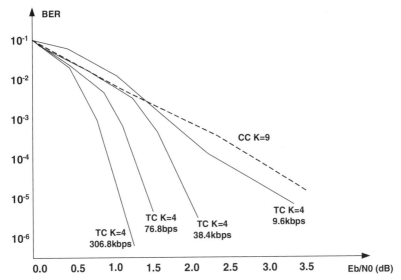

Figure 4.17   A sample turbo code performance comparison for different data rates.

20-ms frame sizes. Note that K in this figure shows the constraint length for the convolutional encoder or each of the two constituent encoders of the turbo code.

The lower the data rate, the smaller the turbo code advantage over convolutional codes and, with a fixed turbo interleaver size, the higher the associated interleaving delay. Therefore, turbo coding is beneficial mostly at higher data rates and for applications with lower sensitivity to delay, for example, high-speed non-real-time packet data application.

4.6   TRANSMIT DIVERSITY

Transmit diversity (TD) refers to a series of technologies considered and adopted for 3G systems to improve the link-level performance. TD introduces spatial diversity into the signal by transmitting through multiple antennas. The spacing of the antennas affects the degree of decorrelation between the "channels" created between the transmitter and the receiver. As in any diversity scheme, the independence of the created "channels" is crucial for achieving the diversity gain.

Typically, antenna spacing of the order of several carrier wavelengths, for example, 10λ to 15λ, leads to uncorrelated fading and results in the best performance. Although, in principle, multiple antennas for TD can be employed at the base station, mobile station, or both, it is usually most cost-effective and practical to employ multiple antenna structures at the base station.

Transmit diversity schemes can be categorized into open- and closed-loop methods.

The open-loop schemes create spatial diversity without utilizing any feedback from the receiver about current channel conditions. These methods do not require the overhead and complexity associated with the signaling and the channel feedback.

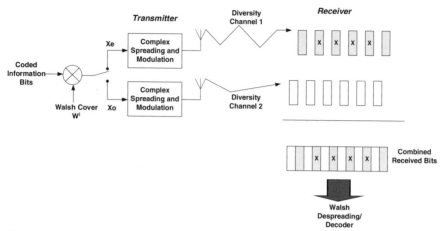

Figure 4.18   Time-switched transmit diversity.

The closed-loop TD schemes, however, utilize the channel feedback to further improve performance gain. These methods can also incorporate adaptive antenna gain adjustments. Our focus in this section is primarily on the open-loop schemes.

The basic transmit diversity concept can be introduced by using the simplest methods, such as *phase-switched transmit diversity* (PSTD) or *time-switched transmit diversity* (TSTD).

The PSTD introduces a known periodically varying phase difference between the symbols transmitted through different antennas to simulate fast fading, whereas in TSTD the transmission is periodically switched among the different antennas according to a known pattern and periodicity.

Figure 4.18 shows a basic two-branch TSTD structure in which the coded information bits after Walsh channelization are divided into even and odd bit positions and alternatively switched between the two diversity branches. Each of the resulting two-bit steams, namely, x_e and x_o corresponding to even and odd bits, has half the data rate of the original coded bit stream and each is carried by one branch only. Both antennas simultaneously transmit the symbols at a reduced power, so that the total power remains unchanged.

Assuming independent fading conditions for channel h1 and channel h2, the coded symbols received will be affected by the channels independently. Therefore, when one stream undergoes severe fading condition, the other may experience a good channel. Thus, as the receiver combines received bits from two antennas to complete a frame, it has a better chance of decoding the frame successfully compared with a non-TD-based system.

This section focuses on the basic concept of TD and the schemes used in cdma2000, which are all open-loop methods using multiple antennas at the base station to enhance the forward link performance. For more information on other TD techniques the reader is referred to [4].

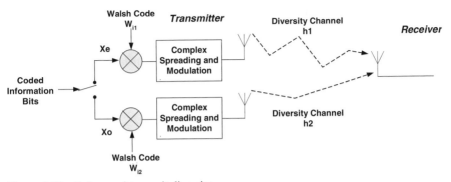

Figure 4.19   Orthogonal transmit diversity.

The two transmit diversity techniques adopted for cdma2000 are orthogonal transmit diversity (OTD) and space-time spreading (STS). They operate based on variations of the basic concept described above. However, they attempt to use some CDMA-specific features to improve the diversity gain.

Orthogonal Transmit Diversity (OTD): OTD is an open-loop TD method in which the interleaved coded symbols are split into even and odd symbol streams x_e and x_o and each of the two streams is spread with a different Walsh code before transmission (see Fig. 4.19).

The length of the Walsh code used in each branch is double the length of the Walsh code that would be used for spreading the original coded symbols when TSTD or no transmit diversity is used. Thus the total number of Walsh codes available is not reduced. In fact, the two Walsh codes W_1 and W_2 can be sibling codes of the half-length Walsh code as previously described.

Let x_e and x_o be the even and odd symbols and let s_1 and s_2 be the corresponding signals after Walsh spreading, respectively; then:

$$s_1 = x_e \cdot W_1$$
$$s_2 = x_o \cdot W_2 \tag{4.4}$$

Where W_1 and W_2 are the two orthogonal Walsh codes used. By applying different Walsh codes the even and odd symbols are effectively transmitted on orthogonal channels with better isolation than the TSTD scheme. Similar to TSTD, the transmit power required on each branch is lower because of lower data rates and higher spreading, thus maintaining the same required total transmit power as non-TD transmissions.

The signal r received at the mobile after demodulation and the x_e and x_o estimates can be calculated as:

$$r = h_1 \cdot s_1 + h_2 \cdot s_2 + n$$
$$\hat{x}_e = r \times W_1$$
$$\hat{x}_o = r \times W_2 \tag{4.5}$$

where h_1 and h_2 are functions representing the channels between the two transmit antennas and the receiver. The $\hat{x}_e$ and $\hat{x}_o$ estimates are obtained by de-spreading the received signal r with each of the two Walsh codes. The estimated even and odd symbols are then combined to form a complete frame, which is then de-interleaved and decoded.

Space-Time Spreading (STS): STS is an open-loop TD technique based on space-time coding theories [3,4] that has been adopted to fit within the cdma2000 framework.

The following briefly describes the structure of transmitted signals and symbol estimation at the receiver. Using the same terminology as in previous sections, one can present the symbols transmitted from the two antennas as:

$$s_1 = x_e \cdot W_1 - x_o^* \cdot W_2$$
$$s_2 = x_o \cdot W_1 + x_e^* \cdot W_2 \qquad (4.6)$$

Where $(\cdot)^*$ represents the conjugate operation. As one can see from these representations; similar to the OTD the transmitted symbols in STS scheme are first de-multiplexed into even and odd symbols and then spread by using two different Walsh codes W_1 and W_2. However, in this case, both even and odd bits contribute to the signals s_1 and s_2 transmitted from each antenna and both Walsh codes are used in a mixed spreading of symbols on each branch.

At the receiver, the original even and odd symbols are estimated by the following equations.

$$r = h_1 \cdot s_1 + h_2 \cdot s_2 + n$$
$$\hat{x}_e = r \times W_1 \times h_1^* + \left(r \times W_2 \times h_2^* \right)^*$$
$$\hat{x}_o = r \times W_1 \times h_2^* + \left(r \times W_2 \times h_1^* \right)^* \qquad (4.7)$$

Like other schemes, the estimated x_e and x_o are then concatenated as the input to the decoding process.

Note that in STS the orthogonality of two disjoint time epochs has been replaced by orthogonality in the spreading code domain. As a result, any symbol of the even and odd streams is exposed to two independent fading channels and therefore benefits from additional temporal and space diversity. The STS has been shown to outperform OTD in most channel conditions with a comparable complexity. However, both schemes are supported, with OTD being a mandatory feature for mobiles.

The details of specific OTD and STS implementation in cdma2000 are presented in Chapter 6.

4.7 NETWORK-CONTROLLED SELECTABLE MODE VOCODER (SMV)

Most CDMA systems use variable-rate speech coding algorithms in conjunction with power control to achieve high voice capacity. These variable rate vocoders exploit the characteristics of the source speech by classifying each voice frame according

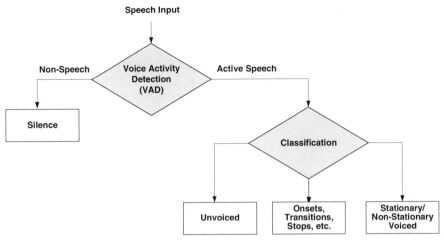

Figure 4.20   Voice activity detection and speech classification.

to its waveform type and energy and select one of multiple compression schemes to achieve a certain average data rate.

Figure 4.20 and Figure 4.21 show the concept of speech activity and the frame classification process. Although the active part of the speech signal is coded at the full or highest rate, the silence intervals are encoded at lower rates. On the other hand, the power control mechanism proportionally lowers the average transmit power for lower vocoder rates. Thus, during the speakers' pauses and silences that constitute approximately 60% of a normal two-way conversation, the user terminals' transmission power levels are reduced. As a result, the overall average interference level is lowered and capacity is increased.

In fact, the system capacity has a direct relationship with the average data rate (ADR) of a vocoder defined as:

$$\text{Average Data Rate:} \quad \text{ADR} = \frac{\sum_{i=1}^{N} n_i R_i}{\sum_{i=1}^{N} n_i} \qquad (4.8)$$

where R_i, $\{i = 1, \ldots N\}$, represent the N vocoder rates and n_i are the percentage of time or frames that rate R_i is used over a long measurement period.

The lower the ADR, the lower the average interference per user and thus the higher the capacity. Therefore, one of the main objectives in a vocoder design for CDMA-based systems is to reduce the ADRs subject to acceptable voice quality.

One of the features of new vocoders designed for 2.5G and 3G systems is that they have multiple modes or encoding options that can be controlled by the network. Each vocoder mode corresponds to a different rate of compression or ADR, and by selecting different modes the network can control the capacity vs. voice quality trade-off.

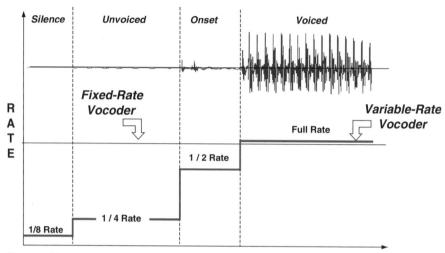

Figure 4.21   Basic variable-rate coding concept.

In the following we describe the two main vocoders used in cdmaONE and cdma2000 systems.

Enhanced Variable-Rate Vocoder (EVRC): The first speech codec designed for CDMA was an 8-kbps variable rate vocoder known by QCELP-8K or the TIA standard number TIA/EIA/IS-96C. This code-excited linear predictive (CELP)-based codec has four output encoding rates compliant with CDMA Rate Set 1 (RS1), which consists of full rate at 8.55 kbps; half-rate at 4 kbps, quarter-rate at 2 kbps, and eighth-rate at 0.8 kbps. This codec was standardized in 1993; however, despite its low ADR of 3.64 kbps, it was not widely deployed. The main reason was that the TIA/EIA/IS-96C codec did not provide a voice quality competitive with those offered by other technologies such as GSM and TIA/EIA/IS-136 (i.e., North American TDMA standard).

In an effort to improve the voice quality, the CDMA standardization group in 1995–1997 developed a new speech codec that employed a lower compression and higher data rates. The new operating rate set is called Rate Set 2 and has four encoding rates of full rate at 13.3 kbps, half-rate at 6.2 kbps, quarter-rate at 2.7 kbps, and eighth-rate at 1.0 kbps. This widely deployed enhanced 2G codec, called TIA/EIA/IS-733, addressed the voice quality issue and achieved quality comparable to that of G.726 at 32 kbps (considered to be toll quality per ITU-T). However, as the average bit rate for TIA/EIA/IS-733 is on the order of 6.6 kbps, system capacity was approximately 33% lower than for TIA/EIA/IS-96C.

Motivated by improved voice capacity, a third speech codec development effort was undertaken by TIA with the objective of achieving voice quality comparable to the TIA/EIA/IS-733 standard but using Rate Set 1. The standard that resulted was the enhanced variable-rate codec (EVRC) or TIA/EIA/IS-127, which operates with an average bit rate of 4.2 kbps [5].

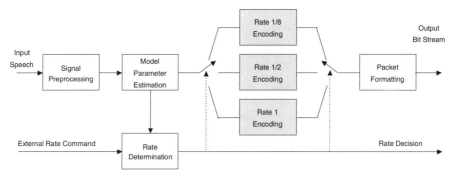

Figure 4.22   EVRC general encoding process.

The EVRC is based on the relaxation code-excited linear prediction (RCELP) paradigm, where, unlike conventional CELP coders, it encodes a preprocessed version of the speech signal that conforms to a linearly interpolated pitch contour (see Fig. 4.22). This pitch contour is obtained by estimating the pitch values at the analysis frame boundaries followed by linearly interpolating the pitch across frames. This approach relaxes the frequent pitch update constraint in low-rate CELP coders.

The encoder also includes a rate determination process that effectively classifies speech frames based on voice activities and their type, for example, voiced or unvoiced. EVRC also integrates a noise reduction module as well as a perceptual postfilter to further improve the perceived voice quality.

EVRC operates based on 20-ms frame sizes where each frame is further divided into shorter subframes. This feature allows the codec to effectively capture the rapid variation of pitch and excitation parameters, which often are not stationary over the extent of the analysis frame. Figure 4.23 shows bit allocations for filter coefficients, pitch, and codebook index for different rates.

The EVRC vocoder has been widely adopted across most CDMA networks because it offers the best combination of quality and capacity available. However, despite its good performance, TIA/EIA/IS-127/EVRC does not efficiently utilize Rate Set 1. The quarter-rate is not used at all, and the half-rate is used only for about 5% of the frames during active speech. Also, the rate determination algorithm (RDA) in TIA/EIA/IS-127 is not very flexible and does not provide the option of operating at various ADRs.

To address these issues, and also to add new capabilities, the CDMA standardization group started working on the next-generation CDMA codecs. These studies resulted in a new multimode network/source-controlled adaptive vocoder called the selectable mode vocoder (SMV or TIA/EIA/IS-893) [6].

Selectable Mode Vocoder (SMV): The SMV coder was mainly developed for replacing the EVRC in CDMA networks by providing improved quality and higher capacity in a multimode framework in which the trade-off between capacity and quality can be made flexibly by network operators.

The following are some of the key design requirements considered for selection of the best SMV codec from a number of proposed candidates:

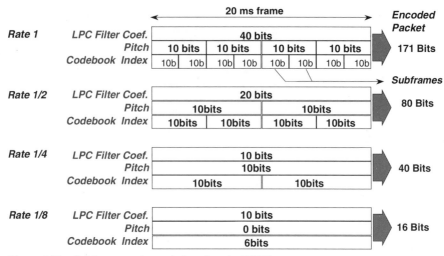

Figure 4.23 Subframes and encoded packets in EVRC.

- To operate within the constraints of the existing Rate Set 1 using a frame length compatible with the 20-ms frame structure of CDMA air interface
- To have computational complexity of 40 WMOPS (million weighted operations per second), compared to about 18 WMOPS used in EVRC
- To operate at a number of average bit rate set points, allowing each service provider to perform its own voice quality/system capacity trade-off
- To offer the choice of multiple modes with better quality at the current rates and the same quality at lower rates than offered by EVRC

The selected SMV vocoder algorithm is based on four encoding schemes operating at the rates of 8.55 kbps, 4.0 kbps, 2.0 kbps, and 0.8 kbps consistent with full-rate, half rate, quarter-rate, and eighth-rate of CDMA RS1, respectively.

The SMV standard as TIA/EIA/IS-893 or Service Option 56 was developed with reduced rate operation (encoding mode of operation) as a network control criterion. Unlike the rate reduction mechanism in EVRC, Service Option 56 does not use a deterministic pattern of Rate 1 and Rate 1/2 to reduce the average encoding rate. Instead, the SMV codec selects the encoding rates based on the characteristics of the input speech, namely, voiced, unvoiced, transient, or stationary, as well as the encoding mode selected.

The SMV algorithm has four network-controlled operating modes:

- Mode 0 or premium mode is designed to achieve better quality than EVRC at the same ADR.
- Mode 1 (standard mode) is designed to achieve similar quality as EVRC at a lower ADR.

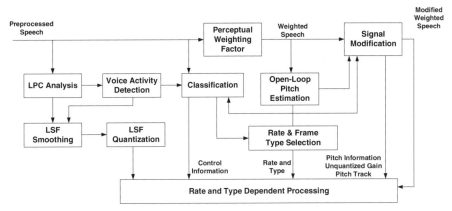

Figure 4.24   The SMV vocoder structure.

- Mode 2 (economy mode) is designed to achieve quality that is slightly worse than EVRC with a significantly lower ADR.
- Mode 3 (capacity-saving mode) is designed for even lower quality with higher capacity.

In addition, Mode 0 and Mode 1 can be configured to operate in a half-rate max (HRM) mode, where the maximum rate allowed is the Rate 1/2. These two modes are sometimes referred to as Modes 4 and 5, respectively.

Under the TIA/EIA/IS-893 standard, which defines the SMV specifications, SMV can instantaneously switch between the modes without any service interruption to allow a trade-off between average data rate (ADR) and speech quality.

Each 20-ms frame of input signal is sampled at the rate of 8 kHz and is processed by one of the four encoders according to a rate-determination algorithm (RDA).

The rate selection is based on the frame characteristics (voiced speech, unvoiced speech, background noise, stationary, etc.), and it is also externally controlled by the SMV operating mode.

The four codecs are based on the excitation-filter approach, in which a linear prediction (LPC) filter is excited with an "excitation vector." Figure 4.24 shows the general structure of an SMV vocoder. Each speech signal over the 20-ms frame is represented by the LPC filter parameters and the excitation parameters, which are transmitted from the encoder to the decoder.

The SMV also classifies each Rate 1 or Rate 1/2 frame as a Type 0 or Type 1 frame (see Fig. 4.25). Type 1 frames are stationary voiced speech, and frames of Type 0 are all other types of speech. The bit allocation for the excitation parameters and the LPC filter parameters differs between Type 1 frames and Type 0 frames.

Another improvement made in SMV is better representation of speech in various background noise environments as part of speech preprocessing with a unique noise reduction and noise classification mechanism. Because of better noise

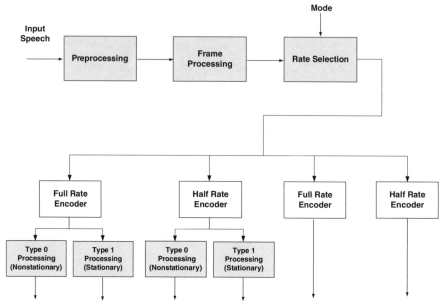

Figure 4.25   Frame classification in SMV.

TABLE 4.1   ADR for Different SMV Modes Assuming 40% Speech Activity

SMV Mode	Rate Distributions (%)				ADR (kbps)	Quality
	Full Rate 8.55 kbps	1/2 Rate 4.0 kbps	1/4 Rate 2.0 kbps	1/8 Rate 0.8 kbps		
Mode 0	40	5	0*	55	3.644	Better than EVRC
Mode 1	22	10	11	57	2.536	EVRC
Mode 2	9	23	11	57	1.951	Better than G.723.1 at 6.3
Mode 3	2	26	10	62	1.91	
Mode 4	0*	38	0*	62	2.02	
Mode 5	0*	31	0*	69	1.8	

0* Implies 0% usage by design.

modeling the SMV also shows significant performance improvements in noisy scenarios such as cars or in office usage environments.

Each mode of SMV uses some or all the rates in RS1, but the coding schemes and the rate selection are different. Table 4.1 shows the usage of different rates and the corresponding ADR for different modes based on a 40% speech activity factor. Figure 4.26 shows the ADRs and performance of SMV compared to the EVRC vocoder. The results show that:

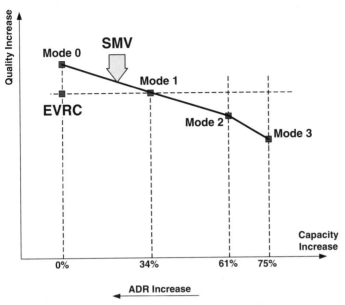

Figure 4.26   SMV capacity improvement over EVRC.

- Mode 0 (premium mode) achieves a statistically higher mean opinion score (MOS) than EVRC with a comparable ADR, that is, it provides better QoS at the same capacity.
- Mode 1 (standard mode) achieves an MOS statistically equivalent to EVRC with a lower ADR of 2.54 kbps, resulting in about 34% increase in the Erlang voice capacity.
- Mode 2 (economy mode) achieves an MOS better than that of G.723.1 (at 6.3 kbps) at an ADR of 1.95 kbps, increasing the effective capacity by 61% over EVRC at slightly lower quality.
- Mode 3 (capacity-saving mode) achieves an MOS statistically equivalent to G.723.1 (at 6.3 kbps) at an ADR of 1.91 kbps, providing about 75% capacity improvement at lower voice quality.

The half-rate modes, Mode 4 and Mode 5, are expected to have MOS performance comparable to G.723.1 (5.3 kbps) at an ADR of 2.02 kbps and 1.8 kbps. respectively. The performance for these modes has not yet been formally evaluated.

One can also see from Table 4.2 that the relative capacity increases, as a result of ADR reduction, are smaller in the reverse link than in the forward link. This is mainly due to the fact that the reverse link average transmit power is a combination of traffic and reverse pilot channel power, where the pilot power cannot be reduced at the lower ADRs.

Various simulations and estimations have suggested that cdma2000 system capacity will be forward link limited, especially when more and more data services

**TABLE 4.2   Forward and Reverse Link Capacity
Improvements with SMV**

SMV Modes	cdma2000 Forward Link	cdma2000 Reverse Link	Erlang-B System Capacity
Mode 0	0%	0%	0%
Mode 1	27%	16%	34%
Mode 2	49%	29%	61%
Mode 3	60%	35%	75%

are offered. Therefore, the higher relative capacity gains in the forward link as a result of SMV can balance some of the inherent imbalance of the cell capacity in forward and reverse links.

In a CDMA system, lower-rate voice bits are repeated and therefore have higher coding gain. This higher coding gain for lower ADR can be used to provide a coverage that is more robust or to increase the cell capacity over the short or long term.

At the edge of a cell where the coverage or link quality is poor, the system can switch the vocoder to a higher mode or a lower ADR to take advantage of a few decibels of coding gain obtained as a result of repetition to maintain a call that may otherwise be dropped.

Another example is for a system that is experiencing a high volume of voice traffic during a busy hour. In this case, the operator can increase the capacity of the high-traffic cells by reducing the ADR by using higher modes for some or all users. With this feature, the system can serve calls that would otherwise have been blocked.

4.8   REFERENCES

1. C. Berrou and A. Glavieux, "Near optimum error correcting coding and decoding: turbo-codes," *IEEE Trans. Commun.*, pp. 1261–1271, Oct. 1996.
2. "Principles of Turbo Codes and Their Application to Mobile Communications," Ushirokawa, Okamura, Kamiya, Vucetic, Invited Paper: IEICE Transaction Fundamentals, July 1998
3. V. Tarokh, H. Jafarkhani, and A. R. Calderbank, "Space-Time Block Codes from Orthogonal Designs," *IEEE Trans. Inform. Theory*, vol. 45, pp. 1456–1467, July 1999.
4. "Multi-antenna Transceiver Techniques for 3G and Beyond", Ari Hottinen, Olav Tirkkonen, Risto Wichman, John Wiley & Sons, 2003.
5. TIA/EIA/IS-127, Enhanced Variable Rate Codec, Speech Service Option 3 for Wideband Spread Spectrum Digital Systems, January 1997.
6. 3GPP2 C.S0030-0 V2.0, Selectable Mode Vocoder Service Option for Wideband Spread Spectrum Communication Systems, 2001.

CHAPTER *5*

CDMA2000 PROTOCOL LAYERS AND CHANNELIZATION

5.1 INTRODUCTION

The radio interface of cdma2000 is designed based on a layered structure that supports a generalized multimedia service model. The air interface channelization and protocols allow concurrent voice and data services along with a quality of service (QoS) control mechanism to balance the varying requirements of services and variation in radio link and system loading conditions.

The layering design generally follows the ISO/OSI reference model, in which Layer 1 is the physical layer and Layer 2 is the link layer, further subdivided into the link access control (LAC) sublayer and the medium access control (MAC) sublayer. Applications and upper-layer protocols corresponding to OSI Layers 3 through 7 utilize the services provided by the LAC, for example, signaling services, voice services, and data services (packet data and circuit data) [1].

As each data unit traverses the protocol stack, each protocol layer or sublayer only processes specific fields of the data unit that are associated with the sublayer-defined functionality. This section provides a short description of each layer of cdma2000 air interface followed by a more detailed presentation of radio and physical channelization.

5.2 PHYSICAL LAYER

The physical layer of cdma2000 air interface is designed to provide a flexible framework for supporting voice and other circuit-switched data as well as bursty packet data bearer services with different QoS requirements. A variety of physical traffic and control channels are defined with different coding, spreading, rate, and frame size options to efficiently carry signaling and user data traffic.

cdma2000 radio channelization is based on transmission masks used in EIA-TIA95A/B, which allows gradual multicarrier deployment of new "3G-CDMA" carriers next to a "2G-CDMA "carrier. The radio and code channelization is designed to allow backward compatibility and overlay of 3G and 2G systems deployed in the same spectrum.

CDMA2000® Evolution: System Concepts and Design Principles, by Kamran Etemad
ISBN: 0-471-46125-3   Copyright © 2004 John Wiley & Sons, Inc.

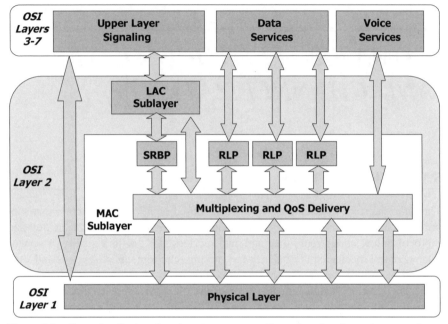

Figure 5.1   General radio interface layering structure. Reproduced under written permission from TIA.

The cdma2000 system supports RF channel bandwidths of SR × 1.25 MHz, where SR is the spreading rate number. Currently SR = 1 and 3 are specified, and SR can be extended to SR = 6, 9, 12.

Within each carrier the cdma2000 physical layer is designed based on spread spectrum and code division multiplex access. A number of fixed- and variable-rate physical channels are defined with new orthogonal variable-length spreading codes and PN codes similar to those used in EIA-TIA95A/B.

The data rates, channel encoding, and modulation parameters supported on the traffic channels are specified by radio configurations (RCs) [2]. In cdma2000 Releases 0, A and B, for spreading rates 1 and 3, there are seven RCs for the reverse link and nine RCs for the forward link. Releases C and D define additional RCs to support high-speed packet data channels, which will be discussed in a later chapter. Collectively, these RCs and their associated physical channels form the basic physical layer structure of cdma2000. Table 5-1 shows the range of data rates for all radio configurations in Release A.

Spreading Rate 1 supports RC1 to RC5 in the forward direction and RC1 to RC4 in the reverse direction. The remaining four RCs are supported in Spreading Rate 3. RC1 and RC2 are the same as Rate Set 1 and Rate Set 2 in IS-95, and they are supported for backward compatibility. Forward and reverse link RCs are chosen such that the FCH base rates for a forward and a reverse link match. For example, if RC3 is used on the reverse link, either RC3 or RC4 can be used on the forward

TABLE 5.1 cdma2000 Release A, Radio Configurations

SR	Forward Link RC	Data Rates (kbps) Base Rate, Max Rate	Reverse Link RC	Data Rates (bps) Base Rate, Max Rate
SR1	RC1	9.6	RC1	9.6
	RC2	14.4	RC2	14.4
	RC3	9.6, 153.6	RC3	9.6, 307.2
	RC4	9.6, 307.2		
	RC5	14.4, 230.4	RC4	14.4, 230.4
SR3	RC6	9.6, 307.2	RC5	9.6, 614.4
	RC7	9.6, 614.4		
	RC8	14.4, 460.8	RC6	14.4, 1036.8
	RC9	14.4, 1036.8		

TABLE 5.2 Reverse Physical Channels

	Channel Type	Max. No. SR1	Max. No. SR3
Common Channels	Reverse Pilot Channel (R-PICH)	1	1
	Reverse Access Channel (R-ACH)	1	NA
	Reverse Enhanced Access Channel (R-EACH)	1	1
	Reverse Common Control Channel (R-CCCH)	1	1
Dedicated Channels	Reverse Dedicated Control Channel (R-DCCH)	1	1
	Reverse Fundamental Channel (R-FCH)	1	1
	Reverse Supplemental Code Channel (RC 1 and 2 only) (R-SCCH)	7	NA
	Reverse Supplemental Channel (RC 3–6 only) (R-SCH)	2	2

link (see Table 5.1). More details on different radio configurations and channel data rates are provided in Chapter 6.

Within a radio channel cdma2000 defines a number of physical channels in forward and reverse directions. Table 5.2 and Table 5.3 show the naming convention used as well as the maximum number of defined physical channels in the reverse and forward links, respectively.

The physical channels are grouped into broadcast, common, and dedicated channels.

- Broadcast channels are used for traffic and signaling transmission in the forward link intended for all users in a cell. Examples of broadcast channels are all pilot channels and the synch channel in the forward link. There is no broadcast channel in the reverse link.

- Common channels are used as shared physical channels, but each transmission on these channels is addressed for a specific user or group of users. Common channels are typically not explicitly allocated. Examples of these

TABLE 5.3 Forward Physical Channels

	Channel Type	Max. No. SR1	Max. No. SR3
Broadcast Channels	Forward Pilot Channel (F-PICH)	1	1
	Forward Auxiliary Pilot Channel (F-APICH)	1	1
	Forward Transmit Diversity Pilot Channel (F-TDPICH)	Not Specified	
	Forward Aux. Trans. Diversity Pilot Channel (F-ATDPICH)	Not Specified	
	Forward Sync Channel (F-SYNCH)	1	1
	Forward Broadcast Channel (F-BCH)	8	8
Common Channels	Forward Paging Channel (F-PCH)	7	NA
	Forward Quick Paging Channel (F-QPCH)	3	3
	Forward Common Control Channel (F-CCCH)	4	4
	Forward Common Power Control Channel (F-CPCCH)	7	7
	Forward Common Assignment Channel (F-CACH)	7	7
Dedicated Channels	Forward Dedicated Control Channel (F-DCCH)	1*	1*
	Forward Fundamental Channel (F-FCH)	1*	1*
	Forward Suppl. Code Channel (F-SCCH) (RC 1 & 2 Only)	7	NA
	Forward Supplemental Channel (F-SCH)	2*	2*

* Per forward Traffic channel.

- channels are paging and forward common control channels in the forward link and access and enhanced access channels in the reverse link.

- Dedicated channels are used for point-to-point transmission, and they need to be allocated to a user on the forward or the reverse link before any data transmission. All traffic channels in the forward and reverse link are dedicated channels.

Note that the reverse pilot channel is considered as a dedicated or common channel when it is transmitted along with a reverse dedicated or a common channel, respectively.

5.3 LINK LAYER (LAYER 2)

The link layer consists of media access control (MAC) and link access control (LAC) sublayers [3,4].

5.3.1 MAC Sublayer

The cdma2000 MAC protocols are designed to allow multiple data service state machines, one for each packet or circuit-switched data application used in an active session. The MAC sublayer also performs multiplexing of data and signaling channels into physical channels and provides some QoS management for each active service.

The MAC sublayer provides the following important functions:

- Best effort delivery of data over the radio link, with a radio link protocol (RLP)
- Multiplexing logical to physical channels and QoS control: enforcing the negotiated QoS levels by managing conflicting and competing service requests coming from one or multiple users
- Providing and monitoring the rules governing access terminal transmission parameters such as power, timing, and rate selection
- Prioritization of access requests on all access channels
- Data rate control on both uplink and downlink traffic channels
- Logical channels are communication paths within the protocol layers on either the base station or the mobile station, and they are classified and labeled based on:
  - Direction: "f" is used for the forward and "r" for the reverse direction
  - Shared mode: "d" is used for dedicated when it is used for one user or "c" for common when it is used for multiple users
  - Information type: "t" is used for user traffic and "s" for signaling traffic
  Thus the following types of logical channel are defined:
  - f-csch/r-csch (forward and reverse common signaling channels, respectively).
  - f-dsch/r-dsch (forward and reverse dedicated signaling channels, respectively).

These logical channels are defined for the purposes of synchronization, broadcast, general signaling, access, and dedicated signaling. Multiple instances of the same logical channel may be deployed.

In principle, the information on a logical channel is carried on one or more physical channels, and multiple logical channels may share the same physical channel.

For example the f-csch carries forward common signaling information, which may be mapped to f-sych, f-pch/f-fcch, or f-bcch physical channels. The mapping between logical and physical channels is performed by the QoS/multiplex function within the MAC sublayer. The logical channel is mapped to the physical channel as shown in Table 5.4.

5.3.2 LAC Sublayer

The cdma2000 LAC sublayer provides reliable, over the air, transmission of signaling services. The LAC sublayer performs the following important functions:

- Delivery of service data units (SDUs) to Layer 3 entity using ARQ techniques to provide reliability.
- Building and validating protocol date units (PDUs) appropriate for carrying the SDUs.
- Segmentation and reassembly (SAR): for segmentation of encapsulated PDUs into LAC PDU fragments of sizes suitable for transfer by the MAC sublayer

TABLE 5.4 Logical to Physical Channel Mapping

Logical Channel	Physical Channel	Purpose
f-csch	F-SYNC	Synch Channel Messaging
	F-PCH/CCCH	Mobile-Directed Common Messaging
	F-BCCH	Broadcast Messaging
r-csch	R-ACH	2G Backward-Compatible, MS Access Messaging
	R-EACH	MS Enhanced Assess Messaging
	R-CCCH	MS Short Data Burst Messaging
f/r-dsch + f/r-dtch	F/R-FCH	L3 Signaling Messages & User Voice and Data Traffic
	F/R-DCCH	L3 Signaling Messages & User Voice and Data Traffic
f/r-dtch	F/R-SCH	User Higher-Speed Data Traffic

and reassembly of LAC PDU fragments into encapsulated PDUs (see Figure 5.2).

- Access control through "global challenge" authentication, such that messages that fail in authentication on a common channel are not delivered to the upper layers for processing.

- Address control for delivery of PDUs based on addresses that identify particular mobile stations.

The LAC sublayer consists of several sublayers each with a different functionality as follows:

Authentication Sublayer: Responsible for executing access control through "global challenge" authentication, message integrity validation, or both. In this regard, some of the Layer 3 functions in IS95B have been transferred to cdma2000 Layer 2. The authentication sublayer is only used on the reverse link common signaling channel (r-csch). For example, messages failing authentication or message integrity check on a common channel may not necessarily need to be delivered to the upper layers for processing. In some cases, the authentication needs to be completed before the data can be passed to higher layers. In other cases, authentication is processed in parallel with the passing of data to the higher layers.

ARQ Sublayer: Provides reliable delivery of SDUs to the Layer 3 peer entity using ARQ techniques, when needed. The ARQ protocols ensure reliability by using a peer-to-peer entity acknowledgment. The IS-2000 ARQ sublayer uses a selective repeat request approach to retransmit only the lost units. The receiving end is required to

- Store the correctly received PDUs
- Send acknowledgments only when requested by the sender
- Detect and discard duplicate packets

The ARQ sublayer of LAC provides two types of delivery services to Layer 3:

Assured Delivery Service where PDUs are repeated autonomously several times at fixed intervals until an acknowledgment from the LAC sublayer of the

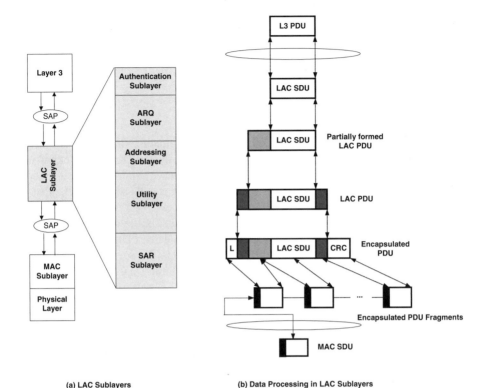

(a) LAC Sublayers   (b) Data Processing in LAC Sublayers

Figure 5.2   IS2000-A LAC sublayer. Reproduced under written permission from TIA.

receiving station is received by the LAC sublayer of the transmitting station. If no such acknowledgment is received after a specific number of retransmissions, the logical channel is dropped. In this mode, the Layer 3 may also request a *confirmation of delivery* from the LAC sublayer.

Unassured Delivery Service where the sent PDUs are not acknowledged by the LAC sublayer of the receiving station, and thus there is no guarantee that they were actually received. When requesting a PDU transfer in unassured mode, Layer 3 may also request the LAC sublayer to increase the probability of delivery of the PDU (for example, by having the PDU sent multiple times in rapid succession and relying on the duplicate detection capabilities of the receiver to achieve unique delivery).

The ARQ sublayer is used in both the forward and reverse links on common and dedicated signaling channels (f/r-csch and f/r-dsch) as well as dedicated media access control channels (f/r-dmch).

Addressing Sublayer: Responsible for address control, for example, adding and removing the addressing-related fields in the messages, to ensure delivery of PDUs based on addresses that identify particular mobile stations. The address control is based on individual mobile station addressing and broadcast addressing. The addressing sublayer is used only on the forward and reverse common signaling channels, and it supports the following address types:

- International mobile subscriber identifier (IMSI)
- Electronic serial number (ESN)
- A combination of IMSI or short IMSI (IMSI_S) and ESN
- Temporary mobile subscriber identifier (TMSI)

Utility Sublayer: Responsible for assembling and validating the PDUs that are appropriate for carrying the SDUs and also discarding those PDUs with unknown MSG_TYPE. Encryption parameters are included whenever required. This sublayer is used on the forward and reverse common signaling channels as well as the dedicated signaling or dedicated MAC channels.

Segmentation and Reassembly (SAR) Sublayer: On the transmit side; the SAR sublayer is responsible for segmentation of encapsulated PDUs into fragments suitable for transfer by the MAC sublayer. It also adds the segment length and number and a CRC field to each segment. There are no limits on the number of segments for an encapsulated PDU. On the receive side, SAR performs reassembly of encapsulated PDU fragments to form the original PDUs.

Note that not all sublayers of LAC need to be implemented for every channel type. Each sublayer function may be enabled or disabled with control plane functions. For example, no addressing or authentication is applied to the dedicated channels. When a sublayer is disabled, it acts as a pass-through or null sublayer.

The interfaces between different LAC sublayers are internal to the system and thus are not specified by the IS-2000 standards. The implementation of these interfaces may vary across vendors and may be different at the base station and the mobile station.

To provide flexible voice services, cdma2000 provides various frameworks to transport encoded voice data in the form of packet data or circuit data traffic. It can also transport voice directly on the physical channels in a manner that is backward compatible with the TIA/EIA-95-B, in which case the LAC and MAC services are null.

Within the data plane, Layer 3 and the LAC sublayer send and receive signaling information on logical channels, thus avoiding the need to be sensitive to the radio characteristics of the physical channels (see Fig. 5.3 and Fig. 5.4).

5.4   LAYER 3 SIGNALING

Layer 3 provides support for a wide range of radio interface signaling related to interoperability of MS and BS, including messages needed for call set up/release, mobility management as well as user identification and authentication. Layer 3 signaling defines parameters and procedures used by the mobile to transition among different call states. In general the mobile station call states include initialization, idle, access and control on the traffic channel states. In IS2000 layer 3 signaling also includes messaging needed to support packet data calls. The details of layer 3 messaging and procedures in these states are discussed in chapters 7 and 8. The layer

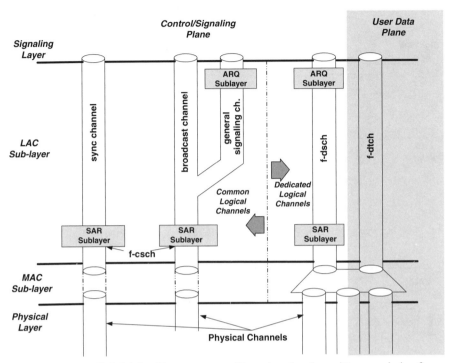

Figure 5.3   Downlink LAC sublayer structure. Reproduced under written permission from TIA.

3 signaling messages in IS2000 are designed to be backward-compatible to TIA/EIA-95-B while supporting upper layer signaling specific to the 3G mode of cdma2000.

In addition to supporting the conventional cellular radio access functions such as broadcast, paging, access and call maintenance, Layer 3 signaling also provides the following radio-related features and capabilities:

- Radio configuration negotiation
- Quick paging operation (to improve battery life)
- Handoff capabilities (i.e., soft handoff, hard handoff, idle and access state handoffs)
- High-speed packet data
- Enhanced access
- Broadcast control operation
- Auxiliary pilot support
- MAC state transitions

For more details on L3 signaling the reader is referred to [5].

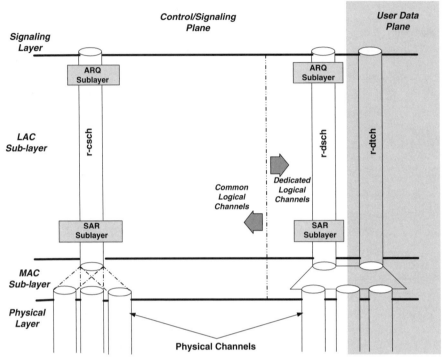

Figure 5.4   Uplink logical channel architecture. Reproduced under written permission from TIA.

5.5   CDMA2000 PHYSICAL CHANNELIZATION

This section describes the physical channels defined in cdma2000 Release A specifications. The overall channelization structure is first introduced, followed by the more detailed description of features and functionality for each physical channel in the reverse and forward links. The physical layer attributes of these channel including coding, spreading, and modulation are discussed in Chapter 6.

5.5.1   Reverse Link Physical Channels

Figure 5.5 shows the reverse physical channels, including both backward-compatible channel and new channels introduced in cdma2000 Release A specifications.

The **Reverse Pilot Channel (R-PICH)** is an unmodulated spread spectrum signal, transmitted by each mobile station, to assist the base station in detecting the corresponding uplink transmission. The R-PICH is a user-dedicated channel, sent before each uplink data transmission as a preamble and during the data transmission as a mean for coherent demodulation of data at the base station. More specifically,

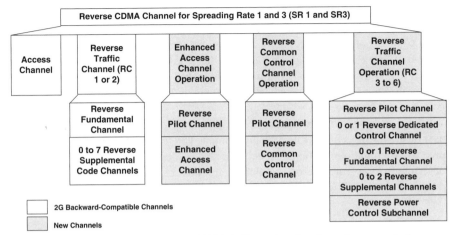

Figure 5.5   Reverse link channels in IS2000-A. Reproduced under written permission from Telecommunications Industry Association.

the R-PICH is sent before and during the R-EACH, R-CCCH, or any reverse traffic channel transmissions.

The **Reverse Power Control Subchannel** allows fast closed-loop power control on the forward link at 800-pbs rate. The mobile station inserts this subchannel into the reverse pilot channel when operating on the R-TCH with RC3 through RC6. Using this channel, the mobile can control the base station transmit power on the forward traffic channels. The reverse power control subchannel bits are time multiplexed with the R-PICH channel.

Although the R-PICH and the uplink coherent demodulation increase the reverse link cell capacity, the reverse power control subchannel allows the base station to more efficiently allocate its power resources and therefore increases the forward link cell capacity.

The **Reverse Access Channel (R-ACH)** is a TIA-EIA 95B backward-compatible or legacy access channel used for uplink common control signaling message exchanges such as call origination, registrations, and responses to pages. The functions and features of this legacy channel are described in Chapter 3.

The **Reverse Enhanced Access Channel (R-EACH)** is the new access channel used by the mobile station to initiate communication with or respond to base station, when no dedicated channel is assigned to the user. R-EACH can be used in the basic access or reservation access mode. In the basic mode, R-EACH is used for short MAC and signaling messages, including origination and page responses, to the base station. In the reservation mode, the R-EACH indicates the mobile's request for transmission of a short data burst on the RCCCH. The R-EACH is a random access channel and supports different data rates and frame sizes. This channel is defined for RC3 and higher.

The **Reverse Common Control Channel (R-CCCH)** is a reservation-based reverse common channel used for short data burst and signaling message transmissions from mobiles to the base station when reverse traffic channels are not in use.

The R-CCCH can be assigned much faster than a dedicated traffic channel; it is power controlled and supports soft handoff. Unlike R-EACH, the transmissions on R-CCCH are not contention based. Mobiles indicate their request for transmission on R-CCCH by sending a short frame on the R-EACH. One could argue that R-CCCH is neither a "common" channel nor a control channel because it is a channel mainly designed to carry user short data bursts and it has to be assigned to a MS before it can be used.

The **Reverse Traffic Channel** is a composite dedicated channel that carries user data as well as a control message channel. The traffic channels may be configured as combinations of fundamental channels or dedicated control channels with or without supplemental or supplemental code channels. These channels are described in the following:

> The **Reverse Fundamental Channel (R-FCH)** is the basic traffic channel that carries voice, low-speed data, and associated signaling messages. The FCH supports the same structure as the legacy reverse traffic channel, and it is backward compatible with IS95A/B. To ensure backward compatibility, FCH supports all transmission rates, frame structures, and signaling multiplex formats of the legacy traffic channel. FCH also supports shorter frames and messages, which are not backward compatible. The FCH can also be used along with supplemental channels or supplemental code channels to offer higher data rates. In this case, the FCH carries all the signaling and MAC-related messages as well as some or all of the retransmitted packets.

> The **Reverse Dedicated Control Channel (R-DCCH)** is a new dedicated control channel that has almost the same set of functions as the FCH defined in addition to the FCH by the cdma2000 specification. The R-DCCH maintains a dedicated link between mobiles and the base station during bursty packet data transmissions to carry power control, handoff, RLP messages, and other MAC and signaling messages. The R-DCCH, however, is not backward compatible with IS95, and although it can be used to support voice, the channel is not designed to efficiently support CDMA vocoder output. A high-speed data channel can be configured as R-FCH + R-SCH or R-DCCH + R-SCH. In some documents, including this book, the term "fundicated channel" is used as a name for either fundamental or dedicated control channels.

> The **Reverse Supplemental Channel (R-SCH)** is a variable-rate packet data channel carrying only high-speed coded information bits. No signaling or control messaging is exchanged over a supplemental channel. Therefore, each R-SCH relies on, and must be accompanied by, a fundamental or a dedicated control channel. Like any dedicated channel, an R-SCH needs to be assigned to a user and transmissions need to be scheduled based on users' packet data buffer size and QoS requirements. The data rate on R-SCH can be changed dynamically by changing the coding and spreading rates.

> The **Reverse Supplemental Code Channel (R-SCCH)** is a fixed-rate data-only channel that can be added to the reverse fundamental channel in IS95B

or RC1 and RC2 of cdma2000 to provide higher transmission rates. Each R-SCCH carries only higher-level user data and has a fixed data rate equal to its associated full-rate R-FCH, that is, 9.6 kbps or 14.4 kbps. Thus high data transmission rates are achieved by aggregation of multiple fixed-rate channels. Up to seven R-SCCHs can be combined with a reverse fundamental channel to provide higher data rate services. This channel was originally introduced in EIA-TIA95B, and it was called a supplemental channel. The cdma2000 system supports these channels only in the backward-compatible RC1 and RC2 modes and calls them "supplemental code channels," not to be confused with the new high-speed and variable-rate supplemental channels.

5.5.2 Forward Link Physical Channels

Figure 5.6 shows the forward physical channels, including both backward-compatible channel and new channels introduced in cdma2000 release A specifications.

The **Forward Pilot Channel (F-PICH)** is the common pilot channel broadcast by each cdma2000 base station based on the same structure and functionality as the legacy pilot channel in IS95A. The F-PICH is an unmodulated spread spectrum signal transmitted across each cell to provide mobile stations with power, timing, and phase references needed for cell selection, timing acquisitions, and handoff decisions. The F-PICH is the first channel searched and acquired by the mobile right after it is turned on. The measured E_c/I_o or signal-to-noise ratio of this channel is the basis for initial cell selection, and its phase and timing are used to achieve chip synchronization with the system. The E_c/I_o from the serving and neighboring base stations are also used by the mobile to evaluate the need for handoff and to identify the handoff candidates.

Optional Pilot Channels: CDMA2000 also supports the following optional pilot channels:

Transmit Diversity Pilot Channel (F-TDPICH): Used on a forward CDMA channel only when one of the transmit diversity schemes is used. In this case, the base station needs to maintain sufficient power on the forward pilot channel to ensure that mobile stations can properly acquire the forward CDMA channel without using energy from the transmit diversity pilot channel.

Auxiliary Pilot Channel (F-APICH): Used to allow intrasector antenna beam selection for more controlled and directional transmission power. The selective beam forming can reduce interference, increase C/I, and therefore improve cell coverage and capacity. Up to four auxiliary pilot channels can be defined and used in each sector of a base station, which correspond to 1–4 subsector beams. When this beam-forming feature is supported by a mobile and the base station, all dedicated traffic channels to the mobile should be transmitted only within the subsector beam defined by the specific auxiliary pilot channel that is in the direction of the mobile.

Auxiliary Transmit Diversity Pilot Channel (F-ATDPICH): When transmit

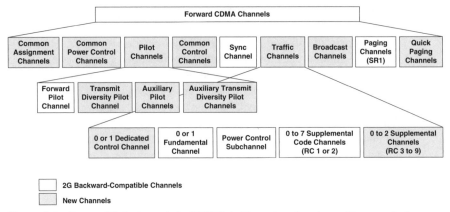

Figure 5.6   Forward link channels in IS2000-A. Reproduced under written permission from Telecommunications Industry Association.

diversity is used on the forward CDMA channel associated with an auxiliary pilot channel, the base station transmits an auxiliary transmit diversity pilot channel.

Forward Sync Channel (F-SYNCH): A backward-compatible broadcast channel transmitted by the base station to provide mobiles with timing information needed for system synchronization as well as network identification. The forward synch channel message is the first channel acquired by the mobile after pilot acquisition. Unlike the pilot, the F-SYNCH carries encoded messages with the same data rate and frame structure as the legacy I95A synch channel.

Forward Paging Channel (F-PCH): A backward-compatible paging channel carrying the forward paging and overhead messages similar to IS95A. It can also carry the new broadcast and common control messages required for cdma2000 operation. F-PCH has the same frame and slot structure as the IS95A paging channel, and it can support both IS95A mobiles and cdma2000 mobiles.

Unlike IS95A, where the paging channel serves as both a common paging and a broadcast channel, in cdma2000 two separate channels are defined and designed to carry broadcast and common channel messaging efficiently.

Forward Broadcast Control Channel (F-BCCH): The new broadcast channel used primarily for sending broadcast and multicast messages as well as overhead parameters. The F-BCCH is not used for paging, and it is typically used together with the forward common control channel.

Forward Quick Paging Channel (F-QPCH): A new channel comprised of a large number of single-bit paging indicators to notify mobiles that are being paged in the next paging channel slot. This channel is introduced with the objective of increasing mobile terminals' battery life by reducing the processing time needed for monitoring the paging messages. Each MS, based on its mobile identification number (MIN), hashes to and monitors a specific indicator symbol on the QPCH, and it would only attempt to decode the next paging channel message if its designated QPCH indicator shows an incoming page for the MS. Because monitoring a single-bit indicator on QPCH uses less energy than monitoring and decoding every paging channel message, the mobile saves its power usage during idle mode and increases its battery life.

Forward Common Control Channel (F-CCCH): Carries the base station's Layer 2 and 3 common channel messages including paging and may also be used to send mobile-terminated short data bursts to users who do not have any dedicated channel assigned to them. The F-CCCH structure is designed to improve packet data performance by reducing Layer 2 signaling latencies with shorter frames and higher data rates. The F-CCCH is not used for sending overhead messages, which are more efficiently transmitted on F-BCCH.

Therefore, the combination of F-BCCH and F-CCCH replaces the legacy paging or the backward-compatible F-PCH, and it provides a more effective and enhanced paging and broadcast messaging for 3G mobiles.

In initial phases of network migration from IS95A to cdma2000, the system may use F-PCH to support both 2G and 3G mobiles. However, as the network moves into an all-3G mode of operation F-PCH may be phased out and replaced by the F-BCCH/F-CCCH combination. In this case, some F-CCCH channels may be defined as the primary channel used for paging, while others are used for other messages such as short data bursts or acknowledgments.

Forward Common Power Control Channel (F-CPCCH): A time multiplex of a number of common power control subchannels, each associated with a reverse common control channel. The F-CPCCH allows power control of mobiles transmitting on R-CCCH as part of reservation access.

Forward Common Assignment Channel (F-CACH): Also a new forward link channel introduced to support enhanced access channel operation in the reverse link. The common assignment channel is specifically designed to provide fast-response reverse link channel assignments to support transmission of random access packets on the reverse link. The F-CACH provides a fast feedback mechanism including the R-CCCH/F-CPCCH allocation to the mobiles requesting access in the reservation access mode. There is a F-CACH/F-CPCCH pair associated with every R-EACH operating in the reservation access mode. F-CACH also implements congestion control. The base station may choose not to support common assignment channels and inform the mobile stations on the broadcast control channel of this choice.

Forward Traffic Channels: The forward physical traffic channel configurations are the same as the reverse link traffic channels. A forward traffic channel may be a combination of a fundamental or a dedicated control channel with or without supplemental or supplemental code channels.

Forward Fundamental Channel (F-FCH): A portion of the forward traffic channel that carries voice, low-rate data, and associated signaling messages. F-FCH support frame structure and messaging that are compatible with the legacy IS95A traffic channel. F-FCH can also carry new signaling messages needed for a 3G mode of operation. Each forward fundamental channel includes a power control subchannel to allow closed-loop fast power control of the reverse traffic channel. The power control bits are punctured into the traffic channel in a manner similar to the IS95A forward traffic channel.

Forward Dedicated Control Channel (F-DCCH): The new physical channel designed mainly to carry lower-rate data and associated signaling and MAC messages involved in data transmission. The DCCH can also be used for voice, but its design is not optimized to efficiently carry variable-rate traffic out of the CDMA vocoder. The functionality of DCCH, which is maintaining a continuous control messaging between the base station and active mobiles during bursty packet data sessions, is very similar to the fundamental channel. Therefore, a traffic channel configuration may include a FCH or a DCCH but not both.

Because many statements about fundamental and dedicated control channels are similar, the term "fundicated" is used when referring to either of these channels in such statements.

Forward Supplemental Channel (F-SCH): The high-speed data channel or "packet pipe" that only carries traffic data bits and is combined with a FCH or DCCH to provide a variable-rate and high-speed traffic channel.

Forward Supplemental Code Channel (F-SCCH): A EIA-TIA95B-compatible channel similar to the reverse supplemental code channel. F-SCCH has a fixed data rate matching the full rate of the corresponding fundamental channel. Because F-SCCH carries only traffic bits, it relies on, and needs to be used together with, a fundamental channel to carry the in-band signaling messages. Up to seven F-SCCHs can be added to a F-FCH, and all codes are aggregated and transmitted to the same user to provide a higher-rate bearer service.

5.6 REFERENCES

1. C.S0001-A, Introduction to cdma2000 Standards for Spread Spectrum Systems, July 2001.
2. C.S0002-A, Physical Layer Standard for cdma2000 Spread Spectrum Systems, July 2001.
3. C.S0003-A, Medium Access Control (MAC) Standard for cdma2000 Spread Spectrum Systems, July 2001.
4. C.S0004-A, Signaling Link Access Control (LAC) Standard for cdma2000 Spread Spectrum Systems, July 2001.
5. C.S0005-A, Upper Layer (Layer 3) Signaling Standard for cdma2000 Spread Spectrum Systems, July 2001.

CHAPTER 6

CDMA2000 PHYSICAL LAYER FEATURES

6.1 INTRODUCTION

The physical layer of cdma2000 is designed to provide a flexible combination of frame structures, channel coding rates, spreading factors, and modulation schemes to efficiently support voice and packet data in mobile radio channels.

This chapter describes some of the most important features of the cdma2000 physical layer for different spreading rates and radio configurations. The supported data rates and frame sizes for all traffic and control channels in the forward and reverse links are described, followed by more detailed specifications of their channel coding, channelization spreading, as well as quadrature spreading and modulation. The detailed specifications of IS2000 release A are documented by 3GPP2 in [1].

Besides spreading rate, cdma2000 channels can also have different rate sets or combinations of data rates supported for the traffic channel. These are captured in the so-called radio configurations (RCs) for different combinations of spreading rates and rate sets.

6.2 SPECTRUM SUPPORT AND RADIO CONFIGURATIONS

The radio channelization and transmission masks of cdma2000 are based on the IS95A system. The cdma2000 radio channels are designed to allow a multicarrier framework in which each carrier can be deployed in the same spectrum as the existing IS95A system or in a spectrum adjacent to IS95A with no need for guard bands.

Initial studies indicated that a multicarrier system ,while allowing gradual migration and deployment of carriers, provides almost the same spectrum efficiency as a direct-spread channel across the entire band. The key attributes of cdma2000 RF channelization are operational spectrum and spreading rates.

Spreading Rates: The total RF channel bandwidth of a multicarrier cdma2000 system is a multiple of 1.25 MHz. This multiple is defined as the spreading rate (SR). The cdma2000 specifications define two spreading rates.

CDMA2000® Evolution: System Concepts and Design Principles, by Kamran Etemad
ISBN: 0-471-46125-3 Copyright © 2004 John Wiley & Sons, Inc.

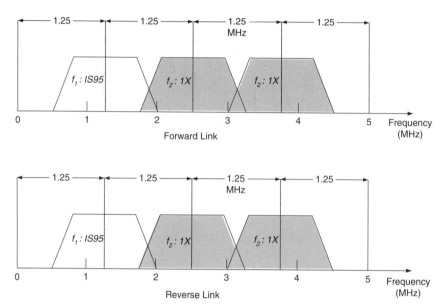

Figure 6.1 Mixed IS95 and cdma2000-1X carriers in the same band. Reproduced under written permission from Telecommunications Industry Association.

- The spreading rate 1 (SR1) or the "1X" system implies a cdma2000 radio channel that uses a single direct-spread carrier with a chip rate of 1.228 Mcps requiring 1.25 MHz of bandwidth (see Fig. 6.1).

- The spreading rate 3 (SR3) or the 3X system is a three-carrier system occupying three times the bandwidth of an IS95A channel or 3.75 MHz of spectrum.

The forward link in SR3 uses three direct-spread carriers each with a chip rate of 1.2288 Mcps. This multicarrier approach in the forward link facilitates the overlay of cdma2000 on 2G IS95A systems in the same or adjacent spectrum according to a gradual migration scheme. The reverse link, however, uses direct-spread spectrum signal over the entire 3X spectrum, that is, 3.75 MHz (see Fig. 6.2). This design provides lower complexity and results in better spreading gain in the reverse link. The effective data rate in the 3X system is three times that in the corresponding 1X system with the same coding and spreading schemes.

The minimum spectrum requirement for cdma2000 deployment is 1.25 MHz plus some guard band, which is usually half the channel size, namely, 0.625 MHz on each side. No guard band is needed between cdma2000 and/or IS95A carriers.

For example, an operator who has 5 MHz of spectrum can deploy three 1X cdma2000 carriers, two 1X carriers, and one IS95A carrier or one 3X cdma2000 carrier. These numbers assume a total of 1.25 MHz (= 2 × 0.625 MHz) for the two guard bands.

Band Classes: cdma2000 defines a large number of band classes reflecting different spectrum allocations around the world. In general, all existing cellular and PCS bands in 800/900 MHz and 1800/1900 MHz as well as all IMT2000 recom-

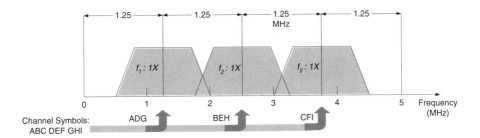

3X Multicarrier Forward Link

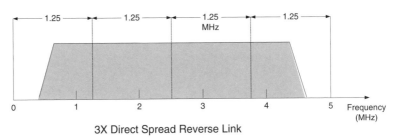

3X Direct Spread Reverse Link

Figure 6.2   3X cdma2000 radio forward and reverse channels. Reproduced under written permission from Telecommunications Industry Association.

mended spectra are supported. The following are supported band classes in Release A of IS-2000:

- Band Class 0 (800-MHz band)
- Band Class 1 (1900-MHz band)
- Band Class 2 (TACS band)
- Band Class 3 (JTACS band)
- Band Class 4 (Korean 1800-MHz PCS band)
- Band Class 5 (NMT 450 band)
- Band Class 6 (1.9/2.1-GHz band)
- Band Class 7 (700-MHz band)
- Band Class 8 (1800-MHz band)
- Band Class 9 (900-MHz band)

For details of each band class the reader is referred to [1].

6.3   RADIO CONFIGURATIONS AND TRAFFIC CHANNEL DATA RATES

The IS-2000A physical layer defines a number of RCs based on different sets of forward and reverse traffic channel formats. Each RC is characterized by physical

TABLE 6.1 Coding, Modulation, and Data Rate Summary for all IS2000-A/B Radio Configurations

SR		Forward Link			Reverse Link	
	RC	Coding/Modulation	Data Rates (kbps) Base Rate, Max Rate	RC	Coding/Modulation	Data Rates (kbps) Base Rate, Max Rate
	RC1	$R = 1/2$ Conv., BPSK prespreading symbols	9.6	RC1	$R = 1/3$, 64-ary orthogonal modulation	9.6
	RC2	$R = 1/2$ Conv., BPSK prespreading symbols	14.4	RC2	$R = 1/2$, 64-ary orthogonal modulation	14.4
SR1	RC3	$R = 1/4$ Conv. or Turbo, QPSK prespreading symbols, TD allowed	9.6, 153.6	RC3	$R = 1/4$ for rates <307.2 kbps, $R = 1/2$ for rates = 307.2 kbps BPSK modulation with a pilot	9.6, 307.2
	RC4	$R = 1/2$ Conv., QPSK prespreading symbols, TD allowed	9.6, 307.2			
	RC5	$R = 1/4$, QPSK prespreading symbols, TD allowed	14.4, 230.4	RC4	$R = 1/4$, BPSK modulation with a pilot	14.4, 230.4
SR3	RC6	$R = 1/6$, QPSK prespreading symbols	9.6, 307.2	RC5	$R = 1/4$ for rates <307.2 kbps $R = 1/3$, For rate = 307.2 & 614.4 kbps, BPSK modulation with a pilot	9.6, 614.4
	RC7	$R = 1/3$, QPSK prespreading symbols	9.6, 614.4			
	RC8	$R = 1/4$ (20 ms) or 1/3 (5 ms), QPSK prespreading symbols	14.4, 460.8	RC6	$R = 1/4$ for rates <1036.8 kbps $R = 1/2$ for 1036.8 kbps, BPSK modulation with a pilot	14.4, 1036.8
	RC9	$R = 1/2$ (20 ms) or 1/3 (5 ms), QPSK prespreading symbols	14.4, 1036.8			

layer parameters such as channel coding, spreading, and modulation schemes combined with a spreading rate, resulting in a set of transmission rate sets. Each transmission rate set includes the base rate of the FCH/DCCH and the associated set of data rates allowed on the SCH.

Nine radio configurations are defined in IS-2000A/B (see Table 6.1). Spreading Rate 1 supports five RCs, namely, RC1 to RC5, in the forward direction and

TABLE 6.2   Comparing RC3 and RC4 in the Forward Link

Data Rate (kbps)	Walsh Code Length	
	RC3, $R = 1/4$	RC4, $R = 1/2$
9.6	64	128
19.2	32	64
38.4	16	32
76.8	8	16
153.6	4	8
307.2	N/A	4

four RCs, namely, RC1 to RC4, in the reverse direction. The remaining four RCs are supported in Spreading Rate 3.

The forward and reverse RCs within a spreading rate are paired such that the fundamental channel base rates for a forward and a reverse link match. For example, in SR1, if RC3 is used on the reverse link, either RC3 or RC4 can be used on the forward link. Similarly, for SR3 either RC6 or RC7 in the forward link is paired with RC5 in the reverse link.

The first two RCs in the forward and reverse links, RC1 and RC2, which are exactly the same as Rate Set 1 and Rate Set 2 of EIA-TIA95A/B, are supported for backward compatibility.

Comparing RC3 and RC4 in the forward link, one can see that although both rate sets are based on 9.6 kbps, RC4 offers greater availability of Walsh codes at the expense of a higher coding rate and relatively higher transmission power per user. For example, at 153.6 kbps, RC3 uses Walsh codes of length 4 with 1/4-rate convolutional coding whereas RC4 uses Walsh codes of length 8 with 1/2-rate channel coding (see Table 6.2). Many of the initial cdma2000 deployments were based on RC3 and, in some cases, RC4.

A similar comparison can be made between RC6 and RC7 or RC8 and RC9 for SR3.

Table 6.3 and Table 6.4 provide more detailed views of the supported data rates for different reverse and forward traffic channels for SR1.

The fundamental channel is a backward-compatible traffic channel that supports variable but low rates according to the same structure as Rate Sets 1 and 2 in the traffic channel in EIA-TIA95A. These channels also support the 1/2, 1/4, and 1/8 rates to carry variable rate vocoder output and in-band associated signaling. In addition to these backward-compatible 20-ms frame structures, the fundamental channel in cdma2000 also supports 5-ms frames to allow faster signaling in support of data services.

Supplemental code channels only carry traffic data in fixed 20-ms frames. The aggregation of a fundamental and up to seven supplemental code channels allows data rates up to 8 times faster than legacy traffic channels, that is, 8×9.6 kbps or 8×14.4 kbps for RC1 and RC2, respectively.

The dedicated control channel introduced in cdma2000 also has a data rate of 9.6 or 14.4 kbps with 20- or 5-ms frames, but it is not a backward-compatible traffic

TABLE 6.3 Data Rates for Reverse Dedicated Channels for SR1

Channel Type	RC	Data Rates (kbps)
Reverse Dedicated Control Channel	RC3	9.6
	RC4	14.4 (20-ms frames) or 9.6 (5-ms frames)
Reverse Fundamental Channel	RC1	9.6, 4.8, 2.4, or 1.2 (20 ms)
	RC2	14.4, 7.2, 3.6, or 1.8 (20 ms)
	RC3	9.6, 4.8, 2.7, or 1.5 (20-ms frames) or 9.6 (5-ms frames)
	RC4	14.4, 7.2, 3.6, or 1.8 (20-ms frames) or 9.6 (5-ms frames)
Reverse Supplemental Code Channel	RC1	9.6 (20 ms)
	RC2	14.4 (20 ms)
Reverse Supplemental Channel	RC3	For 20-ms Frames: 307.2, 153.6, 76.8, 38.4, 19.2, 9.6, 4.8, 2.7, or 1.5
		For 40-ms Frames: 153.6, 76.8, 38.4, 19.2, 9.6, 4.8, 2.4, or 1.35
		For 80-ms Frames: 76.8, 38.4, 19.2, 9.6, 4.8, 2.4, or 1.2
	RC4	For 20-ms Frames: 230.4, 115.2, 57.6, 28.8, 14.4, 7.2, 3.6, or 1.8
		For 40-ms Frames: 115.2, 57.6, 28.8, 14.4, 7.2, 3.6, or 1.8
		For 80-ms Frames: 57.6, 28.8, 14.4, 7.2, 3.6, or 1.8

channel. Unlike FCH, the DCCH does not support factional rates and therefore is not suited for variable-rate vocoders. Both forward and reverse DCCH support discontinuous transmission. The decision to enable or disable the reverse dedicated control channel can be made on a frame-by-frame (i.e., 5 or 20 ms) basis.

The supplemental channel is the high-speed data channel that supports variable rates by using variable-length Walsh spreading and/or variable coding rates. The SCH transmissions only include user data in 20-, 40-, or 80-ms frames.

In RC1 and RC2, the basic traffic channel is bidirectional, formed as a pair of symmetric fundamental channels to support voice, signaling, and low-rate data. Higher data rates can be achieved in each link independently by aggregating up to seven supplemental code channels (SCCH) with FCH. The data rate of each SCCH in RC1 or RC2 is fixed and matches that of the corresponding FCH, that is, 9.6 kbps or 14.4 kbps, respectively.

In RC3 and above, for both links, each traffic channel may be a combination of a fundicated (i.e., DCCH or FCH) and a supplemental channel. The set of data rates on traffic channels for each radio configuration are multiples of the base 9.6-kbps or 14.4-kbps rate used in legacy traffic channels. The overall operation of traffic channel can be based on a fixed or variable data rate.

TABLE 6.4   Data Rates for Forward Dedicated Control and Traffic Channels in SR1

Channel Type	RC	Data Rates (kbps)
Forward Dedicated Control Channel	RC3 or 4	9.6
	RC5	14.4 (20-ms frames) or 9.6 (5-ms frames)
Forward Fundamental Channel	RC1	9.6, 4.8, 2.4, or 1.2 (20 ms)
	RC2	14.4, 7.2, 3.6, or 1.8 (20 ms)
	RC3 or 4	9.6, 4.8, 2.7, or 1.5 (2.0-ms frames) or 9.6 (5-ms frames)
	RC5	14.4, 7.2, 3.6, or 1.8 (20-ms frames) or 9.6 (5-ms frames)
Forward Supplemental Code Channel	RC1	9.6 (20 ms)
	RC2	14.4 (20 ms)
Forward Supplemental Channel	RC3	In 20-ms Frames: 153.6, 76.8, 38.4, 19.2, 9.6, 4.8, 2.7, or 1.5 In 40-ms Frames: 76.8, 38.4, 19.2, 9.6, 4.8, 2.4, or 1.35 In 80-ms Frames: 38.4, 19.2, 9.6, 4.8, 2.4, or 1.2
	RC4	In 20-ms Frames: 307.2, 153.6, 76.8, 38.4, 19.2, 9.6, 4.8, 2.7, or 1.5 In 40-ms Frames: 153.6, 76.8, 38.4, 19.2, 9.6, 4.8, 2.4, or 1.35 In 80-ms Frames: 76.8, 38.4, 19.2, 9.6, 4.8, 2.4, or 1.2
	RC5	In 20-ms Frames: 230.4, 115.2, 57.6, 28.8, 14.4, 7.2, 3.6, or 1.8 In 40-ms Frames: 115.2, 57.6, 28.8, 14.4, 7.2, 3.6, or 1.8 In 80-ms Frames: 57.6, 28.8, 14.4, 7.2, 3.6, or 1.8

The term fixed data rate implies that the data rate of the traffic channel does not change from frame to frame. In variable data rate mode the transmitter can change the data rate among a set of possible choices on a frame-by-frame basis.

The traffic channel with RC3 or higher also supports the flexible data rate mode. In this case the frame format, including the number of information bits, the number of reserved bits, and the number of frame quality indicator bits, is configurable. These frame formats correspond to a range of data rates up to the highest dedicated channel data rate. In this mode, the traffic channel can supports many intermediate rates not listed in Table 6.4.

In SR3, the fundamental or dedicated control channels have the same data rates as in SR1. The supplemental channels, however offer almost three times the data rate of SR1 for the same spreading and coding rates. Table 6.5 and Table 6.6 show the data rates of forward and reverse dedicated channels for different radio configurations with SR3.

TABLE 6.5 Date Rates for Reverse Dedicated Channels for SR3

Channel Type	RC	Data Rates (kbps)
Reverse Dedicated Control Channel	RC5	9.6
	RC6	14.4 (20-ms frames) or 9.6 (5-ms frames)
Reverse Fundamental Channel	RC5	9.6, 4.8, 2.7, or 1.5 (20-ms frames) or 9.6 (5-ms frames)
	RC6	14.4, 7.2, 3.6, or 1.8 (20-ms frames) or 9.6 (5-ms frames)
Reverse Supplemental Channel	RC5	For 20-ms Frames: 614.4, 307.2, 153.6, 76.8, 38.4, 19.2, 9.6, 4.8, 2.7, or 1.5 For 40-ms Frames: 307.2, 153.6, 76.8, 38.4, 19.2, 9.6, 4.8, 2.4, or 1.35 For 80-ms Frames: 153.6, 76.8, 38.4, 19.2, 9.6, 4.8, 2.4, or 1.2
	RC6	For 20-ms Frames: 1036.8, 460.8, 230.4, 115.2, 57.6, 28.8, 14.4, 7.2, 3.6, or 1.8 For 40-ms Frames: 518.4, 230.4, 115.2, 57.6, 28.8, 14.4, 7.2, 3.6, or 1.8 For 80-ms Frames: 259.2, 115.2, 57.6, 28.8, 14.4, 7.2, 3.6, or 1.8

6.4 DATA RATES ON REVERSE COMMON PHYSICAL CHANNELS

The data rates for reverse common physical channels operating with SR1 and SR3 are specified in Table 6.7. These data rates are the same for all radio configurations.

Although the legacy reverse access channel supports a fixed rate of 4.8 kbps with long access slots, the reverse enhanced access and reverse common control channels support multiple rates and frame sizes.

The enhanced access channel can be used in two possible modes: basic access mode and reservation access mode. The enhanced access channel header, which is transmitted on REACH only in the reservation access mode, has a fixed rate of 9.6 kbps. However, the message part of REACH can be transmitted at 9.6, 19.2 or 38.4 kbps in 5-, 10-, or 20-ms frames.

The frame sizes and data rates of reverse common control channels (R-CCCHs) are the same as those of the enhanced access channel, except that on R-CCCH no header is transmitted. R-CCCH can be used to transmit a short data burst at 9.6, 19.2, or 38.4 kbps in 5-, 10-, or 20-ms frame sizes.

The base station specifies the supported rate for REACH and RCCCH channels as part of channel configuration messaging. More details on the operation and structure of RACH, REACH, and RCCCH are provided in Chapter 7.

Note that the reverse pilot channel that is time multiplexed with the reverse power control channel is an unmodulated channel and does not carry any data. The rate and structure of the reverse power control channel are discussed in Chapter 8.

TABLE 6.6 Data Rates for Forward Dedicated Control and Traffic Channels in SR3

Channel Type	RC	Data Rates (kbps)
Forward Dedicated Control Channel	RC6 or 7 RC8 or 9	9.6 14.4 (20-ms frames) or 9.6 (5-ms frames)
Forward Fundamental Channel	RC6 or 7 RC8 or 9	9.6, 4.8, 2.7, or 1.5 (20-ms frames) or 9.6 (5-ms frames) 14.4, 7.2, 3.6, or 1.8 (20-ms frames) or 9.6 (5-ms frames)

Forward Supplemental Channel	RC6	For 20-ms Frames:
		  307.2, 153.6, 76.8, 38.4, 19.2, 9.6, 4.8, 2.7, or 1.5
		For 40-ms Frames:
		  153.6, 76.8, 38.4, 19.2, 9.6, 4.8, 2.4, or 1.35
		For 80-ms Frames:
		  76.8, 38.4, 19.2, 9.6, 4.8, 2.4, or 1.2

	RC7	For 20-ms Frames:
		  614.4, 307.2, 153.6, 76.8, 38.4, 19.2, 9.6, 4.8, 2.7, or 1.5
		For 40-ms Frames:
		  307.2, 153.6, 76.8, 38.4, 19.2, 9.6, 4.8, 2.4, or 1.35
		For 80-ms Frames:
		  153.6, 76.8, 38.4, 19.2, 9.6, 4.8, 2.4, or 1.2

	RC8	For 20-ms Frames:
		  460.8, 230.4, 115.2, 57.6, 28.8, 14.4, 7.2, 3.6, or 1.8
		For 40-ms Frames:
		  230.4, 115.2, 57.6, 28.8, 14.4, 7.2, 3.6, or 1.8
		For 80-ms Frames:
		  115.2, 57.6, 28.8, 14.4, 7.2, 3.6, or 1.8

	RC9	For 20-ms Frames:
		  1036.8, 460.8, 230.4, 115.2, 57.6, 28.8, 14.4, 7.2, 3.6, or 1.8
		For 40-ms Frames:
		  518.4, 230.4, 115.2, 57.6, 28.8, 14.4, 7.2, 3.6, or 1.8
		For 80-ms Frames:
		  259.2, 115.2, 57.6, 28.8, 14.4, 7.2, 3.6, or 1.8

TABLE 6.7 Data Rates of Reverse Common Channels for SR1 and SR3

Channel Type		Data Rates (bps)
Access Channel (SR1 Only)		4800
Enhanced Access Channel	Header	9600
	Data	38,400 (5-, 10-, or 20-ms frames), 19,200 (10- or 20-ms frames), or 9600 (20-ms frames)
Reverse Common Control Channel		38,400 (5-, 10-, or 20-ms frames), 19,200 (10- or 20-ms frames), or 9600 (20-ms frames)

TABLE 6.8 Data Rates for Forward Common and Broadcast Control Channels in SR1 and SR3

Channel Type	Data Rates (bps)
Sync Channel	1200
Paging Channel (for SR1 Only)	9600 or 4800
Broadcast Control Channel	19,200 (40-ms slots), 9600 (80-ms slots), or 4800 (160-ms slots)
Quick Paging Channel	4800 or 2400
Common Power Control Channel	19,200 (9600 bps per I and Q arm)
Common Assignment Channel	9600
Forward Common Control Channel	38,400 (5-, 10-, or 20-ms frames), 19,200 (10- or 20-ms frames), or 9600 (20-ms frames)

6.5 DATA RATES ON FORWARD COMMON PHYSICAL CHANNELS

The data rates for forward common physical channels operating with SR1 and SR3 are specified in Table 6.8. These data rates are the same for all radio configurations.

Note that the forward pilot channel, transmit diversity pilot channel, auxiliary pilot channels, and auxiliary transmit diversity pilot channels are not listed in Table 6.8 because they are unmodulated spread spectrum signals with no frame structure and they do not carry any data.

The forward pilot channel is transmitted at all times by the base station on each active forward CDMA channel, unless the base station is classified as a hopping pilot beacon. In that case, the hopping pilot beacons change frequency periodically to simulate multiple pilot beacons transmitting pilot information, resulting in discontinuous transmissions on a given forward CDMA Channel.

If transmit diversity is used, the base station uses a transmit diversity pilot. In this case the base station should continue to use sufficient power on the forward pilot channel to ensure that all mobile stations can acquire and estimate the forward CDMA Channel without using energy from the transmit diversity pilot channel. If transmit diversity is used associated with an auxiliary pilot channel, the base station transmits an auxiliary transmit diversity pilot.

The synch channel and paging channels are backward compatible and have the same coding, modulation, and frame structure as in EIA-TIA95.

The broadcast control channel transmits information at a data rate of 19,200, 9600, or 4800 bps, which correspond to slot durations of 40, 80, and 160 ms, respectively. The base station may support discontinuous transmission on the broadcast control channel, in which case the decision to enable or disable transmission is made on slot-by-slot basis.

The quick paging channel is an uncoded, spread, and on-off-keying (OOK) modulated spread spectrum signal, which consists of paging indicators, configuration change indicators, and broadcast indicators, structured in 80-ms slots. The quick paging channel has a fixed data rate of 4800 or 2400 bps. For more details on the QPCH structures and usage, see Chapter 7.

TABLE 6.9   Common Power Control Subchannels for Spreading Rate 1

Rate (bps)	Duration (ms)	Power Control Subchannels (N) per I and Q Arms	Power Control Subchannels ($2N$)
800	1.25	12	24
400	2.5	24	48
200	5.0	48	96

The common power control channel consists of several time-multiplexed sub-channels, each used to power control the transmissions on a reverse common control channel. There are $2N$ common power control subchannels (1 bit per subchannel) in one common power control group of the common power control channel. The sub-channels are divided equally between the I arm and the Q arm of the common power control channel. The data rate on each of the I or Q arm is 9600 bps.

In a 20-ms frame, there are 16 common power control groups for an 800-bps power control update rate, 8 common power control groups for a 400-bps power control update rate, and 4 common power control groups for a 200-bps power control update rate. The start of the first power control bit in the first common power control group aligns with the beginning of the 20-ms frame. Table 6.9 shows the number of common power control subchannels for power control update rates of 800, 400, and 200 bps per subchannel.

The base station transmits messages on the forward common assignment channel (F-CACH) at a fixed data rate of 9600 bps in 5-ms frames. The F-CACH supports discontinuous transmission, and the base station can make the decision to enable or disable the F-CACH on a frame-by-frame basis.

The forward common control channel is an encoded, interleaved, spread, and modulated spread spectrum signal by the base station to transmit mobile station-specific messages. The messages on F-CCCH can be transmitted at a variable data rate of 9600, 19,200, or 38,400 bps in 20-, 10-, or 5-ms frames, with only 6 combinations allowed as shown in Table 6.8. Note that these rates and frame sizes are the same as those supported in R-CCCH and R-EACH. Although the data rate of the forward common control channel is variable from frame to frame, the data rate transmitted to a mobile station in a given frame is predetermined and known to that mobile station. The transmissions on forward common control channels are structured based on F-CCCH slots that are each 80 ms in duration. For more details on the operation and messages of F-CCCH see Chapter 7.

6.6   REVERSE AND FORWARD LINK CHANNEL CODING

The channel coding in cdma2000 physical channels includes cyclic redundancy codes (CRC) used as frame quality indicators followed by convolutional or turbo encoding.

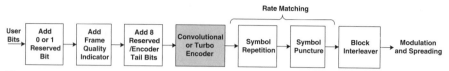

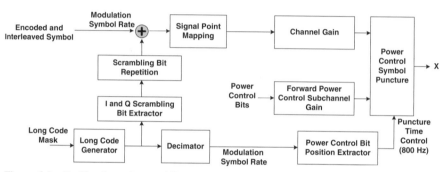

Figure 6.3 Overall channel coding structure for encoded channels. Reproduced under written permission from TIA.

Figure 6.4 Traffic channel scrambling, power control symbol puncturing in forward traffic channel. Reproduced under written permission from TIA.

Channel-encoded symbols are repeated and punctured to match a nominal value of data rate. The rate-matching process is applied to those channels, which support variable coding rate or flexible data rates.

A block interleaver is also used to mitigate the fading effects through randomization of errors in time. This general structure is shown in Figure 6.3, and it applies to all encoded physical channels in cdma2000.

In Release A of cdma2000, all reverse physical channels except for the reverse pilot channel are encoded and they follow the same general coding structure as described above.

The channel coding for encoded forward physical channels is also similar to the reverse link and consistent with the general structure shown in Figure 6.3. The exceptions are the sync channel and the paging channel, which do not use the frame quality indicator.

Block interleaving is used on the sync channel, the paging channels, the broadcast channel, the common assignment channel, the forward common control Channel, and all the forward traffic channels.

Note that in the forward link all the pilot channels, the quick paging channel, and the common power control channel are uncoded.

Note that the traffic channel processing also includes power control bit puncturing and scrambling. Figure 6.4 shows the scrambling, power control symbol puncturing, and demultiplexer structure for the forward traffic channels with RC3 through RC9. The channel-encoded and interleaved output symbols are scrambled by the long code.

The power control symbols are punctured on the forward fundamental channel and forward dedicated control channel only.

6.6.1 Convolutional Encoders

The cdma2000 forward and reverse link use convolutional encoders for all control channels and low-rate traffic channels. Convolutional encoding involves the modulo-2 addition of selected taps of a serially time-delayed data sequence. The length of the data sequence delay is equal to $K - 1$, where K is the constraint length of the code. All convolutional codes in cdma2000 have a constraint length of 9 and thus require 8 tail bits so that the convolutional encoder, on initialization, always starts with the all-zero state.

Figure 6.5 shows different convolutional encoders of 1/2, 1/3, and 1/4 rates used in cdma2000 reverse and forward links.

6.6.2 Turbo Encoder

For high-rate data transmission on supplemental channels turbo codes are used, which have shown to provide better coding gains than convolutional encoders at higher data rates. The basic concepts of turbo encoding and iterative decoding are discussed in Chapter 4.

The turbo encoder structure used in cdma2000 reverse link is shown in Figure 6.6. The turbo encoder employs two systematic, recursive, convolutional encoders connected in parallel, with a turbo interleaver preceding the second recursive convolutional encoder. Each of the two recursive convolutional encoders is called a constituent encoder of the turbo code.

Figure 6.7 describes the structure of each constituent encoder. The input to this encoder consists of the data, frame quality indicator (CRC), and two reserved bits. Each constituent encoder generates and adds its six tail bits. The system uses a common constituent encoder for all different code rates, and the outputs of constituent encoders are punctured and repeated to achieve the desired output symbol rate.

For the total number of N_{turbo} input bits the turbo encoder generates N_{turbo}/R encoded data output symbols followed by $6/R$ tail output symbols, where R can be a code rate of 1/2, 1/3, 1/4, or 1/5. The transfer function for each constituent code is

$$G(D) = \left[1 \quad \frac{n_0(D)}{d(D)} \quad \frac{n_1(D)}{d(D)} \right] \qquad (4.1)$$

Where $d(D) = 1 + D^2 + D^3$, $n_0(D) = 1 + D + D^3$, and $n_1(D) = 1 + D + D^2 + D^3$. Let X, Y_0, Y_1, X', Y_0', Y_1' denote the output of the encoder with the X output first. The puncturing patterns used for different coding rates are shown in Table 6.10. In this table in each puncturing pattern, a "0" or a "1" means that the symbol is deleted from or included in the output sequence, respectively. Also note that for each rate the puncturing table is read first from top to bottom and then from left to right and the entire pattern corresponds to one information bit period. For example for rate

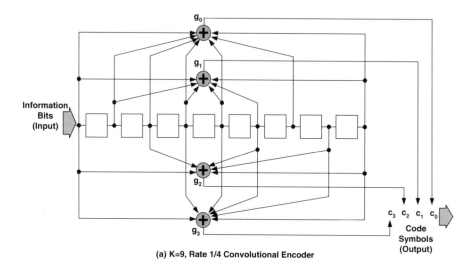

(a) K=9, Rate 1/4 Convolutional Encoder

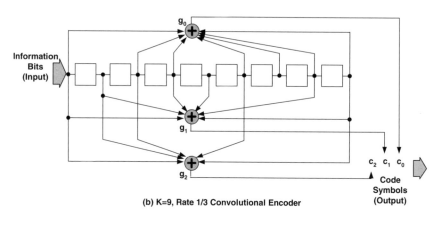

(b) K=9, Rate 1/3 Convolutional Encoder

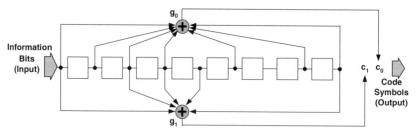

(c) K=9, Rate 1/2 Convolutional Encoder

Figure 6.5 Convolutional encoders in cdma2000 reverse link.

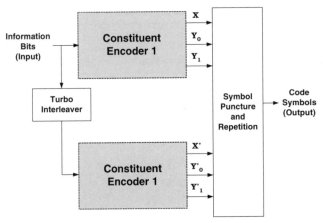

Figure 6.6   The structure of turbo encoder used in cdma2000.

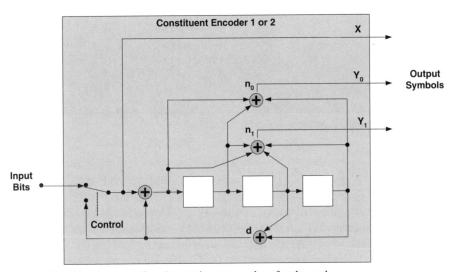

Figure 6.7   The structure of each constituent encoder of turbo code.

TABLE 6.10   Puncturing Patterns for Different Coding Rates

Output	Code Rate			
	1/2	1/3	1/4	1/5
X	11	11	11	11
Y_0	10	11	11	11
Y_1	00	00	10	11
X'	00	00	00	00
Y'_0	01	11	01	11
Y'_1	00	00	11	11

TABLE 6.11 Forward Error Correction in the Reverse Physical Channels

Channel Type	FEC	Coding Rate	
		Spreading Rate 1	Spreading Rate 3
Reverse Pilot Ch.	—	—	—
Access Ch.	Conv.	1/3	NA
Enhanced Access Ch.	Conv.	1/4	1/4
Reverse Common Control Ch.	Conv.	1/4	1/4
Reverse Dedicated Control Ch.	Conv.	1/4	1/4
Reverse Fundamental Ch.	Conv.	1/3 (RC1) 1/2 (RC2) 1/4 (RC3 and 4)	1/4
Reverse Supplemental Code Ch.	Conv.	1/3 (RC1) 1/2 (RC2)	NA
Reverse Supplemental Ch.	Conv.	1/4 (RC3, $N \leq 3048$) 1/2 (RC3, $N > 3048$) 1/4 (RC4)	1/4 (RC5, $N \leq 3048$) 1/3 (RC5, $N > 3048$) 1/4 (RC6, $N \leq 9192$) 1/2 (RC6, $N > 9192$)
	Turbo ($N \geq 360$)	1/4 (RC3, $N \leq 3048$) 1/2 (RC3, $N > 3048$) 1/3, 1/4, or 1/5 (RC4)	1/4 (RC5, $N \leq 3048$) 1/3 (RC5, $N > 3048$) 1/4 (RC6, $N \leq 9192$) 1/2 (RC6, $N > 9192$)

N is the number of channel bits per frame.

1/2 the symbols output by the turbo encoder for even-indexed and odd-indexed data bit periods is XY_0 and XY_0', respectively.

Note that regardless of coding rate, the original input bits X are always transmitted as indicated by "11" in the table, effectively resulting in a systematic code. Also, the information bits are not repeated, as indicated by "00" in the X' row.

The overall forward error correction (FEC) scheme, that is, the convolutional and turbo encoder structure used in the forward link, is the same as that used in the reverse link, with very minor differences. Table 6.11 and Table 6.12 show the channel coding type and rate for different reverse and forward link physical channels, respectively.

6.7 REVERSE LINK SPREADING AND MODULATION

The uplink-encoded bits out of the channel encoder and interleaver for each physical channel are spread with specific orthogonal channelization codes and subsequently combined and passed through quadrature spreading and modulation processes. This section provides more details of the spreading and modulation process in the reverse link.

TABLE 6.12 Forward Error Correction in the Forward Physical Channels

Channel Type	FEC	Coding Rate	
		SR1	SR3
Sync Ch.	Conv.	1/2	1/2
Paging Ch.	Conv.	1/2	N/A
Broadcast Control Ch.	Conv.	1/4 or 1/2	1/3
Quick Paging Ch.	None	—	—
Common Power Control Ch.	None	—	—
Common Assignment Ch.	Conv.	1/4 or 1/2	1/3
Forw. Common Control Ch.	Conv.	1/4 or 1/2	1/3
Forw. Dedicated Control Ch.	Conv.	1/4 (RC3 or 5) 1/2 (RC4)	1/6 (RC6); 1/3 (RC7); 1/4 (RC8, 20 ms), 1/3 (RC8, 5 ms); or 1/2 (RC9, 20 ms), 1/3 (RC9, 5 ms)
Forw. Fundamental Ch.	Conv.	1/2 (RC1, 2, or 4) 1/4 (RC3 or 5)	1/6 (RC6); 1/3 (RC7); 1/4 (RC8, 20 ms), 1/3 (RC8, 5 ms); or 1/2 (RC9, 20 ms), 1/3 (RC9, 5 ms)
Forw. Supplemental Code Ch.	Conv.	1/2 (RC1 or 2)	N/A
Forw. Supplemental Ch.	Conv.	1/2 (RC4) 1/4 (RC3 or 5)	1/6 (RC6)
	Turbo ($N \geq 360$)	1/4 (RC3) 1/2, 1/3, 1/4 or 1/5 (RC4) 1/3, 1/4, or 1/5 (RC5)	1/3 (RC7) 1/4 (RC8) 1/2 (RC9)

N is the number of channel bits per frame.

6.7.1 Reverse Link Orthogonal Channelization Spreading

In a 2G CDMA system the mobile transmits only one physical channel at a time because either an access channel or a reverse traffic channel is used depending on the mobile state. In the cdma2000 reverse link, however, the mobile transmits multiple physical channels simultaneously. To differentiate the multiple physical channels received from a mobile the system needs to use orthogonal codes for uplink channelization.

Therefore, when transmitting on a reverse pilot channel, the enhanced access channel, the reverse common control channel, or the reverse traffic channel with Radio Configurations 3 through 6, the mobile station uses orthogonal spreading. Table 6.13 specifies the Walsh functions applied to each reverse physical channel.

TABLE 6.13 Walsh Function Allocations for Reverse
Physical Channels

Channel Type	Walsh Function
Reverse Pilot Channel	W_0^{32}
Enhanced Access Channel	W_2^8
Reverse Common Control Channel	W_2^8
Reverse Dedicated Control Channel	W_8^{16}
Reverse Fundamental Channel	W_4^{16}
Reverse Supplemental Channel 1	W_1^2 or W_2^4
Reverse Supplemental Channel 2	W_2^4 or W_6^8

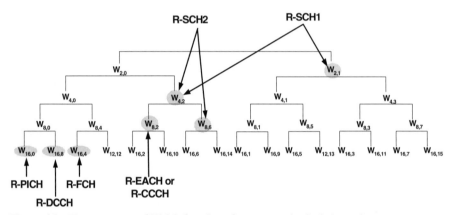

Figure 6.8   Tree structure of Walsh functions for reverse physical channels.

In this table the Walsh function W_n^N represents a Walsh function of length N that is serially constructed from the nth row of an $N \times N$ Hadamard matrix with the 0th row being Walsh function 0, the first row being Walsh function 1, etc. Walsh function time alignment is such that the first Walsh chip begins at the first chip of a frame. The Walsh function spreading sequence repeats with a period of $N/1.2288\,\mu s$ for Spreading Rate 1 and with a period of $N/3.6864\,\mu s$ for Spreading Rate 3. For more details on variable-length Walsh codes, see Chapter 4.

The variable-length Walsh code assignment to reverse channels in their tree structure is also shown in Figure 6.8. Note that the variable-length Walsh codes is used to provide orthogonal channelization among physical channels of different rates. Thus the longer Walsh codes are assigned to lower-rate channels, for example, the fundamental channel, and shorter codes are assigned to higher-rate channels such as supplemental channels. This specific Walsh allocation ensures that any two channels that are likely to be transmitted at the same time are orthogonal. However, note that because of this allocation R-EACH or R-CCCH cannot be used while SCH2 is being transmitted. Also, R-EACH and R-CCCH share the same code and cannot be transmitted together.

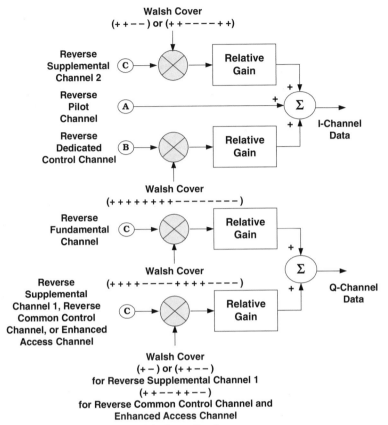

Figure 6.9   Reverse link Walsh channelization.

When a mobile station only supports one reverse supplemental channel, it should support Reverse Supplemental Channel 1. Reverse Supplemental Channel 1 should use Walsh function W_2^4 when possible.

After orthogonal spreading the reverse physical channels are grouped into two disjoint sets, one creating the I-channel data and one forming the Q-channel data.

In each set a relative gain is applied to each physical channel before symbols are summed together (see Fig. 6.9). This form of data modulation is sometimes referred to as "dual-channel QPSK," because each channel symbol contributes to one of the I or Q data channels.

6.7.2   Reverse Link Quadrature Spreading and Modulation

After the orthogonal Walsh spreading, the channel-encoded and interleaved symbols are modulated and are direct spread before transmission. The spreading chip rate is $N \times 1.2288\,\text{Mcps}$. Figure 6.10 shows the uplink spreading and the modulation operation for Radio Configurations 3 through 6.

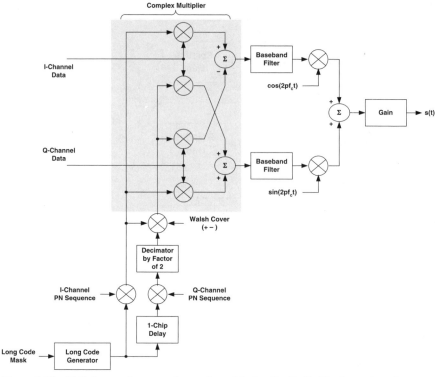

Figure 6.10   Uplink complex spreading and modulation for Radio Configurations 3
through 6.

For Radio Configurations 1 and 2 the access channel and the reverse traffic
channel have the same quadrature spreading as EIA-TIA95, as described in
Chapter 3.

For the enhanced access channel, the reverse common control channel, and
the reverse traffic channel with Radio Configurations 3 through 6, the I-channel data
and Q-channel data are multiplied by a complex spreading sequence before baseband filtering as shown in Figure 6.10.

The in-phase spreading sequence is formed by a modulo-2 addition of the
I-channel PN sequence and the I long code sequence. The quadrature-phase
spreading sequence is formed by the modulo-2 addition of the following three terms:
the W_1^2 Walsh function, the modulo-2 addition of the I-channel PN sequence and the
I long code sequence, and the decimated-by-2 output of the modulo-2 addition of
the Q-channel PN sequence and the Q long code sequence.

For Spreading Rate 1 the I and Q long code for Spreading Rate 1 has a chip
rate of 1.2288 MHz. The Q long code for Spreading Rate 1 is the I long code delayed
by one chip.

For Spreading Rate 3, the I long code consists of three multiplexed component sequences, each having a chip rate of 1.2288 Mcps, as shown in Figure 6.11:

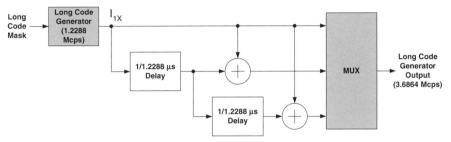

Figure 6.11   Long code generator for Spreading Rate 3. Reproduced under written permission from TIA.

- The first component sequence is the I long code for Spreading Rate 1 (I_{1X}).
- The second component sequence is the modulo-2 addition of I_{1X} and I_{1X} delayed by one chip time, that is, by $1/1.2288\,\mu s$.
- The third component sequence is the modulo-2 addition of I_{1X} and I_{1X} delayed by two chip times, that is, by $2/1.2288\,\mu s$.

The three component sequences are multiplexed to form the I long code for Spreading Rate 3, which has a chip rate of 3.6864 Mcps. The Q long code for Spreading Rate 3 is the I long code delayed by one chip.

Remember that in the EIA-TIA95 the long code masks used for access and traffic channels are different. Similarly, the mask used for generating the I long code for Spreading Rate 1 (or equivalently, the first component sequence of the I long code for Spreading Rate 3) varies depending on the channel type on which the mobile station is transmitting.

6.8   SPREADING AND MODULATION IN THE FORWARD LINK

The downlink-encoded bits out of the channel encoder and interleaver for each physical channel are spread with specific orthogonal channelization codes and subsequently combined and passed through quadrature spreading and modulation processes. This section provides more details of the spreading and modulation process in the forward link.

6.8.1   Forward Link Orthogonal Channelization Spreading

Each physical channel transmitted on the forward CDMA channel is spread with an orthogonal Walsh function or a quasi-orthogonal function (QOF) at a fixed chip rate of 1.2288 Mcps to provide code domain channelization. The maximum length of the assigned Walsh codes is 128 for SR1 and 256 for SR2.

Before applying Walsh or QOF spreading, the DEMUX functions distribute the scrambled and punctured symbols X sequentially from the top to the bottom,

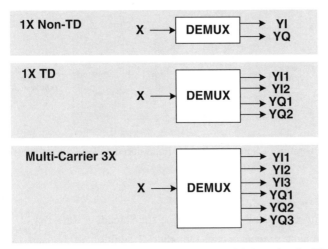

Figure 6.12   De-multiplexing symbols to I and Q branches in 1X, with and without TD, and in 3X multicarrier.

forming pairs of quadrature modulation symbols. Three cases are shown in Figure 6.12.

- 1X Non-TD: When operating in Spreading Rate 1 (1X) without transmit diversity (TD), each modulation symbol in a quadrature pair is spread by the appropriate Walsh function or QOF and is then spread by a quadrature pair of PN sequences at a fixed chip rate of 1.2288 Mcps.

- 1X TD: With transmit diversity, each modulation symbol in two quadrature pairs is spread by the appropriate Walsh function or QOF and a quadrature pair of PN sequences to a chip rate of 1.2288 Mcps. Then the two quadrature pairs are transmitted on two separate antennas.

- Multicarrier 3X: When operating in the Spreading Rate N multicarrier mode, each modulation symbol in N quadrature pairs is spread by the appropriate Walsh function or QOF and a quadrature pair of PN sequences to a chip rate of 1.2288 Mcps. Then the N quadrature pairs are transmitted on N adjacent 1.25-MHz carriers.

The QOFs are used in the cdma2000 forward link to increase the number of code channels, and they may be utilized when the system reaches its Walsh code capacity. Figure 6.13 shows the channelization spreading in the forward link.

The QOFs are created by changing the symbol polarities of a repeated sequence of an appropriate Walsh function according to a nonzero sign multiplier QOF mask (QOF_{sign}) and then applying a nonzero rotate enable Walsh function ($Walsh_{rot}$).

The repeated sequence of the Walsh function is multiplied by the repeated sequence of the sign multiplier QOF mask converted to bipolar (± 1) form. The resulting sequence is also multiplied by the repeated sequence of masks with symbols 1 and j, where j is the complex number representing a 90° phase shift, cor-

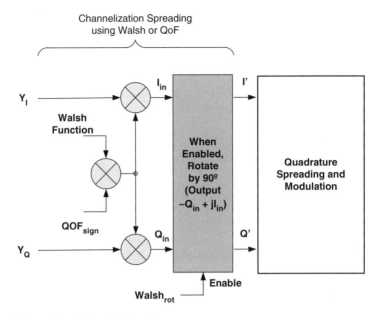

Walsh function = ±1 (mapping: '0' → +1, '1' → −1)
QOF_{sign} = ±1 sign multiplier QOF mask (mapping: '0' → +1, '1' → −1)
$Walsh_{rot}$ = '0' for no rotation or '1' for 90°-rotation-enable Walsh function
For RC1 and RC2 the QOF is null with QOF_{sign} = +1 and $Walsh_{rot}$ = '0'.

Figure 6.13   Channelization spreading in the forward link using Walsh codes and QOF.

responding to the $Walsh_{rot}$ values of 0 and 1, respectively. More details of QOF_{sign} and $Walsh_{rot}$ are specified in [1].

Note that the quasi-orthogonal Walsh functions are used only with Radio Configurations 3 through 9.

The assignment of code channels is such that each code channel is orthogonal or quasi-orthogonal to all other code channels in use. This orthogonality needs to be maintained despite variable-length spreading. Therefore, the binary tree structured Walsh codes in the forward link need to be assigned based on similar rules mentioned in Chapter 4 and used in the reverse link.

A code channel that is spread with Walsh function n from the N-ary orthogonal set ($0 \leq n \leq N - 1$) is represented as W_n^N. The Walsh function spreading sequence repeats with a period of ($N/1.2288$) μs. Table 6.14 and Table 6.15 show the Walsh code assignment to forward physical channels for SR1 and SR3. In cases where the Walsh codes are defined with parameters n and N, these parameters are specified by the base station.

Note that if an auxiliary pilot channel is used with an auxiliary transmit diversity pilot channel, the auxiliary pilot channel is assigned a code channel W_n^N and the auxiliary transmit diversity pilot channel is assigned a code channel $W_{n+N/2}^N$, where $N \leq 512$ and $1 \leq n \leq N/2 - 1$.

For the case of the supplemental channel the value of n is explicitly specified by the base station, whereas the value of N is indirectly specified by the base station

TABLE 6.14   Walsh Code Allocation for the Forward Common Channels in IS2000A

Code Channel	SR1	SR3
Pilot Ch.		W_0^{64}
TD Pilot Channel		W_{16}^{128}
Auxiliary Pilot Ch.	W_n^N, where $N \leq 512$, and $1 \leq n \leq N - 1$	
Synch Ch.		W_{32}^{64}
Paging Ch.		W_1^{64} to W_7^{64}
Quick Paging Ch.	W_{80}^{128}, W_{48}^{128}, and W_{112}^{128} in that order	W_n^{256}, where $1 \leq n \leq 255$
Broadcast Ch.	If rate 1/2 coded: W_n^{64}, where $1 \leq n \leq 63$ If rate 1/4 coded: W_n^{32}, where $1 \leq n \leq 31$	W_n^{128}, where $1 \leq n \leq 127$
Forward Common Control Ch.	If rate 1/2 coded: W_n^N, where $N = 32$, 64, and 128 for the data rate of 38.4, 19.2, and 9.6 kbps, respectively, and $1 \leq n \leq N - 1$ If rate 1/4 coded: W_n^N, where $N = 16$, 32, and 64 for the data rate of 38.4, 19.2, and 9.6 kbps, respectively, and $1 \leq n \leq N - 1$	W_n^N, where $N = 64$, 128, and 256 for the data rate of 38.4, 19.2, and 9.6 kbps, respectively, and $1 \leq n \leq N - 1$
Forward Common Power Control Ch.	For non-TD mode: W_n^{128}, where $1 \leq n \leq 127$ For OTD or STS mode: W_n^{64}, where $1 \leq n \leq 63$	W_n^{128}, where $1 \leq n \leq 127$
Forward Common Assignment Ch.	If rate 1/2 coded: W_n^{128}, where $1 \leq n \leq 127$ If rate 1/4 coded: W_n^{64}, where $1 \leq n \leq 63$ value of n is specified by the base station	W_n^{256}, where $1 \leq n \leq 255$

by explicitly specifying the radio configuration, the number of bits in the frame, and the frame duration.

Also, for supplemental channels in RC3-RC5 for QPSK symbol rates of 4800 sps and 2400 sps, the Walsh function is transmitted two times and four times per QPSK symbol, respectively. Similarly, in RC6-8 for QPSK symbol rates of 7200 sps and 3600 sps, the Walsh function is transmitted two times and four times per QPSK symbol, respectively.

6.8.2   Forward Link Quadrature Spreading and Modulation

After the orthogonal channelization spreading, each code channel is spread in quadrature as shown in Figure 6.14. Similar to the reverse link, the I and Q data chan-

TABLE 6.15 Walsh Code Allocation for the Forward Dedicated Channels in IS2000A

Code Channel	SR1	SR3
Fundamental Or Dedicated Control Ch.	For RC3 and RC5: W_n^{64}, where $1 \le n \le 63$ For RC4: W_n^{128}, where $1 \le n \le 127$	For RC6 and RC8: W_n^{128}, where $1 \le n \le 127$ For RC7 and RC9: W_n^{256}, where $1 \le n \le 255$
Supplemental Ch.	For RC3–5: W_n^N, where $N = 4, 8, 16, 32, 64,$ and 128 for the maximum assigned QPSK symbol rate of 307.2, 153.6, 76.8, 38.4, 19.2 and 9.6 kbps, respectively, and $1 \le n \le N - 1$	For RC6–9: W_n^N, where $N = 4, 8, 16, 32, 64, 128,$ and 256 for the maximum assigned QPSK symbol rate of 921.6, 460.8, 230.4, 115.2, 57.6, 28.8 and 14.4 kbps, respectively, and $1 \le n \le N - 1$

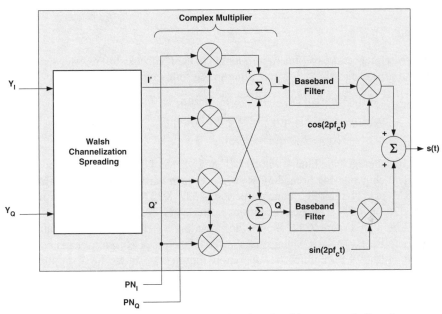

Figure 6.14   Downlink spreading and modulation for 1X without transmit diversity.

nels go through a complex multiplier as follows:

$$I = I \times PN_I - Q' \times PN_Q$$

$$Q = I' \times PN_Q + Q' \times PN_I$$

where the input I' and Q' signals are cross-multiplied to provide a more balanced distribution of power on the final quadrature I and Q branches before RF modulation is applied. The quadrature pilot PN spreading sequences, namely, PN_I and PN_Q, are the same short codes used in EIA-TIA95A/B.

For Spreading Rate 1 and each carrier of Spreading Rate 3, the quadrature

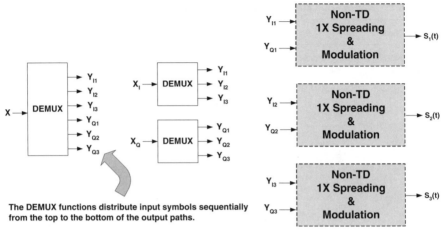

Figure 6.15 Symbol de-multiplexing onto multiple carriers in a 3X system.

pilot PN sequences are two maximum-length linear feedback shift register sequences $i(n)$ and $q(n)$ of length $2^{15} - 1$ according to the following linear recursions:

$$i(n) = i(n-15) \oplus i(n-10) \oplus i(n-8) \oplus i(n-7) \oplus i(n-6) \oplus i(n-2)$$
$$q(n) = q(n-15) \oplus q(n-12) \oplus q(n-11) \oplus q(n-10) \oplus q(n-9)$$
$$\oplus q(n-5) \oplus q(n-4) \oplus q(n-3) \quad (4.2)$$

where $i(n)$ and $q(n)$ are binary valued ("0" and "1") and the additions are modulo-2.

The chip rate for Spreading Rate 1 and each carrier of Spreading Rate 3 is 1.2288 Mcps. The pilot PN sequence period is $32768/1228800 = 26.666\ldots$ ms, and exactly 75 pilot PN sequence repetitions occur every 2 s.

Although Figure 6.14 shows the downlink spreading and modulation for Spreading Rate 1 (1X), it can also be viewed as the process used for each carrier in Spreading Rate N for multicarrier operation. Figure 6.15 shows the de-multiplexing of input data symbols into three quadrature pairs, where each pair (Y_{Ii}, Y_{Qi}) undergoes the same but independent 1X Walsh spreading, quadrature spreading and modulation as described above.

To provide better link performance on the forward link, cdma2000 also supports transmit diversity on the forward CDMA channel. Two variations of transmit diversity may optionally be implemented by the base station manufacturers. These two techniques are the orthogonal transmit diversity (OTD) and space time spreading (STS). The basic concept of transmit diversity in general and the specific OTD and STS methodologies are described in Chapter 4.

Figure 6.16 shows the spreading and modulation structure of the forward CDMA channel when OTD is used. In this case, the input symbols are de-multiplexed into two pairs and then each symbol is repeated once based on two orthogonal patterns (i.e., ++ and +−) to create two orthogonal paths. Each of the two paths then follows the same spreading and modulation process as a non-TD 1X system.

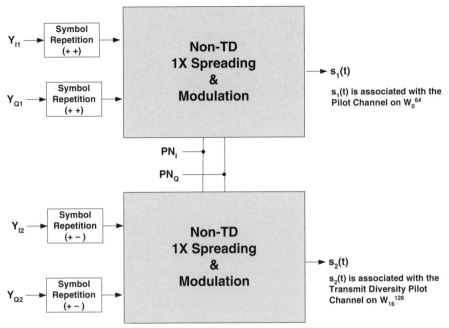

Figure 6.16   Forming transmission symbols for orthogonal transmit diversity (OTD).
Reproduced under written permission from Telecommunications Industry Association.

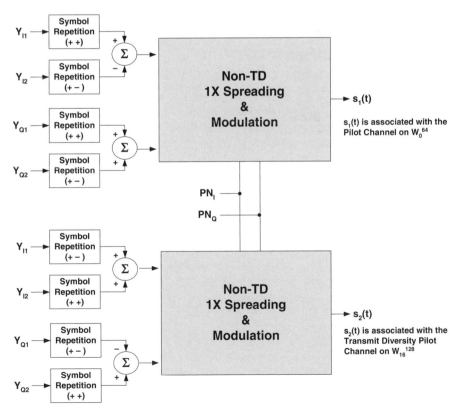

Figure 6.17   Forming transmission symbols for space time spreading (STS).

In the case of the STS scheme, shown in Figure 6.17, the process is similar to OTD except that the input symbols are repeated and combined differently.

Note that whenever transmit diversity is used the base station needs to define a transmit diversity pilot identified by Walsh W_{16}^{128} to be used and transmitted on the second transmit branch.

6.9 REFERENCE

1. C.S0002-A, Physical Layer Standard for cdma2000 Spread Spectrum Systems, July 2001.

CHAPTER 7

IS2000 CALL PROCESSING

7.1 INTRODUCTION

One of the key features of each air interface standard is its Layer 2 and Layer 3 messaging structure and usage of various control channels. From the moment the user terminal is turned on until it is ready to transmit or receive traffic frames, there are many procedures that must be followed and parameters to be set to ensure proper operation of the system. In this chapter our focus is on call processing and control channel structure in the context of mobile station states. For more detailed information about mobile station states, call processing procedures, and signaling messages the reader is referred to 3GPP2 specifications [1–4].

During normal operation, an IS2000 mobile station may occupy one of the following states (see Fig. 7.1):

- Mobile station initialization state
- Mobile idle state
- System access state
- Mobile station control on the traffic channel state

After power-up, the mobile first enters the "mobile station initialization state," where it selects and acquires a serving system. On exiting the initialization state, after the mobile has fully acquired the system and its timing, the mobile station enters the mobile station idle state. In the idle state, the mobile monitors messages on the paging or forward common control channels.

Any of the following three events will cause the mobile to transition from the idle state to the system access state.

- The mobile receives a paging message that requires an acknowledgement or response.
- The mobile originates a call.
- The mobile performs registration.

In the access state, the mobile sends messages to the base station on the access channels. When the mobile is directed to the traffic channel, it enters the mobile station control on the traffic channel state, where the mobile communicates with the

CDMA2000® Evolution: System Concepts and Design Principles, by Kamran Etemad
ISBN: 0-471-46125-3   Copyright © 2004 John Wiley & Sons, Inc.

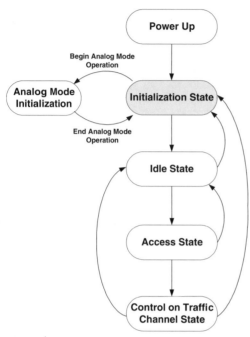

Figure 7.1   Mobile station call processing states.

base station with forward and/or reverse traffic channels. On call termination, the mobile returns to the initialization state.

The following sections describe the mobile station procedures in initialization, idle and access states. The mobile station control on the traffic channel state is discussed in Chapter 8.

7.2   MOBILE STATION INITIALIZATION STATE

In the initialization state, the mobile station first selects a system to use. If the selected system is a CDMA system, the mobile proceeds to acquire the pilot and then synchronize to the system. If the selected system is an analog system, the mobile station begins analog mode operation.

The mobile station initialization state consists of four substates as follows (see Fig. 7.2):

- *System Determination Substate*: involves selecting which system to use
- *Pilot Channel Acquisition Substate*: involves acquiring the pilot channel of a CDMA system
- *Sync Channel Acquisition Substate*: involves obtaining system configuration and timing information for the selected CDMA system
- *Timing Change Substate*: involves timing synchronization with the selected CDMA system.

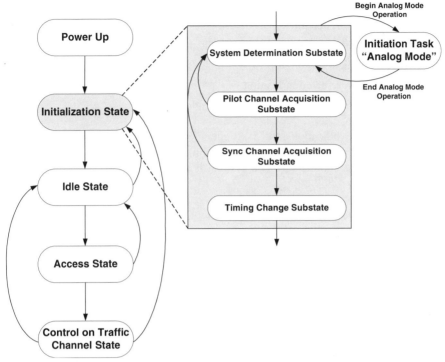

Figure 7.2   Initialization state.

These substates are described in more detail below.

System Determination Substate: On entering the initialization state, the mobile station first selects which system to use. Because all CDMA cellular phones have dual-mode capability, the mobile station can decides whether to operate in the CDMA or analog mode. The preference for selecting the mode can be set in the handset by the operator or the user.

If the mobile is a multiband mobile, capable of operating in different frequency bands, it would choose the best serving system based on roaming preferences set by the operator.

Depending on various events and signaling triggers, the mobile station may enter the system determination substate with different indications, such as:

- Power-up indication: when the mobile powers up
- Acquisition failure indication: when unable to acquire a acceptable pilot channel within the allotted time
- New system indication
- CDMA available indication
- Reselection or redirection indication
- Encryption failure indication

- Protocol mismatch indication
- System lost indication
- Access denied or account blocked indication
- Release indication
- Registration rejected indication
- Wrong system or wrong network indication

Depending on the indications, that is, triggers for entering the system determination substate, the mobile may need to adjust certain parameters before entering the next substate. Once the mobile selects a system, it will attempt to acquire the selected system as follows:

- If the selected system is an analog system, the mobile station starts the analog initialization process.
- If the selected system is a CDMA system, the mobile station enters the pilot channel acquisition substate.

Pilot Channel Acquisition Substate: In this substate, the mobile station attempts to acquire the pilot channel of the selected CDMA system by tuning its receiver to the appropriate CDMA channel number and code channel. If the mobile station acquires the pilot channel within a specified time frame, the mobile station with enter the *sync channel acquisition substate*; otherwise, it turns back to the *system determination substate* with an acquisition failure indication.

Sync Channel Acquisition Substate: In this substate, the mobile station receives and processes the *sync channel message* to obtain system configuration and timing information. On entering the sync channel acquisition substate, the mobile station tunes its receiver to the prespecified sync code channel and looks for a valid synch channel message.

On receiving a valid synch channel message within acceptable time window showing supportable protocol revision, the mobile station reads and stores all the parameters in the message that are defined in the mobile protocol revision and ignores the rest.

Some of the key parameters carried by the synch channel are:

- Protocol revision level and minimum protocol revision level supported
- System identification (SID) and network identification (NID)
- Paging channel or broadcast channel support as well as their data rates and code channels
- Timing information, such as
  - System time (SYS_TIME)
  - Pilot PN sequence offset index (PILOT_PN) and log code state
  - Number of leap seconds that have occurred since the start of system time
  - Offset of local time from system time and daylight saving time indicator

If the mobile station does not receive a valid sync channel message within a preset time window or if it does not meet the minimum protocol revision level supported by the base station, the mobile station re-enters the system determination substate with an acquisition failure or protocol mismatch indication, respectively.

If the synch channel message is valid and the mobile's protocol revision is supported by the network, the mobile saves all relevant information from the synch channel and moves to the next substate.

Timing Change Substate: In this substate, the mobile station synchronizes its long code timing and system timing to those of the CDMA system, using the $PILOT\_PN_s$, LC_STATE_s, and SYS_TIME_s values obtained from the received sync channel message. The mobile then initializes some of its internal parameters and enters the mobile station idle state.

7.3 MOBILE STATION IDLE STATE

In the idle state, the mobile station monitors the forward broadcast and common channels including, the legacy paging channel or the new forward common control channel/primary broadcast control channel. The mobile may also monitor the quick paging channel, if supported. While monitoring these channels, the mobile station can receive directed or overhead messages and it may receive or initiate calls.

The mobile may also initiate a registration or other message transmission or it may cancel a priority access channel assignment (PACA) call while in idle state. If the mobile station declares a loss of the paging channel or the forward common control channel/primary broadcast control channel, it should enter the system determination substate of the mobile station initialization state with a system lost indication.

While in the mobile station idle state, the mobile station performs the following procedures:

- Monitoring forward common and broadcast channels
  - Paging channel monitoring or
  - Forward common control channel monitoring
  - Forward common control channel monitoring
  - Primary broadcast control channel monitoring
  - Quick paging channel monitoring (if supported)
  - Mobile station page match operation
- Registration
- Mobile station origination operation
- Mobile station order and message processing operation
- Mobile station message transmission operation
- Mobile station power-down Operation
- Mobile station PACA cancel operation

- Response to overhead information operation
- Idle handoff procedures
- System reselection procedures

Some of these procedures are described in the following sections.

7.3.1 Monitoring the Forward Common and Broadcast Channels

On the basis of the information received during initialization and on entering the idle state the mobile station sets the code channel, data rate, and code rate for the primary broadcast control channel, if supported; otherwise, it sets the code channel and data rate for the paging channel. The mobile also determines whether the slotted mode and/or transmit diversity mode are to be used. Once these key parameters are set, the mobile starts the common channel supervision processing.

The proper supervision and monitoring of forward common channels is managed by a timer. The mobile station performs the following:

- Sets a timer (T30m) whenever it begins to monitor F-PCH or F-CCCH/F-BCCH
- Resets the timer for whenever it receives a message on these channels, whether addressed to the mobile station or not
- Disables the timer when it is not monitoring these channels

If the timer expires, the mobile station declares a loss of F-PCH or F-CCCH/F-BCCH and returns to the system determination substate.

A mobile station that monitors the F-PCH or F-CCCH/F-BCCH only during certain assigned slots is referred to as operating in the slotted mode. During the slots in which the paging channel or the forward common control channel is not being monitored, the mobile station can stop or reduce its processing for power and battery life conservation. The slotted mode may only be used in the mobile station idle state. A mobile station that monitors the F-PCH or F-CCCH/FBCCH at all times or in all slots is referred to as operating in the nonslotted mode.

The broadcast control channel is divided into 40-, 80-, or 160-ms slots called the broadcast control channel slots. The primary broadcast control channel is used for control messages. Support for the primary broadcast control channel is mandatory for mobile stations. The primary broadcast control channel will operate with the forward common control channels with or without the quick paging channels.

After a mobile station acquires and synchronizes with a new base station that supports a primary broadcast control channel (BCCH), the mobile station monitors the primary BCCH to receive overhead information. Once the mobile station has received the updated overhead information from the primary BCCH, the mobile station may begin to monitor a forward common control channel or a quick paging channel, if it is supported.

While monitoring the primary BCCH, if the mobile station determines that the CONFIG_MSG_SEQ has changed, it should monitor the primary broadcast control channel to receive updated overhead messages.

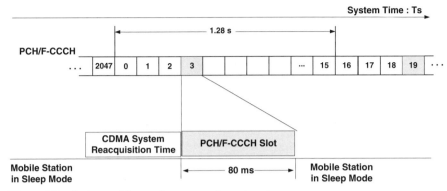

Figure 7.3   Paging and forward common channel cycles.

The paging channel and forward common control channels are divided into 80-ms slots called F-PCH slots and F-CCCH slots, respectively. The F-PCH and F-CCCH protocols provide for scheduling the transmission of messages for a specific mobile station in certain assigned slots. However, the support of this feature is optional and may be enabled by each mobile station.

A mobile station operating in the slotted mode generally monitors the paging channel or the forward common control channel for one or two slots per slot cycle, but it may optionally monitor additional slots to receive broadcast messages and/or broadcast pages.

The minimum-length slot cycle is 1.28 s, consisting of 16 slots of 80 ms each. Hence, any slot cycle T is measured in units of 1.28 s, given by $T = 2^{\text{SLOT\_CYCLE\_INDEX}}$, and there are $16 \times T$ slots in a slot cycle. SLOT_NUM is the F-PCH/F-CCCH slot number, calculated from the systems time T_s in units of 20-ms frames divided by 4 to reflect 80-ms slot size and representing modulo 2048. Therefore, SLOT_NUM is always a number between 0 and 2045.

For each mobile station, the starting time of its slot cycle is a fixed but randomly selected number called PGSLOT. Figure 7.3 shows an example for F-PCH/F-CCCH slot structure assuming SLOT_CYCLE_INDEX = 0 or cycle length of 1.28 s and PGSLOT = 3. In this case the mobile station's slot cycle begins when SLOT_NUM = 3 and repeats every 16 slots. Thus the mobile station begins monitoring the F-PCH/F-CCCH at the start of the slot in which SLOT_NUM = 3, 19, 35, etc.

The broadcast pages and messages can be transmitted on the broadcast control channel or the paging channel.

7.3.2   Monitoring Broadcast Messages on F-PCH

The F-PCH can be used to send broadcast messages or broadcast pages. A broadcast message is a *data burst message* that has a broadcast address type, whereas a broadcast page is a record within a *general page message* that has a broadcast

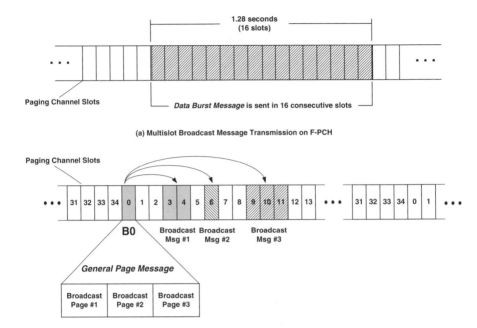

Figure 7.4 Broadcast message transmission methods on F-PCH.

address type. A paging channel slot that is identified to be monitored by the mobile to receive either a broadcast page or a broadcast message is referred to as a broadcast slot.

The paging channel protocol provides two methods for the transmission of broadcast messages.

- **Multislot Broadcast Message Transmission:** In this method, a broadcast message is sent in a sufficient number of assigned slots such that it may be received by all mobile stations that are operating in the slotted mode. Figure 7.4a shows an example in which the maximum slot cycle index is equal to 0 and the broadcast message fits in a single slot. In this case, the data burst message is transmitted in 16 consecutive slots.

- **Periodic Broadcast Paging:** In this method, mobile stations configured to receive broadcast messages monitor a specific broadcast slot (the first slot of a broadcast paging cycle, e.g., B0). If all of the broadcast messages to be transmitted fit within one slot it will be sent in B0. Alternately, one or more broadcast pages may be transmitted in B0 as part of the general page message, each associated with and pointing to a subsequent broadcast slot. For each broadcast page, an associated broadcast message may be transmitted in the associated subsequent broadcast slot. Figure 7.4b shows an example of periodic broadcast paging when the broadcast index is set to 1. A general page message containing three broadcast pages is transmitted in the first slot of the broadcast paging cycle. For each of the three broadcast pages, a data burst message is transmitted in a subsequent slot.

7.3.3   Monitoring Broadcast Messages on F-BCCH and F-CCCH

The broadcast control channel/forward common control channel protocol enables mobile stations operating in the slotted mode or in the nonslotted mode to receive broadcast messages on the broadcast control channel. A broadcast message is a data burst message that has a broadcast address type. An enhanced broadcast page is a record within a general page message or a *universal page message* that has a broadcast address type and includes broadcast message scheduling information.

Similar to the paging channel protocol, the broadcast control channel/forward common control channel protocol provides two methods for the transmission of broadcast messages.

- **Multislot Broadcast Message Transmission:** According to this method, an enhanced broadcast page is sent in a sufficient number of assigned slots on the F-CCCH such that it can be received by all mobile stations operating in the slotted mode. Figure 7.5a shows an example for the case where the maximum slot cycle index is equal to 0 and the enhanced broadcast page is transmitted in 16 consecutive slots.

- **Periodic Broadcast Paging:** In this method, the base station transmits an enhanced broadcast page in an assigned F-CCCH slot, or in a broadcast slot, to inform mobile stations that a broadcast message will be transmitted in a specified broadcast control channel slot. The enhanced broadcast page identifies the broadcast control channel and the slot the mobile station is to monitor to receive the broadcast message. According to this method, mobile stations monitor a specific broadcast slot (the first slot of a broadcast paging cycle, e.g., B0; see Fig. 7.5). One or more enhanced broadcast pages may be transmitted in B0 and/or in the subsequent slot as part of a general/universal page message.

A mobile station configured to receive broadcast messages should support reception of broadcast messages transmitted with both multislot and periodic enhanced broadcast paging.

7.3.4   Monitoring the Quick Paging Channel

The quick paging channel (QPCH) is used by the base station to quickly and efficiently notify the mobile about an upcoming configuration change, broadcast, or page message on the associated F-PCH or F-CCCH/F-BCCH. Mobiles can save battery power because they monitor a single bit on QPCH and they start monitoring the normal paging channel only if the corresponding bit is turned on. The amount of mobile battery savings is dependent on the number of pages sent in the sector, the number of QPCHs, and their rate. Support for QPCH is optional.

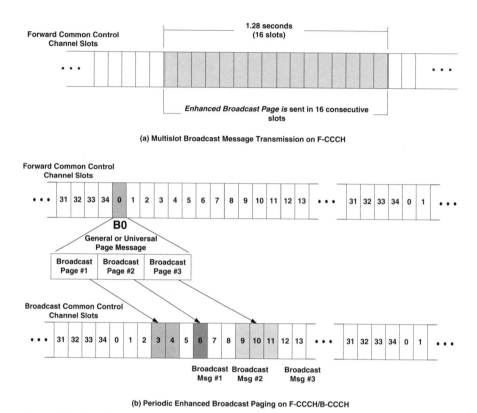

(a) Multislot Broadcast Message Transmission on F-CCCH

(b) Periodic Enhanced Broadcast Paging on F-CCCH/B-CCCH

Figure 7.5   Broadcast message transmission methods on F-CCCH/B-CCCH.

The QPCH is used for transmission of paging indicators, configuration change indicators, and broadcast indicators in specific slots assigned to the mobile stations.

The mobile is informed via an extended system parameters message about the existence and number of QPCHs in the sector. Because there is a fixed number of paging indicators on QPCHs, the indicators are shared by groups of mobiles. Paging indicators are assigned to mobiles based on their hashing algorithm. When one mobile is paged, every mobile with the same assigned paging indicator also wakes up to monitor the paging message on the paging channel.

The QPCH is divided into 80-ms slots called QPCH slots. Each QPCH slot consists of simple binary indicators, which can be detected with very little processing, to reduce the battery power consumed by the mobile during the idle state.

The mobile station's assigned QPCH slots are offset from its assigned paging channel slots or its assigned forward common control channel slots by 100 ms. Two paging indicators are transmitted in each QPCH slot to direct the mobile either to the paging channel or to the F-CCCH in the next frame (20 ms later). A standard hashing function is used to determine the position of the mobile station's two assigned paging indicators within the assigned QPCH slot. To provide some time

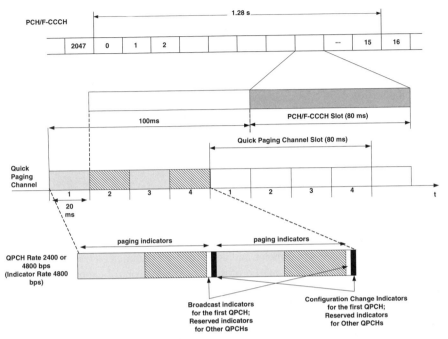

Figure 7.6   Quick paging channel slots.

diversity, the hashing algorithm is devised such that the two paging indicators for a mobile station are separated in time and they fall within the first and third quarter slot or the second and fourth quarter slot as shown in Figure 7.6.

The configuration indicators are enabled by the base station after a change in configuration parameters, so that the mobile knows that configuration messages have been updated and it starts reading the new configuration messages.

The QPCH has a data rate of 2400 or 4800 bps. At the end of each half-slot of QPCH, there are $2n$ broadcast and $2n$ configuration change indicators and $4n$ reserved indicators, where $n = 1$ for a 2400-bps rate and $n = 2$ for 4800-bps rate of QPCH.

A mobile station supporting the QPCH, during its idle mode monitors paging indicators on the QPCH as follows:

- When the mobile station checks assigned paging indicators, it performs the following:
  - If the mobile station detects that one of the paging indicators is set to "OFF," the mobile station need not detect another paging indicator and it can go back to "sleep mode."
  - If the mobile station does not detect that at least one of the paging indicators is set to "OFF," the mobile station will monitor its assigned paging channel or forward common control Channel slot immediately following its assigned QPCH slot.

- When a mobile station performs an idle handoff to a base station whose PCH or F-CCCH/primary BCCH has recently been monitored, it may monitor one or more configuration change indicators. Configuration change indicators are scheduled every 40 ms on the first QPCH.

- A mobile station configured to receive broadcast messages should also monitor the broadcast indicators in the mobile station's assigned QPCH. If the mobile station determines that broadcast indicators are not set to "OFF," the mobile station should receive its assigned broadcast slot on the forward common control channel or paging Channel immediately following its assigned QPCH broadcast slot.

7.3.5 Response to Overhead Information

A mobile station in the idle state should always monitor the broadcast or paging channel to read and update its stored configuration and access parameters. Overhead messages include access parameter and configuration parameter messages.

The overhead messages on the primary broadcast control channel are:

- Enhanced access parameters message
- ANSI-41 system parameters message
- MC-RR parameters message
- Universal neighbor list message
- Private neighbor list message
- Extended CDMA channel list message
- Extended global service redirection message
- User zone identification message
- ANSI-41 RAND message

The overhead messages on the paging channel are:

- Access parameters message
- System parameters message
- Extended system parameters message
- Neighbor list message
- Extended neighbor list message
- General neighbor list message
- Private neighbor list message
- CDMA channel list message
- Extended CDMA channel list message
- Global service redirection message
- Extended global service redirection message
- User zone identification message

When a mobile station receives an overhead message, it should update all of its related stored information accordingly. Associated with the set of configuration messages sent on each paging channel or primary broadcast control channel is a configuration message sequence number (CONFIG_MSG_SEQ). When the contents of one or more of the configuration messages change, the configuration message sequence number is incremented.

The mobile station stores the most recently received configuration message sequence number (CONFIG_MSG_SEQ$_s$) contained in any message.

Also, for each of the configuration messages received, the mobile station stores the corresponding configuration message sequence number based on the type of the message. For example, when a neighbor list message is received, the NGHBR_LST_MSG_SEQ$_s$ will be set to CONFIG_MSG_SEQ$_s$ of the current message.

Thus the mobile station can examine the stored values of the configuration message sequence numbers by type to determine whether the specific configuration parameters stored by the mobile station are current.

Access parameters messages or enhanced access parameters messages are independently sequence-numbered by the ACC_MSG_SEQ field. The mobile station stores the most recently received access parameters message or enhanced access parameters message sequence number (ACC_MSG_SEQ$_s$). The stored configuration and access parameters are considered to be current only if

- All stored configuration/access message sequences for all types of messages are equal to the most recent configuration/access message sequence numbers received and
- No more than T_{31m} seconds have elapsed since the mobile station last received a valid message on the PCH or F-CCCH/primary BCCH for which the parameters were stored.

If the configuration parameters are not current, the mobile station should process the stored parameters on receipt of the configuration messages.

Paging channels, broadcast control channels, and forward common control channels are considered to be different if they are

- Transmitted by different base stations,
- Transmitted on different code channels, or
- Transmitted on different CDMA channels

Configuration and access parameters from one paging channel or primary broadcast control channel should not be used while monitoring a different F-PCH or primary BCCH/F-CCCH. The only exception is for registration and authentication parameters while the mobile station is performing an access probe handoff or access handoff, to be discussed later.

7.3.6 Idle Handoff

An idle handoff occurs as a mobile station moves from one cell to another cell, while it is in the idle state. In the idle state, the mobile station continuously searches for the strongest pilot channel signal on the CDMA frequency assignment, in which it monitors the paging channel or the forward common and broadcast control channels. During this search, if the mobile station detects a pilot channel signal from another base station that is sufficiently stronger than that of the current base station, the mobile station determines that an idle handoff should occur.

To properly detect and select the best serving base station, the mobile station in the idle state groups the pilot channels into the following four disjoint sets and it updates these sets frequently.

- Active Set: The pilot offset of the forward CDMA channel whose forward common/broadcast channel, F-PCH or F-CCCH/F-BCCH, is being monitored
- Neighbor Set: The offsets of the pilot channels that are likely candidates for idle handoff. The members of the neighbor set are specified in one of the following neighbor list messages:
  - On the F-PCH: neighbor list message, extended neighbor list message, and general neighbor list message
  - On the primary BCCH: universal neighbor list message
- Remaining Set: The set of all possible pilot offsets in the current system and on the current CDMA frequency assignment, excluding the pilots in the neighbor set and the active set. These pilot offsets are integer multiples of PILOT_INC$_s$.
- Private Neighbor Set: The offsets of the pilot channels for the private systems that are likely candidates for idle handoff. The members of the private neighbor set are specified in the private neighbor list message. These pilots are optionally specified by the system to allow hierarchical cell structure for the proper reselection of public- vs. private layer networks.

The mobile should also be capable of searching for pilots in frequencies and band classes other than its current frequency. If a pilot in the neighbor set or in the private neighbor set is on a different frequency assignment from that of the mobile station, this target frequency should be included in the search criteria.

The mobile uses four search windows: SRCH_WIN_A, SRCH_WIN_N, SRCH_WIN_R, and SRCH_WIN_PRI_NGHBR$_s$ to search for those pilots contained in active, neighbor, remaining, and private sets, respectively.

If the mobile determines that one of the neighbor set or remaining set pilot signals is sufficiently stronger than the pilot of the active set, the mobile performs an idle handoff.

When multiple idle handoff candidates are available, the mobile station should select a candidate that supports the primary broadcast control channel, if possible.

In most cases, after the selection of a new base station, the mobile station operates in nonslotted mode until it has received at least one valid configuration message or mobile station-addressed page on the new F-PCH or F-CCCH/primary BCCH. After completion of the idle handoff , the mobile station should discard all unprocessed messages received from the old base station's channels.

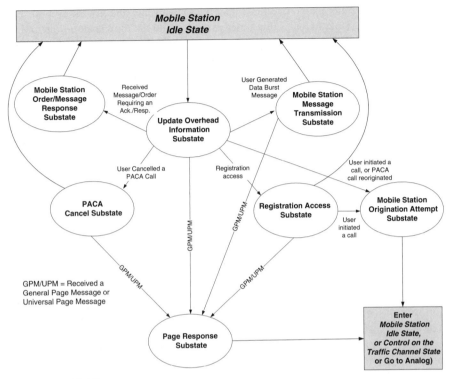

Figure 7.7 Mobile access state.

7.4 SYSTEM ACCESS STATE

In the system access state, the mobile station monitors the F-PCH or F-CCCH/ primary BCCH channel for any updates on overhead messages and it may transmit mobile-originated or response messages to the base station on the access or enhanced access channel.

The system access state consists of several substates as shown in Figure 7.7 and described in the following.

Update Overhead Information Substate: On entering the access state from the idle state, and before sending any access probe, the mobile station should first monitor the primary broadcast control channel or paging channel until it receives the current overhead messages. The mobile station should update its access and system parameters, including neighbor list parameters, based on overhead messages received from the target base station.

If the mobile station receives a (extended) global service redirection message, which directs the mobile station away from the new base station, the mobile station should not access the new base station and should look for alternative strong candidates for handoff.

On entering the update overhead information substate, the mobile station sets the system access state timer. If this timer expires while the mobile station is in this substate, the mobile station enters the system determination substate of the mobile station initialization state with a system lost indication.

Mobile Station Origination Attempt Substate: In this substate, the mobile station sends an origination message to the base station using access procedures.

Registration Access Substate: In this substate, the mobile station sends a registration message to the base station. The mobile enters this state only if it has not already registered with the network, for example, as the mobile is turned ON by the user.

Page Response Substate: In this substate, the mobile station sends a page response message in response to a mobile-station-addressed page from a base station.

Mobile Station Order/Message Response Substate: In this substate, the mobile station sends a response or acknowledgment to messages received from the base station using the access procedures. For example, the mobile may send an authentication challenge response message in response to an authentication challenge message sent by the base station.

Mobile Station Message Transmission Substate: In this substate, the mobile station sends a data burst message or a device information message to the base station. Support of this state is optional.

PACA Cancel Substate: PACA (priority access channel assignment) is a service in which priority is given to mobile station originations. In the PACA cancel substate, the mobile station sends a PACA cancel message to the base station.

Once the mobile station receives confirmation of any message sent in an access substate, it sends a response and re-enters the mobile station idle state.

After receiving the page response or origination message, the base station may send the mobile a channel assignment message on the F-PCH or F-CCCH to start setting up a call. In this case, the mobile will move to the control on traffic channel or conversation state.

7.4.1 Enhanced Random and Reservation Access Procedures

The random access in IS2000-A can be performed by using the legacy reverse access channels (R-ACH) or reverse enhanced access channels (R-EACH).

When legacy access channels are used the random access timing and protocols are the same as those in IS95A. The legacy access channel structure and procedures as well as the concepts of access probes and open-loop access power control are described in Chapter 3. In this section, our focus is on the new access channels and protocols introduced in IS2000-A.

The R-EACH procedures in IS2000 are introduced to offer more flexibility and efficiency along with lower latencies and higher data rates for short data burst and common signaling transmissions on the reverse link [2,5].

Figure 7.8 shows the frame structure of R-EACH and reverse common control channels (R-CCCH) including the slot preambles and data segments. The enhance-

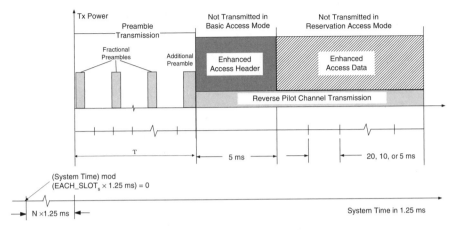

(a) Enhanced Access Channel Preamble, header and message part.

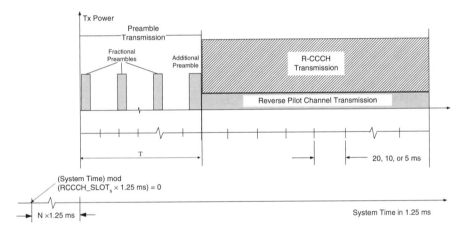

(b) Reverse Common Control Channel Preamble and Data Transmission

Figure 7.8   Reverse enhanced access and common control channel structure.

ments made in the design of R-EACH include shorter slot sizes, variable frame sizes, multiple data rates, gated preamble, and a new mode of operation to allow fast and contention-free reservation access.

The R-EACH may be used in one of two modes: the basic bAccess mode (BA) mode and the reservation access (RA) mode. The mode of operation for each R-EACH as well as the maximum data rates and message sizes permitted for transmission on that channel are defined as system parameters by the base station.

Before initiating a transmission, the mobile station randomly selects a R-EACH corresponding to the applicable access mode and forms the enhanced access probe (EAP) accordingly. The SRBP entity at the mobile MAC sublayer uses an algorithm to select an access mode before each Layer 2 encapsulated PDU transmission. The selection is mainly based on the length of the Layer 2 encapsulated PDU and configuration parameters from the base station. Whereas the BA mode is mostly suited for very short MAC and signaling messages with delay constraint, RA provides a more reliable mechanism to send extended short messages. The following describes the operation of R-EACH in each of the two modes.

Basic Access Mode: The BA mode uses a slotted Aloha-like approach for contention resolution and has a protocol structure similar to the legacy, that is backward-compatible, access protocol on the R-ACH. As shown in Figure 7.9, each basic EAP consists of a preamble immediately followed by the enhanced access data. The enhanced access data contain the Layer 2 encapsulated PDU similar to the message capsule of R-ACH. The mobile also transmits the reverse pilot channel along with the enhanced access data to allow coherent demodulation at the base station.

Following the same access attempt, subattempt, and probe sequencing procedure as R-ACH, a mobile station using a R-EACH in the BA mode structures its transmissions in sequences of basic enhanced access probes (B-EAPs). After transmitting each B-EAP the mobile reverts to monitoring the F-CCCH for an L3 acknowledgment. The probe sequence protocol for R-EACH follows the same access attempt structure and timing including the random back-off times, persistence delay, and open-loop power control described for the R-ACH channel. The difference is mainly in the preamble gating, timing scale, and frame structure within an access probe.

Note that in the BA mode, there is no power control or access control feedback from the base station during the mobile's transmission. Therefore, similar to the legacy access, in the BA mode the entire enhanced access data is transmitted in every probe of a sequence in a contention mode until the L3 acknowledgment from the base station is received.

Reservation Access Mode: When operating in the RA mode, the mobile sends its request for a R-CCCH reservation in the form of a reservation access enhanced

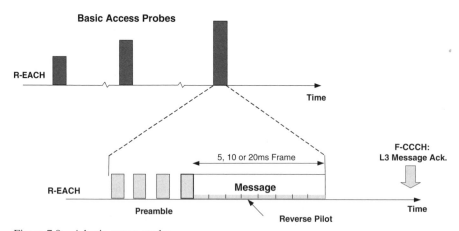

Figure 7.9   A basic access probe.

access probe (RA-EAP) on the R-EACH. Each RA-EAP consists of a preamble followed by an enhanced access header, but it does not include the L2 message.

The transmission of the preamble may be gated, with total preamble duration, multiples of 1.25 ms, and duty cycle of gating being defined as part of the access parameters.

The enhanced access channel header is a short, 5-ms, coded frame with a fixed data rate and includes:

- A request mode indicator showing whether a reservation or power control access request is being used
- A short hash ID, pseudo-randomly selected by the mobile, that serves as a temporary and locally significant identification number for the user's request
- A rate word, indicating the requested data rate-frame size combination
- A soft handoff request indicator and the pilot code offset of the handoff candidate base station. This field is included only if soft handoff on R-CCCH is supported by the base station.

Reservation access data, including the L2 message, are sent on the R-CCCH only on receiving a permission from the base station in the form of an *early acknowledgment channel assignment message* (EACAM) (see Fig. 7.10). The base station echoes the HASH ID, used by the mobile in the request header, as an acknowledgment ID within the EACAM to address the target mobile.

For every R-EACH operating in the RA mode, there is an associated forward common assignment channel (F-CACH) and the corresponding forward common power control subchannel (F-CPCSCH). The EACAM transmitted on the F-CACH not only serves as an early MAC layer acknowledgment to the mobile request, it also assigns an R-CCCH and its common power control subchannel of F-CPCCH to the mobile.

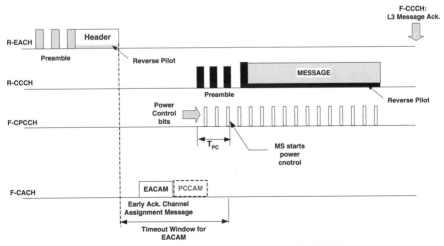

Figure 7.10   An RA-EAP followed by data transmission on R-CCCH.

Once the mobile station transmits an RA-EAP on the next R-EACH slot boundary, it starts monitoring the F-CACH to receive an EACAM. If the mobile station does not receive an EACAM with its hash identifier within this time, it waits for a random amount of back-off time and sends another RA-EAP at a higher power.

If the hash identifier in the EACAM matches the one used by the mobile, the mobile station begins transmitting the R-CCCH preamble followed by enhanced access data on the next R-CCCH slot boundary. The mobile station initially applies an open-loop power estimate to its transmission. After a delay of $RA\_PC\_DELAY_s$ from the beginning of transmission of the R-CCCH preamble, the mobile station begins closed-loop power control operation at 800 bps based on the power control bits received on the assigned F-CPCSCH.

Whereas the presence of the power control subchannel is an indication to the mobile that its data are received by the base, any long interruption of these power control bits would show a problem with the link and a failed access. In this case, the mobile should stop its transmission immediately and reattempt the access.

The power control subchannels on F-CPCCH are time multiplexed, and their number depends on the power control rate. For example, if 1.25-ms common power control group (CPCG) is used, 24 mobile stations can be power controlled, each at a rate of 800 bps, and if 2.5 ms CPCG is used, 48 mobiles stations can be power controlled, each at a rate of 400 bps. The common power control rate is one of the system parameters.

Based on the transmission rate requested in the R-EACH header, the length of the burst to be transmitted cannot exceed $ACC\_MAX\_DURATION_s$ seconds. If the length of the burst exceeds the maximum duration, the mobile station should not proceed to the reservation mode and may need to request for a dedicated traffic channel.

In some cases the base may allow the mobile to transmit its data at a lower rate than the mobile had originally requested in its reservation access header. In this case, the mobile needs to send its packet at the reduced data rate, if the packet is still within $ACC\_MAX\_DURATION_s$ seconds. If the message size in time exceeds the maximum allowed on the reservation R-CCCH channel, the mobile would need to request for a dedicated traffic channel.

In this mode of access, as the R-CCCH channel may only be used on reservation basis, the entire reservation access data is sent free of contention and with closed-loop power control.

Reservation Access with Soft Handoff: In the RA mode only, if the mobile receives significant pilot signal power from a neighboring base station, and if the base station supports soft handoff on R-CCCH, the mobile can indicate its soft handoff request in the RA-EAP header. If soft handoff is requested, the mobile should also include the pilot code offset of the candidate base station in the enhanced access header. This offset will be used by the base station to identify the neighbor that needs to be involved in the soft handoff process. However, the base station may choose not to grant soft handoff on a case-by-case basis. Figure 7.11 shows a reservation access scenario with soft handoff granted.

The mobile indicates this request in the enhanced access header along with the strongest neighbor's PN code offset. If the request is granted, with soft handoff

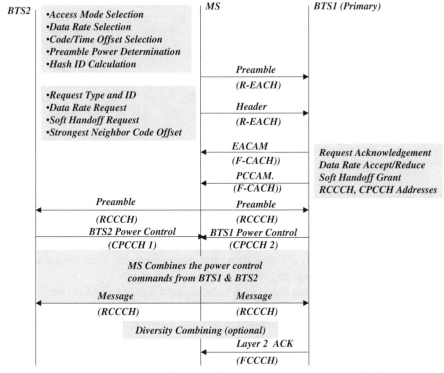

Figure 7.11   Reservation access diagram with power control and soft handoff.

allowed, the primary base station sends an extended common channel assignment message, which includes a power common control assignment message (PCCAM), through which the allocated R-CCCH and F-CPCCH channels of both base stations are specified.

In this case, both base stations involved in soft handoff receive and decode the enhanced access data sent on the assigned R-CCCH and they both power control the mobile. During the transmission of data on R-CCCH, the mobile follows both base stations' power control commands.

The message received by BTS1 and BTS2 may be combined to achieve diversity gain. If the two base stations are two sectors of the same cell site, the diversity combining can be done at the base station; otherwise, a selection combining can be performed at the base station controller. The details of diversity combining at the base station or at the network are subject to implementation. The network sends a Layer 3 message acknowledgment to the mobile through the primary base station.

The mobile station transmits the enhanced access Header on the R-EACH at a fixed data rate of 9600 bps in a 5-ms frame. The enhanced access data to be sent on the R-EACH or RCCCH, however, can be sent with one of six possible combinations of data rates and frame sizes, shown in Table 7.1. Regardless of its frame

TABLE 7.1 Data Rates and Frame Sizes on R-EACH and R-CCCH

RATE_WORD	Data Rate (kbps)	Frame Length (ms)	No. of Information Bits
"000"	9.6	20	172
"001"	19.2	20	360
"010"	19.2	10	172
"011"	38.4	20	744
"100"	38.4	10	360
"101"	38.4	5	172
"110–111"		Reserved	

size, an enhanced access channel frame begins only at the enhanced access slot boundary.

A complete message typically spans over several slots. The maximum message durations permitted on REACH and RCCCH are specified by the base station. Because the system allows multiple data rates, the system sets a maximum on the message duration in seconds and not in number of bits or bytes.

Frame Sizes and Data Rates: All transmissions on the R-EACH or R-CCCH begin at a microslot boundary, synchronized with system time, with microslot sizes defined and broadcast by the system as part of the access parameters. The smallest microslot size is 1.25 ms.

The slot duration is a parameter that may be specified independently for each R-EACH supported. The mobile station should initiate transmission of a R-EACH probe and subsequent probes in the probe sequence on a slot boundary.

Enhanced access data rates and frame sizes permitted are determined by the R-EACH- or R-CCCH-specific parameter RATE_WORD provided by the base station (see Table 7.1).

The base station also defines ACC_MAX_DURATION$_s$ as the maximum message duration that a mobile station is permitted to transmit on the R-EACH (in basic access mode) or R-CCCH (in reservation access mode).

In the basic access mode the enhanced access data can be sent at any of six different combination of frame sizes and data rates defined for R-EACH, if supported by the base station. If more than one data rate is allowed, the base station must perform blind rate determination.

In the reservation access mode, the mobile station transmits the enhanced access header on the R-EACH at a fixed data rate of 9600 bps in 5-ms frames. In this header, the mobile requests one of the data rates and frame sizes allowed on the R-CCCH. The base station also indicates, explicitly in the EACAM, the data rate and frame size to be used by the mobile. Therefore, in this mode there is no need for rate and frame size determination at the base station.

7.4.2 Access State Handoffs

In IS95A, once a mobile station (MS) enters the system access state, it is not allowed to hand off to another base station. However, during the call setup for mobiles in the "handoff regions," the propagation and interference conditions may change so quickly that committing to the same base station throughout the process may result in call setup failures. The call setup failures in the handoff regions may be the results of lost access and/or channel assignment messages or the delay in bringing the MS into soft handoff (SHO) on the traffic channel.

To improve the robustness of messaging during the access state, the IS-95-B/IS-2000 standards remove the earlier limitation of the IS-95-A standard, which required the mobile station to commit to a single PN during call setup.

A mobile station in the access state continues monitoring pilots and maintaining PCH/F-CCCH active, neighbor, and remaining sets similar to the idle state. The mobile also assists the base station in the traffic channel assignment process by monitoring and reporting the pilot strength of the pilot in the mobile station's PCH/F-CCCH active set.

When access state handoff is supported, the mobile station also monitors and reports other pilots on the same frequency by creating an ACCESS_HO_LIST and reporting it in every access channel message. The ACCESS_HO_LIST is defined as a set of strongest pilots, exceeding T_ADD, to which access handoff is allowed.

The *access handoff list* is similar to the pilot strength measurement message sent during traffic channel operation, in which the MS reports certain pilots that are strong and likely candidates for soft handoff. The ACCESS_HO_LIST is created by the mobile immediately before sending the first access probe after entering the system access state and can be updated before each successive probe.

The EIA-TIA95B and IS2000 standards define the following access state handoff procedures to be used based on the timing of the handoff (see Fig. 7.12).

- Access entry handoff
- Access probe handoff
- Access handoff
- Channel assignment into soft handoff

Support for access state handoffs is optional for the infrastructure. These access state handoffs are controlled by some parameters in the extended system parameters message or the mobility management and radio resource (MM-RR) message, which

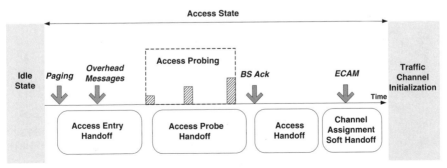

Figure 7.12   Access state handoffs.

is broadcast on the paging or broadcast control channel, respectively. These parameters can be enabled/disabled on a per neighboring cell basis, for every neighbor specified in the neighbor list message or the extended neighbor list message. Using different parameters in these messages, the serving base station specifies for the MS the neighbors' PNs to which any of these access state handoffs are s allowed.

The following briefly describes each of the access state handoff procedures:

Access Entry Handoff (AEHO) is the process of switching the reception of the paging, or F-CCCH/F-BCCH, channel from one base station to another when the mobile station is transitioning from the mobile station idle state to the system access state. The AEHO allows an IS-95B/IS-2000 mobile to send a page response message to and receive subsequent paging channel messages from a cell other than the cell from which the original page message was received. Figure 7.13 shows this process for a mobile station that is paged by BTS1 while it is moving the cell covered by BTS2.

This feature increases the probability that the mobile is always on the paging/access Channel associated with the relatively strongest RF signal and therefore has improved chances of completing the call setup sequence.

Access Probe Handoff (APHO) is the handoff that occurs while the mobile station is performing an access attempt (the period between the access probes) in the system access state. The APHO allows the mobile to perform a handoff of the paging and access channels before receiving a Layer 2 ACK to an access channel message. The mobile stops transmitting access probes on one access channel and resumes transmitting access probes on the target cell's access channel.

Access Handoff (AHO) is the act of transferring reception of the paging channel from one base station to another, after a successful access attempt and before

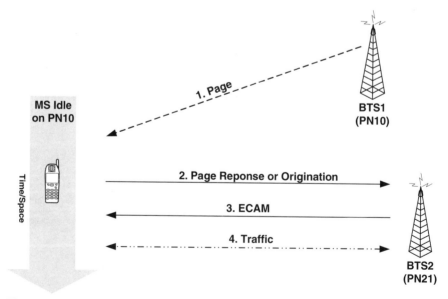

Figure 7.13   Access entry handoff.

receiving an extended channel assignment message (CAM or ECAM). The AHO feature allows the mobile to listen to the paging channel from an alternative stronger cell, after a successful access attempt with the old cell, while the mobile is still in the page response or the origination attempt substate. In this case, the mobile may receive the ECAM from a base station other than the one that received mobile's page response (see Fig. 7.14). The likely target BSs are listed in the access handoff list in any access channel message (origination/page response) from the MS.

Channel Assignment into Soft Handoff (CASHO) allows the assignment of multiple traffic channels to a mobile during the call setup phase to reduce the chance of a failed mobile call attempt for both mobile origination and termination. The ECAM allows a cell to assign multiple traffic channels to the mobile station. A set of PNs, called the assignment set, will be sent to the MS in the ECAM (see Fig. 7.15).

7.4.3 Registration

Another important part of call processing is the mobile's registration. Registration is the process by which the mobile station notifies the base station of its location, status, identification, slot cycle, and other characteristics.

The information provided through the registration process is used by the base station to efficiently page the mobile station when establishing a mobile station-terminated call. For operation in the slotted mode, the mobile station supplies the SLOT_CYCLE_INDEX parameter so that the base station can determine which slots

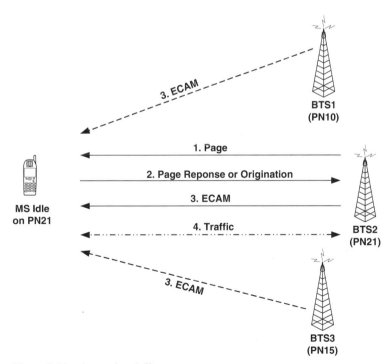

Figure 7.14   Access handoff.

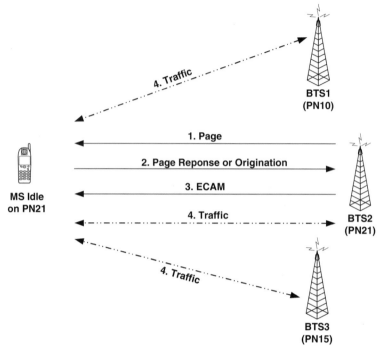

Figure 7.15   Channel assignment into soft handoff.

TABLE 7.2   CDMA Mobile Protocol Revisions

MOB_P_REV	Standard Rev.
MOB_P_REV 1	J-STD-008, 1.8 to 1.9 GHz
MOB_P_REV 3	IS95A, 800 MHz
MOB_P_REV 4, 5	TIA/EIA-95B
MOB_P_REV 6	cdma2000, Rev. 0 (1 × RTT)
MOB_P_REV 7	cdma2000, Rev. A

the mobile station is monitoring. The mobile station supplies the station class mark and the protocol revision number so that the base station knows the capabilities of the mobile station. Table 7.2 shows the mobile protocol revisions up to Release A.

The IS2000A specifications support 11 different forms of registration:

- Power-up registration, when the MS powers on, switches from using a different frequency block, a different band class, or an alternative operating mode or from using the analog system

- Power-down registration, when the MS powers off if previously registered in the current serving system

- Timer-based registration, when a prespecified timer expires

- Distance-based registration, when the distance between the current BS and the BS in which the MS last registered exceeds a threshold
- Zone-based registration, when the MS enters a new zone
- Parameter change registration, when some of the MS's stored parameters change or when it enters a new system
- Ordered registration, which is initiated by the BS through an order message
- Implicit registration, when the MS successfully sends an origination message, reconnect message, or page response message. In these cases the base station can infer the MS's location, and thus this is considered an implicit registration. This type of registration does not involve the exchange of any registration messages between the BS and the MS.
- Traffic channel registration. Whenever the BS has registration information for a MS that has been assigned to a traffic channel, the BS can notify the MS that it is registered.
- User zone registration, when the MS registers when it selects an active user zone
- Encryption/message Integrity re-sync required registration. The MS registers when the new extended encryption is turned on or when message integrity is supported and the MS determines that it can not decrypt or validate any messages from the BS.

The first five forms of registration, user zone registration, and encryption/message integrity re-sync required registration, as a group, are called autonomous registration.

Autonomous registration and parameter change registration can be enabled or disabled, and their parameters can be communicated in the system parameters message on the paging channel or the ANSI-41 system parameters message on the primary broadcast control channel. Enabling or disabling autonomous registrations depends on the roaming status. Parameter change registration is independent of roaming status.

The base station can also obtain registration information by sending the status request message to the mobile station on either the f-csch or the f-dsch. The mobile station can be notified that it is registered through the mobile station registered message.

7.5   REFERENCES

1. C.S0002-A, Physical Layer Standard for cdma2000 Spread Spectrum Systems, July 2001.
2. C.S0003-A, Medium Access Control (MAC) Standard for cdma2000 Spread Spectrum Systems, July 2001.
3. C.S0004-A, Signaling Link Access Control (LAC) Standard for cdma2000 Spread Spectrum Systems, July 2001.
4. C.S0005-A, Upper Layer (Layer 3) Signaling Standard for cdma2000 Spread Spectrum Systems, July 2001.
5. "Enhanced Random Access and Reservation Scheme in CDMA2000," K. Etemad, *IEEE Personal Communications Magazine*, April 2001

CHAPTER *8*

TRAFFIC CHANNEL OPERATION

8.1 INTRODUCTION

One of the main design objectives of IMT2000 systems and IS2000 specifically is to allow efficient support of high-speed data traffic along with conventional or enhanced voice services. To achieve this objective in a mobile cellular network, IS2000 traffic channels are designed to allow flexible and dynamic rate allocation for high-speed packet data as well as voice traffic transmissions [1–4]. Different combinations of high- and low-speed traffic channels are dynamically allocated to users and applications according to operator-defined priority rules and for admission control and scheduling. Depending on the implementation, the radio resource management may support different levels of user-level or application-level QoS control.

A IS2000 mobile station communicates with the serving base stations with the forward and reverse dedicated traffic channels while in *mobile station control on the traffic channel state*.

The mobile station control on the traffic channel state consists of the following substates:

- *Traffic Channel Initialization Substate:* In this substate, the mobile station verifies that it can receive the forward traffic channel and begins transmitting on the reverse traffic channel.

- *Traffic Channel Substate:* In this substate, the mobile station exchanges traffic channel frames with the base station based on a specific service configuration. During this substate one or more concurrent call control instances can be activated.

- *Release Substate:* In this substate, all call control instances are disconnected and the associated physical channels are released.

In addition to user data the traffic channel also carries signaling and MAC messages to allow closed-loop power control, radio link protocol for packet data, as well as handoff. The traffic channel type, frame sizes and data rates as well as the assignment and usage depend on application type, user class, and channel conditions.

This chapter describes on the operation of traffic channels focusing on traffic channel service negotiation, voice and data transmission on FCH/DCCH and SCH channels, as well as admission control and rate allocation issues. Other areas of focus in this chapter are traffic channel power control and handoff procedures.

CDMA2000® Evolution: System Concepts and Design Principles, by Kamran Etemad
ISBN: 0-471-46125-3   Copyright © 2004 John Wiley & Sons, Inc.

8.2 TRAFFIC CHANNEL SERVICE CONFIGURATION AND NEGOTIATION

During the traffic channel operation the mobile station and base station use a common set of attributes, referred to as service configurations, for building and interpreting traffic channel frames.

The mobile station can request a default service configuration associated with a service option at call origination and can request new service configurations during traffic channel operation.

Similarly, it is possible for the base station to request a default service configuration associated with a service option when paging the mobile station and to request new service configurations during traffic channel operation.

The service configuration can also be determined as part of a service negotiation process between the base station and the mobile station. If the mobile station requests a specific service configuration that is not acceptable to the base station, the base station can reject the requested service configuration or propose an alternative service configuration. The mobile station may then accept or reject the base station's proposed service configuration or propose yet another service configuration. This service negotiation process ends when service configuration is mutually agreed upon, or when either the mobile station or the base station rejects a service configuration proposed by the other.

The service configuration parameters include both negotiable and nonnegotiable parameters. The nonnegotiable parameters are those sent from the base station to the mobile stations only.

The set of negotiable service configuration parameters consists of the following:

- *Forward and Reverse Multiplex Options* are parameters that define how the information bits of the forward and reverse traffic channel frames, respectively, are divided into different types of traffic, such as signaling traffic, primary traffic, and secondary traffic. A multiplex option combined with a radio configuration specifies the frame structures and transmission rates.

- *Forward/Reverse Traffic Channel Configurations* define the radio configurations and other necessary attributes for the forward or reverse traffic channels.

- *Forward/Reverse Traffic Channel Transmission Rates* are the transmission rates actually used for the forward and reverse traffic channels, respectively. These transmission rates can include all or a subset of the transmission rates supported by the radio configuration associated with the corresponding traffic channel multiplex option.

Note that the multiplex option, channel configuration, and transmission rates used for the forward and reverse traffic channel can be different.

Service option connections include all the services in use on the traffic channel. A mobile station may have zero, one, or multiple simultaneous service option connections. When there is no service option connection the mobile station may send null frames or signaling traffic on the reverse traffic channel. A service option connection has the following attributes:

- *Service Option type*, which formally defines the way in which traffic bits are processed by the mobile station and base station
- *Service Option Connection Reference*, which provides a means for uniquely identifying the service option connection, especially in cases in which there are multiple service option connections in use
- The *Forward/Reverse Traffic Channel traffic type(s)* used to support the service option. A service option may require the use of a particular type of traffic, such as primary or secondary, or it may accept more than one traffic type. This attribute also defines whether the service option requires only a one-way traffic channel or it needs to be supported on the forward and reverse traffic channels simultaneously.

The set of nonnegotiable service configuration parameters includes the following:

- *Reverse Pilot Gating Rate*, which controls the way the reverse pilot channel is gated.
- *Forward and Reverse Power Control Parameters*, which define the mode of power control and the associated parameters.
- *Logical to Physical Mapping*, a table of logical to physical mapping entries, consisting of service reference identifier, logical resource, physical resource, forward flag, reverse flag, and priority.

The mobile station can request a default service configuration associated with a service option at call originatio, and can request new service configurations during traffic channel operation.

8.3   VOICE AND DATA TRANSMISSION ON TRAFFIC CHANNELS

This section describes data and voice transmission on the forward and reverse traffic channels as well as the system considerations in their allocation and release processes. More details can be found in the 3GPP2 specification [2,3].

Voice and Data Transmission in RC1 and RC2: The traffic channels in RC1 and RC2 of IS2000 are the fundamental and supplemental code channels (FCH and SCCH), which are designed to be backward compatible with IS95A/B.

In these radio configurations the FCH is used for all voice and associated signaling traffic. FCH can also be used to send low-speed circuit or packet data in forward and reverse directions. The FCH may be combined with up to seven supplemental code channels to provide higher data rates. Note that the SCCH in IS2000 is the same as the supplemental channel of EIA-TIA95B and the new name is adopted to differentiate this channel from the high-speed and variable-rate supplemental channel of IS2000. Each supplemental code channel (SCCH) carries only user data and has a fixed rate equal to the corresponding FCH full rate, namely, 9.6 kbps for RC1 and 14.4 kbps for RC2. The SCCH can only be allocated along with an FCH.

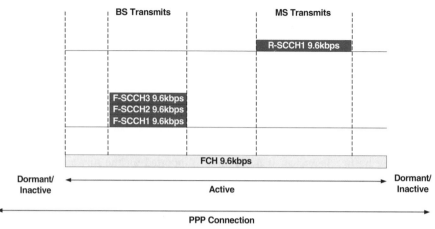

Figure 8.1 Code channel aggregation during data bursts to provide higher data rates in EIA-TIA95B and IS2000.

Whereas the FCH is allocated and established between the mobile station and the base station, the additional SCCHs can be assigned independently and dynamically for the duration of each forward or reverse data burst and released afterward. The SCCHs are requested and allocated using the supplemental request message (SCRM) and supplemental channel assignment message (SCAM) carried on the FCH, respectively.

This concept of code channel aggregation is simplified in an example shown in Figure 8.1 where the mobile is exchanging traffic and signaling frames with the base station on an FCH in RC1. In this example, the base station assigns 3 forward SCCHs to the mobile in an SCAM message and sends a data burst at $(3 + 1) \times 9.6$ = 38.4 kbps by aggregating the FCH with the 3 F-SCCHs each with a rate of 9.6 kbps. The mobile then requests for a 19.2-kbps data burst by sending a SRCM message to the base station, and after receiving the corresponding SCAM message it sends a data burst at 19.2 kbps using the assigned reverse SCCH and the FCH.

The SCCHs are not expected to be used for IS2000 mobiles because SCHs provide more efficient and more flexible framework for high-speed data transmission.

Note that in the backward-compatible modes with RC1 and RC2, the packet data MAC functions are similar to EIA-TIA95B and they are based on the concept of packet data service over an underlying circuit switched-based call model. In this MAC the mobile with packet data service option has two states, namely, the active and the dormant state. In the active state a traffic channel is assigned to the mobile, and a link layer along with a point-to-point protocol (PPP) connection is established between the IWF/PDSN and the mobile. In the dormant state no physical traffic channel is assigned to the call; however, the state of the user's registration for packet data service and the logical PPP connection is maintained until the mobile terminates the session.

Voice and Date Transmissions in RC3 and Higher: For RC3 and higher the IS2000 uses FCH or DCCH as the basic traffic channel along with variable-rate SCHs when needed for high-speed data burst transmissions.

The variable-rate circuit-switched voice traffic is supported with legacy Rate Set 1 and Rate Set 2 vocoders as well as the new selectable mode vocoder (SMV). The SMV is used to provide better voice quality at the same capacity or at a higher capacity with a quality comparable to legacy vocoders. For details on the SMV see Chapter 4.

All voice traffic is carried on the FCH channel, using backward-compatible frame structure and associated signaling. Note that even in RC3 and higher, although DCCH can in principle be used to carry voice traffic, FCH is generally the preferred channel for this purpose. This is mainly because of FCH's backward-compatible design and its support for rate reduction according to the data source activity that minimizes cochannel interference. The set of available FCHs can dynamically be allocated to 2G or 3G mobiles. Although the frame structure of FCH is the same as the traffic channels of IS95A, a number of new messages and vocoder options are introduced to be addressed to and used by 3G mobiles.

For data traffic in RC3 and higher the system may assign either an FCH or a DCCH as the basic traffic channel. These bidirectional channels carry low-rate user data and their retransmissions as part of the radio link protocol (RLP) as well as the in-band associated signaling traffic. The in-band signaling and MAC messages include the power control and handoff-related messages as well as RLP acknowledgments.

The FCH/DCCH for data is established in the same way as the voice traffic channel after exchanging signaling messages on paging and access channels. The FCH/DCCH needs to be set up before a mobile's data session enters an "active" state, where it can exchange traffic frame with the base stations on a dedicated traffic channel (see Fig. 8.2).

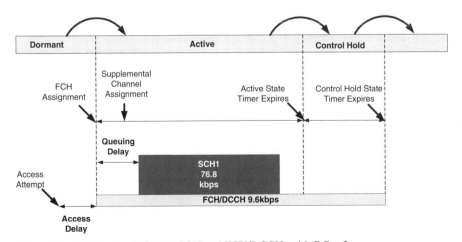

Figure 8.2   Data transmission on SCH and FCH/DCCH, with RC > 2.

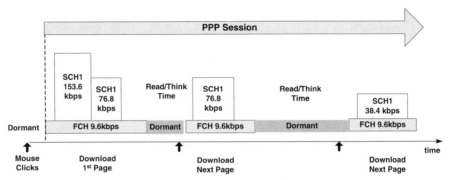

Figure 8.3   Multiple traffic channel assignments and releases within the same PPP session.

This setup occurs at the beginning of a data session or when the user returns from a "dormant" state. A users' data session is in dormant state if no dedicated channel is used by the mobile. The dormant-to-active transition may happen because of a mobile or network request for data exchange. As shown in Figure 8.3, the mobile station may enter and exit the active state several times during a packet data session, under the same PPP logical connection.

8.3.1   Forward SCH Assignment

A forward supplemental burst assignment, indicated as part of an *extended supplemental channel assignment message* (ESCAM), specifies the start time, data rate, and the duration of the transmission. The ESCAM also defines the code channel, that is, Walsh or QOF index, and the active set for the assigned channel, identified by FOR_SCH_ID.

The duration or the forward supplemental assignment interval is specified by the parameter FOR_SCH_DURATION. Whereas a duration value of "1111" indicates infinite duration, setting this parameter to "0000" indicates that the mobile station should stop processing the forward supplemental channels at the specified start time.

Depending on the amount of data buffered for transmission and the allocated data rate, the duration of transmission on F-SCH allocation can span over multiple 20-ms frames.

The system defines a minimum and a maximum transmission time on the SCH. Note that although in principle the SCH is the system's preferred channel for packet data transmission, for small packet sizes it may be more efficient to carry the traffic on a low-rate FCH/DCCH channel.

For example, the minimum and maximum durations may be set to 160 ms (8 frames) and 2.56 s, respectively. In this case, assuming a minimum SCH data rate of 19.2, all packets shorter than $8 \times 20 \text{ ms} \times 19.2 \text{ kbps} = 3072$ bits are transmitted on FCH only and they do not require an SCH allocation. For any packet longer than this minimum size the system allocates the F-SCH.

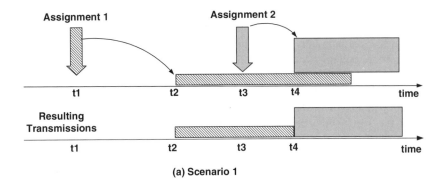

(a) Scenario 1

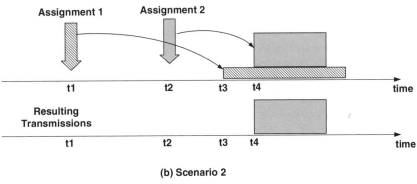

(b) Scenario 2

Figure 8.4   A new F-FCH may be assigned while the previous one is still in use.

A mobile may receive a new forward supplemental assignment while processing the F-SCH based on the last assignment. The time start of the second assignment may be before or after the end of the first assignment. In any case, the new
assignment overwrites the previous assignment.

If the new assignment starts after the previous assignment ends, the result is
clear. There are two scenarios in which the mobile station receives a new forward
supplemental assignment before the end of the first assignment. These scenarios are
shown in Figure 8.4.

- In Scenario 1, Assignment 2 starts after the start time, but before the end, of
  Assignment1, that is, when the mobile has already started the processing of
  F-SCH according to Assignment1. In this case, Assignment 2 takes effect
  immediately at its start time and Assignment 1 is ignored thereafter.
- In Scenario 2, Assignment 2 starts before the mobile starts processing the supplemental channel according to Assignment 1. In this case, the second assignment simply replaces the first assignment.

The F-SCH transmission may be continued beyond the initially assigned duration if more data are buffered and if the resources are safely available. However, the number of consecutive continuations may be limited to prevent one or a small number of users from monopolizing system resources for a long period of time.

In general, the FCH/DCCH allocation (for both voice and data calls) takes precedence over the supplemental channel allocation. Therefore, if resources needed for setting up an FCH/DCCH are unavailable because of their utilization by an existing SCH, the system may terminate transmission on the SCH early to release its resources to the incoming FCH/DCCH.

Data transmission on F-SCH is terminated normally when the transmission time spans over the entire duration assigned to the mobile in the corresponding ESCAM message. The F-SCH transmissions may, however, be terminated earlier than specified in ESCAM because of base station power overload, increased interference, or the need to free up base station resources to allocate FCHs for new or soft handoff voice or data calls. In this case the BS may ask the MS to stop processing F-SCH by assigning a zero interval in the ESCAM, that is, by setting FOR_SCH_DURATION to "0000," at the specified start time. Alternatively, and based on the reevaluation of resource availability at the time of termination, the BS may assign a new F-SCH at a lower rate to the mobile. In either case, a low-rate connection on the underlying FCH or DCCH is typically maintained to avoid any service interruption. Transmission rate of the F-SCH may subsequently be increased if the system determines that a significantly larger amount of resources have become safely available. Figure 8.5 shows a scenario in which the F-SCH1 data rate is reduced to release some Walsh codes, power, or channel elements needed for a new FCH to support an incoming voice or data call.

The radio resource management should allow sufficient margin to reduce the probability of early F-SCH termination.

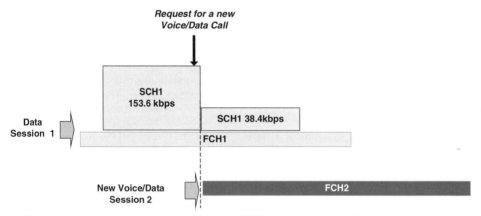

Figure 8.5   Lowering the rate on an ongoing SCH Transmission to allow new traffic channels.

8.3.2   Reverse SCH Assignment

On the reverse link, the mobile requests for a R-SCH channel by sending an enhanced SCRM message on the R-FCH or the R-DCCH. The base station allocates the R-SCH by sending a ESCAM message on the F-FCH/F-DCCH specifying the rate, duration, and start time for the data burst.

The duration of R-SCH allocation can span one or more 20-ms frames depending on the amount of data in the mobile's transmit buffer. Similar to the forward link, the system defines a minimum burst duration for R-SCH. Any data bursts shorter than this minimum length can be transmitted at lower rates on the R-FCH or the R-DCCH. On the reverse link, the system lets the mobile station decide whether to send data over R-FCH and R-SCH simultaneously.

Unlike the single-leg F-SCH handoff operation, the operation of the R-SCH may be on the full active set, that is, on all soft handoff legs of FCH. Therefore, the rate of the assigned R-SCH is determined by the minimum burst rate that can be supported simultaneously by all base stations in the active set.

In a normal R-SCH transmission termination, the mobile station completes transmission of data in its buffer and requests the R-SCH release by sending a SCRM message specifying a zero amount of data to be transmitted. However, the R-SCH may be terminated early by the system before the mobile requests such termination. The early termination of an R-SCH assignment may be caused by approaching reverse link overload conditions or because of the need to free up base station resources, for example, channel elements, to allow processing of the new voice or handoff call requests. Also, as the R-SCH can be in soft handoff with multiple base stations, it may need to be terminated early to adjust, and typically lower, the data rate such that the new handoff leg can support it.

When R_SCH early termination does occur, the MS may be assigned a new R-SCH at a lower rate indicated by a ESCAM based on the reevaluation of resources available at the time of termination. Note that the R-SCH termination does not necessarily mean service or connection interruption, as traffic channel frames may still be exchanged on the FCH/DCCH at low rates.

8.3.3   Admission Control and Traffic Channel Allocation

The traffic channel allocation in IS2000 relies on priority-based admission control for voice and data call requests as well as radio resource management to efficiently and uniformly utilize base station resources across different frequency carriers.

Call admission algorithms are designed to maximize the number of simultaneous active users while protecting the system from overload. The system defines a set of admission thresholds and load distributions to provide an acceptable level of service to all existing and incoming users.

The radio resource management is tightly related to the admission control because the priority-based admission control not only applies to new call arrivals but also impacts the channel assignments and scheduling for ongoing voice or data calls.

In IS2000 the call admission control and resource control depend greatly on the system implementation; however, the following considerations are typically made in most cases.

- High- vs. Low Priority Applications: Whereas in most systems the voice has higher call admission priority than packet data traffic, the system can also be configured to give define alternative priority rules if desired. The specific priority rules supported by the system are implementation dependent. For example, when the highest priority is given to voice, a new voice call or a soft handoff leg request with no available systems resources will immediately seize resources by terminating a high-speed transmission on a SCH or by pushing the existing active low-rate packet data connections into the dormant state.

- Load Balancing Among Carriers: Another criterion for call admission is load balancing across different frequency channels of a multicarrier system. The load balancing ensures fair utilization of channel resources as well as uniform distribution of interference across carriers.

- 2G vs. 3G Call Admission: In IS2000 the resource control also needs to differentiate different generations of mobile stations and their capabilities and take those into consideration for channel assignments. In a backward-compatible network with a mix of legacy and 3G-capable mobiles the system needs to support both 2G- and 3G-type voice and data traffic. The base station may be configured to carry 2G- and 3G-type traffic on separate carriers or dynamically share each of, or some of, its carriers between the two types of traffic. In each case the strategy for balancing the 2G and 3G traffic must be optimized to ensure that while each carrier is fairly utilized the interfrequency or intergeneration handoff can be performed seamlessly. The specific scheduling algorithm depends on the implementation and can be optimized based on different criteria. However, most scheduling schemes consider the following factors in determining the rate and duration of transmission on the F-SCH:

- User class, application class, and amount of data buffered for transmission to the mobile

- Any estimate of user's current channel condition

- Resources required for an assignment
  - Fractional power
  - Number of channel elements
  - Length of Walsh code
  - Backhaul capacity

- Current resource availability based on
  - The current forward link power utilization or loading in terms of fraction of total power consumed by other applications and users
  - Channel element, backhaul, Walsh function, and other hardware and system resource availability
  - An additional margin to ensure resource availability for urgent assignment such as voice and/or handoff users.
  - Scheduling policy
  - Mobile's soft handoff state: if F-SCH supports soft handoff. Note that in the forward link the high-speed and scheduled F-SCH has a smaller active set

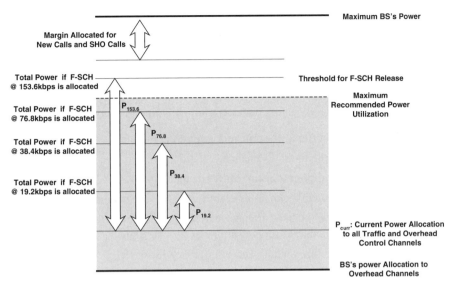

Figure 8.6   Power-based downlink rate determination.

than the FCH and it may only support single-leg handoffs. See the following sections for details on traffic channel soft handoff.

Figure 8.6 schematically shows the concept of rate determination based on power resource utilization. The same approach is typically taken for other resources such as Walsh code and channel elements. In this example, the base station adds the power P_{Ri}, required for an F-SCH assignment at rate R_i, to the total power P_{curr} currently used by other channels including the overhead channels to estimate the total power P_{post} used after such assignment. The base station may allocate the SCH at rate R_i if the estimated $P_{post} = P_{Ri} + P_{curr}$ does not exceed the recommended power usage at the base station.

For each data rate R one can define a parameter called the power factor, PF_R as the product of data rate R and its required E_b/N_o. The higher the data rate PF_{Ri} on an SCH, the higher power P_{Ri} required by that channel.

In this example, the base station, based on the current power allocations and thresholds, can only allocated up to a 76.8-kbps rate to the F-SCH channel. The base station may allocate the maximum rate on SCH that meets this criterion or a lower rate based on other considerations.

Also note that Figure 8.6 shows a margin threshold for new or handoff calls as well as a threshold for releasing an ongoing F-SCH transmission if such levels are reached.

The same evaluation needs to be performed on the Walsh code, channel elements, and backhaul resources before the feasible rate can be determined.

The higher the data rate on an F-SCH, the shorter the Walsh code needed for that channel and the larger the number of equivalent FCH channel it occupies. More specifically, an F-SCH at 19.2-, 38.4-, 78.6-, and 153.6-kbps rate will take 2×,

4×, 8×, and 16× codes than an F-FCH or voice channel at 9.6 kbps requires, respectively.

Similarly, from a channel elements (CE) perspective, in most initial implementations transmitting on a high data rate a F-SCH channel requires a larger number of CEs than the low-rate channel.

Similar to F-SCH, the reverse link SCH assignment is subject to various scheduling and resource allocation consideration including the following:

- Amount of data buffered for transmission by the mobile
- Any estimate of the mobile's channel condition
- The mobile's maximum transmit power available for R-SCH
- Channel element and backhaul availability at the base station
- The current loading, that is, uplink interference and noise floor, and interference budgets available to support a new SCH, based on
  - Current reverse link loading and interference levels
  - Estimated rise of reverse link interference caused by the R-SCH after it is assigned
  - Maximum allowed interference rise and loading to allow for a margin that reduces the need for early termination of the R-SCH assignments because new or handoff call arrival

If the mobile is in soft handoff, all the above considerations need to be made by all base stations in the active set such that the assigned rate and duration on the R-SCH are feasible on all soft handoff legs.

Note that a similar rate determination approach can be used for the reverse SCH. The main difference is that the key parameter to be considered is the rise in received interference in the uplink rather the required power or Walsh code that is used for the downlink. In this case, under the same radio channel conditions, the higher the data rate on an assigned R-SCH the larger the interference rise and loading increase caused by that assignment. Therefore, the system can measure the current uplink loading and estimate the expected loading increase as a result of each data rate allocation and assign the highest rate that meets the maximum loading limit recommended for that base station.

Note that, similar to forward link, the rate determination also depends on the base station's CE and backhaul capacity availability. However, in this case the same receiver CE at the base station can be used to process a high- vs. low-rate SCH.

8.4 TRAFFIC CHANNEL HANDOFF

Similar to IS95A, the mobile station in IS2000 supports the following handoff procedures while in the mobile station control on the traffic channel state:

- Soft Handoff: a handoff in which the mobile station commences communications with a new base station without interrupting communications with the old base station. Soft handoff can only be used between CDMA channels having identical frequency assignments. Soft handoff provides diversity of forward traffic channels and reverse traffic channel paths on the boundaries between base stations

- CDMA-to-CDMA Hard Handoff: a handoff in which the mobile station is transitioned between disjoint sets of base stations, different band classes, different frequency assignments, or different frame offsets
- CDMA-to-Analog Handoff: a handoff in which the mobile station is directed from a CDMA traffic channel to an analog voice channel

8.4.1 Traffic Channel Soft Handoff

The traffic channels in IS2000 support soft handoff based on the same concepts and procedures as EIA-TIA95A. However, the soft handoff messaging and triggers are improved to enhance the system performance. This section describes some of these enhancements as they relate to fundamental and supplemental channel operation. These changes mainly involve enhanced messages, new dynamic soft handoff thresholds for fundamental channels, and the reduced active sets for supplemental channels.

Soft Handoff Messaging: In the following we describe the soft handoff scheme in IS2000 starting with a quick review of the key messaging involved in the process.

The base station indicates the PN offset of pilots from neighboring base stations in a broadcast *neighbor list message.* The base station also sends an (*extended*) *neighbor list update message*, which contains the latest combined neighbor list for the pilots in the active set.

The mobile informs the base station as to the relative strengths of the pilots that should be considered for adding or removing from the active set, by sending:

- *Pilot Strength Measurement Message* (PSMM), for IS95A cellular/PCS mobiles
- *Extended Pilot Strength Measurement Message* (EPSMM), for IS95B and 3G1X voice mobiles

These messages include information on the E_c/I_t, arrival time, and handoff drop timer for each of the pilots received by the mobile.

If the primary sector/cell decides that a change in the mobile active set is warranted, it will specify the new desired active set in:

- *Extended Handoff Direction Message*, for IS95A cellular/PCS mobiles
- *General Handoff Direction Message*, for IS95B and 3G1X voice mobiles
- *Universal Handoff Direction Message*, for 3G1X data mobiles

These messages include parameters such as HDM sequence number, CDMA channel frequency assignment, PN offset for updated active set, Walsh Code associated with each pilot in the active set, search window size for the active and candidate sets, and handoff parameters (e.g., T_ADD, T_DROP, . . .).

On establishing the new Active Set, the mobile will respond to the cell with a *handoff completion message*, which conveys a positive acknowledgment and also includes the PN offset for all active set pilots.

8.4.2 Dynamic Thresholds for Soft Handoff

As indicated in many CDMA deployments, optimizing the soft handoff areas has a major impact on system performance and efficiency. Although creating the soft handoff links at the right time and location improves link integrity and capacity, putting too many users in soft handoff will negatively impact system capacity and network resources.

On the forward link, excessive handoff can significantly reduce the capacity as each user in soft handoff takes a relatively large portion of the base station's limited power resources. On the reverse link, the impact of excessive soft handoff is less dramatic but it still takes extra system resources, such as channel elements and backhaul capacity, increasing the cost and complexity of the network.

In many cases adjusting fixed handoff thresholds at the base station cannot solve the problem. For example, areas in a cell where all pilots are weak require lower handoff thresholds to create more diversity by allowing more base stations in soft handoff (see Fig. 8.7). Meanwhile, in other areas of the same cell, where one or two strong pilots are dominant, higher handoff thresholds are desirable to limit the number of base stations in soft handoff. In this case, engaging multiple base stations in soft handoff takes extra network resources without providing additional diversity gains.

To address this issue, IS2000 defines new soft handoff parameters and messages, most of which were introduced in EIA-TIA95B but were not widely implemented until IS2000 Release 0/A. The new handoff mechanism involves dynamic thresholds to optimize the active set management, resulting in better handoff performance and resource utilization.

The main criterion for adding pilots to, or removing them from, the active set is the combined received power from exiting pilots in the active set.

If this combined power is low, the system lowers the dynamic pilot-add threshold to accept more pilots to the active set to be demodulated and provide higher

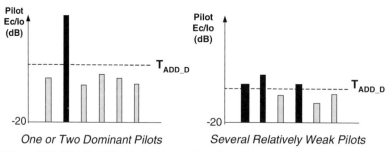

Figure 8.7   The T-ADD threshold must be adjusted based on received pilots' strengths.

diversity gain. However, when the combined power of existing active pilots is high, for example, when there are one or two strong dominant pilots, the system raises the threshold to avoid adding additional pilots to the active set. At that point, adding more pilots of weaker power would only take more network resources without providing improvement in the link quality.

The same concept is used to remove a pilot from the active set if that pilot does not significantly contribute to the overall received power and removing that pilot does not change the combined link quality.

The dynamic soft handoff thresholds are captured in three main parameters, namely, SOFT_SLOPE, ADD_INTERCEPT, and DROP_INTERCEPT. The following describes the handoff process in IS2000 and EIA-TIA95B based on these parameters.

When $SOFT\_SLOPE_s$ = "0" or if $P\_REV\_IN\_USE_s \leq 3$, the new parameters have no effects as the IS2000 soft handoff falls back to a process similar to EIA-TIA95A. Figure 8.8 illustrates an example of messaging triggers in this case, as the strength of a pilot P0 in the candidate set gradually rises above the strength of each of the two pilots, P1 and P2, of the active set.

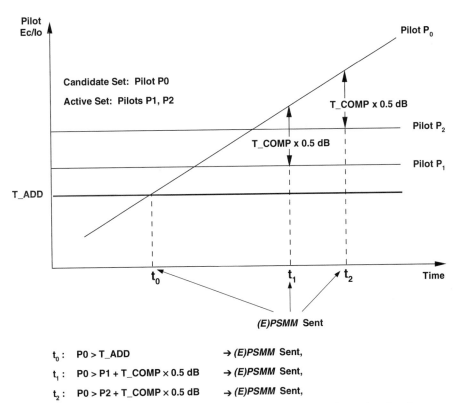

Figure 8.8   PSMM messaging triggers with fixed SHO thresholds. Reproduced under written permission from TIA.

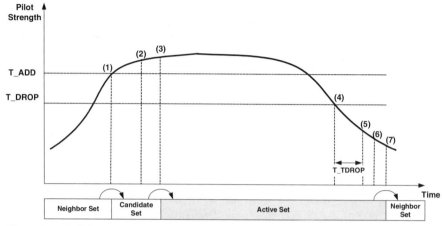

Figure 8.9 Soft handoff steps when SOFT_SLOPE = 0.

The mobile station sends an autonomous (extended) pilot Strength measurement message once any of the following events occur:

- The strength of a neighbor set or Remaining set pilot is found to be above T_ADDs or

- The strength of a candidate set pilot exceeds the strength of an active set pilot by T_COMPs × 0.5 dB.

Figure 8.9 and the following steps show the messages exchange and pilot set transitions during a typical handoff process if $P\_REV\_IN\_USE_s \leq 3$ or $SOFT_SLOPE_s = $ "0."

1. Pilot strength exceeds T_ADD. Mobile station sends a pilot strength measurement message, as a request to add to the BS, and transfers the pilot to the candidate set.

2. Base station sends an extended handoff direction message, a general handoff direction message, or a universal handoff direction message.

3. Mobile station transfers the pilot to the active set and sends a handoff completion message.

4. Pilot strength drops below T_DROP. Mobile station starts the handoff drop timer.

5. Handoff drop timer expires. Mobile station sends a pilot strength measurement message as a request to remove to the BS.

6. Base station sends an extended handoff direction message, a general handoff direction message, or a universal handoff direction message.

7. Mobile station moves pilot from the active set to the neighbor set and sends a handoff completion message.

One can see that these steps are the same as those in IS95A, except for new handoff-related messages in EIA-TIA95B and IS2000.

The main effect of the dynamic threshold is when P_REV_IN_USE$_s$ > 3, that is, the mobile is EIA-TIA95B and IS2000 compliant, and when SOFT_SLOPE$_s \neq$ "0." In this case the system defines two new thresholds as follows:

A dynamic add threshold $T_{\text{ADD\_D}}$ is defined as:

$$T_{\text{ADD\_D}} = \frac{\text{SOFT\_SLOPE}_s}{8} \times 10 \times \log_{10} \sum_{i \in A} \text{PS}_i + \frac{\text{ADD\_INTERCEPT}_s}{2} \qquad (8.1)$$

where $10 \times \log_{10} \sum_{i \in A} \text{PS}_i$ is combined pilot E_c/I_o and the summation is performed over all pilots currently in the active set.

Similarly, a dynamic drop threshold $T_{\text{DROP\_D}}$ is defined as:

$$T_{\text{DROP\_D}} = \frac{\text{SOFT\_SLOPE}_s}{8} \times 10 \times \log_{10} \sum_{i \in A} \text{PS}_i + \frac{\text{DROP\_INTERCEPT}_s}{2} \qquad (8.2)$$

where the summation is performed over all pilots currently in the active set except for the pilot under consideration to be removed. Figure 8.10 graphically shows the difference between the effects of static and dynamic add thresholds on the active set management. In this figure, the shaded area corresponds to pilots above T_ADD that would not be reported to be added to the active set with the new soft handoff thresholds.

Figure 8.11 illustrates the messaging triggered by a pilot of the candidate set P0 as its strength gradually rises above the strength of each pilot of the active set, for example, P1 and P2, for the case where P_REV_IN_USE$_s$ > 3 and SOFT_SLOPE$_s \neq$ "0." Note that in this case the mobile station reports that a candi-

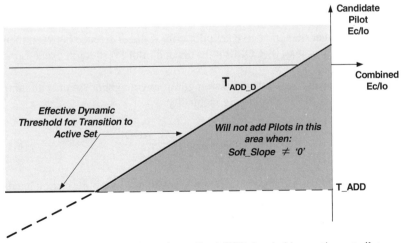

Figure 8.10 The effect of dynamic vs. fixed SHO threshold on active set pilots.

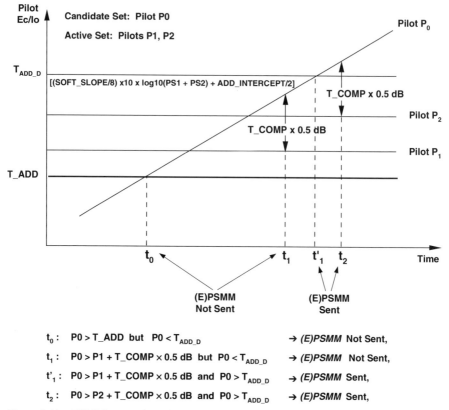

t_0 : P0 > T_ADD but P0 < T_{ADD_D} → *(E)PSMM* Not Sent,

t_1 : P0 > P1 + T_COMP × 0.5 dB but P0 < T_{ADD_D} → *(E)PSMM* Not Sent,

t'_1 : P0 > P1 + T_COMP × 0.5 dB and P0 > T_{ADD_D} → *(E)PSMM* Sent,

t_2 : P0 > P2 + T_COMP × 0.5 dB and P0 > T_{ADD_D} → *(E)PSMM* Sent,

Figure 8.11 PSMM messaging triggers with dynamic SHO thresholds. Reproduced under written permission from TIA.

date set pilot is stronger than an active set pilot only if the difference between their respective strengths is at least T_COMP × 0.5 dB and pilot P0 strength exceeds the dynamic add thresholds T_{ADD_D}.

The EPSMM is also sent when the strength of any neighbor set or remaining set pilot is found to satisfy the following inequality:

$$10 \times \log_{10} PS > \max\left(T_{ADD\_D}, -\frac{T\_ADD_s}{2} \right) \qquad (8.3)$$

Figure 8.12 provides the step-by-step pilot transitions as a mobile station hands off from base station 1 with pilot P1 to base station 2 with pilot P2.

1′. Pilot P2 strength exceeds T_ADD. Mobile station transfers the pilot to the candidate set.

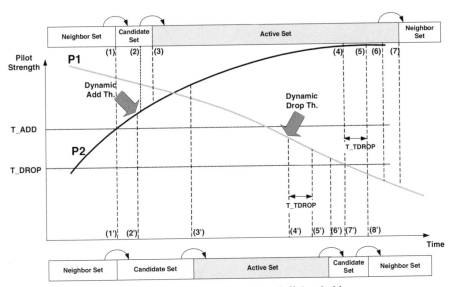

Figure 8.12 Pilot set transitions with dynamic soft handoff thresholds.

2′. Pilot P2 strength exceeds [(SOFT_SLOPE/8) × 10 × $\log_{10}(PS_1)$ + ADD_INTERCEPT/2]. Mobile station sends a pilot strength measurement message.

3′. Mobile station receives an extended handoff direction message, a general handoff direction message or a universal handoff direction message, transfers the pilot P2 to the active set, and sends a handoff completion message.

4′. Pilot P1 strength drops below [(SOFT_SLOPE/8) × 10 × $\log_{10}(PS_2)$ + DROP_INTERCEPT/2]. Mobile station starts the handoff drop timer.

5′. Handoff drop timer expires. Mobile station sends a pilot strength measurement message.

6′. Mobile station receives an extended handoff direction message, a general handoff direction message or a universal handoff direction message, transfers the pilot P1 to the candidate set, and sends a handoff completion message.

7′. Pilot P1 strength drops below T_DROP. Mobile station starts the handoff drop timer.

8′. Handoff drop timer expires. Mobile station moves the pilot P1 from the candidate set to the neighbor set.

Comparing the active sets, shown by the shaded areas at the top and the bottom of Figure 8.12, one can see the difference between the legacy and the new soft handoff regions. By combining static and dynamic thresholds (versus static thresholds only) the system can limit the location and duration of soft handoff to cases in which diversity is most beneficial.

8.4.3 Soft Handoff for Supplemental Channels

In the forward link, the data transmission on a high-rate traffic channel to each user in the handoff areas usually takes significant resources, such as power, Walsh space, and channel elements at the base stations. These resources need to be shared by all users served by the base station. In fact, it is the combination of power, channel elements, and Walsh code availability that determines base station capacity at any instance of time.

Therefore, it is often undesirable to operate a high-speed forward traffic channel in the soft handoff mode, requiring multiple base stations to simultaneously allocate major parts of their resources to a mobile. However, to ensure link integrity throughout the handoff process, it is typically necessary to maintain the in-band or associated signaling channel in soft handoff.

In IS2000 each high-speed traffic channel is a code-multiplexed channel including a fundicated channel that carries low-rate data and in-band signaling and a supplemental channel that carries high-rate user data on a different code.

In principle, different physical channels dedicated to a mobile station can have different soft handoff active sets. In IS2000, whereas the full active set is defined and maintained for the fundamental channel, a reduced active set, as a subset of the full active set, can be used for high-speed supplemental channels. As a result, in some soft handoff scenarios while the fundamental channel is in soft handoff with multiple stations, the supplemental channel may or may not be in soft handoff at all.

The higher the data rate on a forward supplemental channel, the shorter the Walsh code, the higher the allocated power, and in many implementations the higher the number of channel elements needed for the traffic channel. Therefore, the higher the data rate, the higher the cost of simulcasting the same data from multiple base stations, and therefore the smaller the reduced active set should be. In many cases, the optimal power and resource allocation strategy for SCH is to allocate all the resources on the better link, while the underlying signaling channel remains in soft handoff.

Figure 8.13 shows this scenario as the mobile moves from BS1 to BS2. Initially the mobile is connected to BS1 on both forward fundamental F-FCH and forward supplemental F-SCH channels. As the mobile realizes the strength of the pilot from BS2 and brings that to the active set, it starts receiving a second fundamental channel from BS2. The F-SCH, however, does not go into soft handoff, and it switches from BS1 to BS2 only when BS2 becomes the stronger link. During the handoff, the F-SCH may switch back and forth, a few times, between the two base stations without causing a session drop as the link integrity is maintained by the FCHs that are in soft handoff.

Soft handoff on the R-SCH has less impact on the system resources. In fact, the reverse SCH can be in soft handoff with the full active set of its fundamental channel. This is because of the fact that in the reverse link the power and Walsh codes for R-SCHs are allocated by each mobile separately and therefore such allocations will not directly affect capacity.

However, the R-SCH handoff still has an impact on the backhaul and channel element resources at the base station.

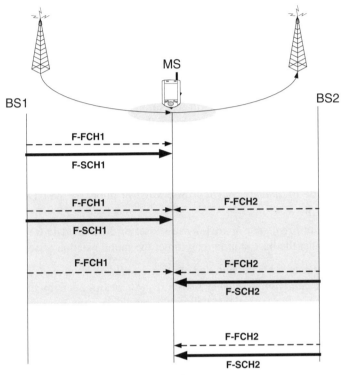

Figure 8.13   Handoff for supplemental channel.

8.4.4   CDMA-to-CDMA Hard Handoff

A CDMA-to-CDMA hard handoff takes place when a mobile station is transitioned between disjoint sets of base stations, different frequency assignments, or different frame offsets.

In initial deployments of multicarrier CDMA networks, the interfrequency hard handoff, which is a break-before-make process, was one of the main sources of dropped calls. Therefore, during the deployment of such networks, different vendors developed different ad hoc solutions to improve handoff performance. These solutions were primarily based on pilot beacons and/or round-trip delay measurements, which, once reported to the base station, could indicate the need for inter-carrier handoff.

To further improve CDMA-to-CDMA hard handoff performance, in EIA-TIA95B and IS2000 a new process is incorporated in the standard that defines a measurement-based mobile-assisted handoff across carriers. This process is described in the following.

A CDMA mobile station is capable of receiving only one carrier at a given time. Therefore, to make any measurement of a candidate frequency in support of interfrequency handoff the mobile should "leave" the currently serving frequency for a period of time.

The base station may direct the mobile station to search for pilots on a different CDMA frequency to detect the presence of CDMA channels and to measure their strengths. The mobile station reports the results of the search to the base station with the *candidate frequency search report message*.

Depending on the pilot strength measurements reported in the candidate frequency search report message, the base station may direct the mobile station to perform an interfrequency hard handoff.

The pilot search parameters are expressed in terms of the following sets of pilots on the CDMA candidate frequency:

- *Candidate Frequency Neighbor Set:* a list of pilots on the CDMA candidate frequency

- *Candidate Frequency Search Set:* a subset of the candidate frequency neighbor set that the base station may direct the mobile station to search

The measurements may be initiated with a *candidate frequency search control message* or the *candidate frequency search request message* sent by the base station. In these messages, the base station defines the candidate frequency search set as well as the periodicity and duration of each search. Table 8.1 shows the range of the search periods.

Based on the direction from the base station, the mobile station may perform a single or a periodic search of the candidate frequency search set. After each search the mobile station reports the results to the base station in the candidate frequency search report message. The mobile station may measure all pilots in the candidate frequency search set in one visit to the candidate frequency, or it may visit the candidate frequency several times in a search period, each time measuring all or some of the pilots in the candidate frequency search set. The mobile station procedures are defined such that all measurements are subject to a "freshness" test to ensure that despite multiple visits all reported pilot measurements are valid. Thus the mobile station measures the strength of all pilots in the candidate frequency search set in one or more visits, and in each visit it performs the following (see Fig. 8.14):

TABLE 8.1 Search Window Sizes

SEARCH_PERIOD$_S$	Search Period (s)	SEARCH_PERIOD$_S$	Search Period (s)
0	0.48	8	30
1	0.96	9	40
2	2	10	50
3	2.96	11	60
4	4	12	80
5	4.96	13	100
6	10	14	150
7	20	15	200

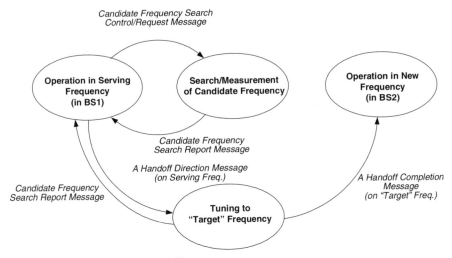

Figure 8.14   Interfrequency handoff process.

- It stops processing, that is, transmitting or receiving, any traffic channel in use on the serving frequency and stops or locks any timer and counter associated with those channels.

- It stores the CDMA band class, frequency assignment, and pilot detection thresholds and search window parameters based on its current configuration and sets them to corresponding values for the candidate frequency.

- It measures the mean input power on the candidate frequency and the serving frequency in dBm/1.23 MHz serving. If the mean input power of the candidate frequency is weaker than a threshold, the mobile may end the search for pilots in the current visit to the candidate frequency; otherwise, it will measure E_c/I_o for all or some of the pilots in its candidate frequency search set.

- It restores the CDMA band class, frequency assignment, and pilot detection threshold of the original serving frequency.

- It retunes to the serving frequency and resume processing of the traffic channels using the serving frequency active set. The processing of supplemental channels may be resumed only if their assignment has not expired.

After receiving the mobile's measurements, the base station may direct the mobile station to perform a CDMA-to-CDMA hard handoff by sending an extended handoff direction message, a general handoff direction message, or a universal handoff direction message.

At the action time specified by any of these handoff direction messages, the mobile station must disable its transmitter, reset the fade timer, suspend counters for FER measurement, and tune to the assigned forward traffic channel. The mobile station should acquire the pilots in the new active set and begin monitoring the assigned forward traffic channel within a specified time.

After the action time, on receiving sufficient signal quality on the assigned forward traffic channel over a prespecified period of time, the mobile station should reenable its transmitter and transmit the traffic channel preamble, followed by a handoff completion message or extended handoff completion message.

Hard handoff process may include the option to return to the serving frequency upon a failure, in which case the mobile should set up a handoff timer with a specified expiration time. If the mobile station declares the handoff attempt to be unsuccessful, it should restore the configuration to what it was before the handoff attempt and send a candidate frequency search report message.

The configuration parameters to be restored include service configuration, list of calls stored, protocol revision level, service negotiation type, long code mask, search window sizes, pilot add and drop detection thresholds, frame offset, nominal power setting, power control step, CDMA band class, frequency assignment, active set, and code channel list.

The mobile should then resume its operation on the traffic channels so long as the applicable timers for scheduled channels are not expired.

Intergeneration and Intersystem Handoff: The IS2000 system provides the ability to hand off voice and data calls to and from the IS95/2G system. To support this intergeneration handoff the IS2000 infrastructure needs to support the backward-compatible channels and the mobile stations need to be both IS95- and IS2000 capable. The procedures for IS95-to-IS2000 handoff and vise versa are straightforward extensions of the hard handoff procedures defined in IS95 and IS2000.

In the case of IS95-to-IS2000 handoff, sometimes referred to as "hand up," the backwards-compatible IS2000 system sends an IS2000 signaling message to the IS95/IS2000-capable MS directing the MS to perform an IS95-to-IS2000 hard handoff.

A voice or data handoff from IS95 to IS2000, if it does not require a change of service option, can be executed with no, or minimal, service interruption.

Similarly, the IS2000-to-IS95 handoff, sometimes referred to as "hand down," can occur with minimal service interruption if the currently connected IS2000 service options can be successfully mapped into IS95 services. Otherwise, connected service options are terminated as gracefully as necessary. For data service the hand-down process typically does not require reconnections of protocol layers above RLP, resulting in only a minor service interruption in the worst case.

The intersystem handoff between two IS2000 networks is also supported according to the recently developed IS-41E standard. However, in initial deployments in which this standard was not implemented, packet data calls could not continue into the new system and voice calls had to be downgraded to IS95 Rate Set 1 or Rate Set 2 before an intersystem handoff could occur.

8.5 TRAFFIC CHANNEL POWER CONTROL

The traffic channel power control in IS2000 involves a number of enhancements to forward and reverse link power control in IS95A. These enhancements include faster power control in the forward link, better channel measurement in the reverse link, and independent power control of fundamental and supplemental channels.

8.5.1 Forward Traffic Channel Power Control

The forward traffic channel power control in IS2000, for RC3–RC6, is based on a fast and closed-loop control process similar to the IS95A reverse link power control. The new process uses a reverse power control subchannel at a rate of 800 bps, which depending on the mode of power control can be split into two subchannels to control forward fundamental and supplemental channels. This major change in the forward link power control, compared with the slow power control of IS95A, is expected to improve the power efficiency of the system and therefore increase the forward link capacity.

The reverse power control subchannel applies to Radio Configurations 3 through 6 only. The mobile station maintains both the inner power control loop and the outer power control loop for forward traffic channel power control on the forward fundamental channel, the forward dedicated control channel, and the forward supplemental channel, when assigned (see Fig. 8.15).

The outer power control loop estimates the set point value based on E_b/N_t to achieve the target frame error rate (FER) on each assigned forward traffic channel. These set points are communicated to the base station, either implicitly through the inner loop or explicitly through signaling messages. The differences between set points help the base station derive the appropriate transmit levels for the forward traffic channels that do not have inner loops.

The inner power control loop compares the E_b/N_t of the received forward traffic channel with the corresponding outer power control loop set point, if one is maintained for that channel, to determine the value of the forward power control (FPC) bit to be sent to the base station on the reverse power control subchannel.

The mobile station may also transmit the erasure indicator bits (EIB) or the quality indicator bits (QIB) on the reverse power control subchannel on the command of the base station.

Erasure Indicator Bit (EIB) is a bit used in the RC2 reverse traffic channel frame structure to indicate an erased forward fundamental channel frame. In RC3, -4, -5, and -6 the EIB is used in the reverse power control subchannel to indicate frame erasure(s) and/or nontransmission on the forward Fundamental channel or forward dedicated control channel.

Quality Indicator Bit (QIB) is a bit used in the RC3, -4, -5, and -6 reverse power control subchannel to indicate signal quality on the forward dedicated control channel. When the forward fundamental channel is present, this bit is set the same as the EIB.

Reverse Power Control Subchannel Structure: The reverse link power control subchannel is time multiplexed with the reverse pilot, and it is transmitted on the reverse pilot channel (see Fig. 8.16). Each 1.25 ms power control group on the reverse pilot channel contains $1536 \times N$ PN chips, where N is the spreading rate number ($N = 1$ for SR1 and $N = 3$ for SR3).

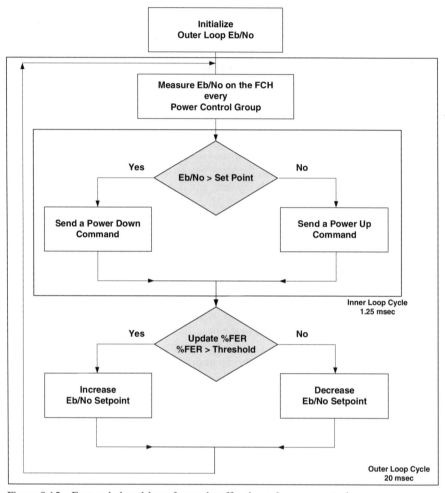

Figure 8.15   Fast and closed-loop forward traffic channel power control.

The pilot signal is sent in the first $1152 \times N$ PN chips, and the reverse power control subchannel is transmitted in the following $384 \times N$ PN chips in each power control group on the reverse pilot channel.

The reverse power control subchannel carries FPC bits, EIB, or QIB generated by the mobile. Each of the $384 \times N$ PN chips on the reverse power control subchannel is a repetition of the FPC bit, EIB, or QIB.

Power control bits, signifying power-up or power-down, are multiplexed with reverse link pilot at a rate of 800 Hz. The power control bit is punctured in the 20 ms frame every 1.25 ms, which results in a power control frequency of 800 bps.

The reverse pilot channel can be transmitted with the gated transmission mode enabled or disabled. When the reverse pilot gating is enabled, the mobile can trans-

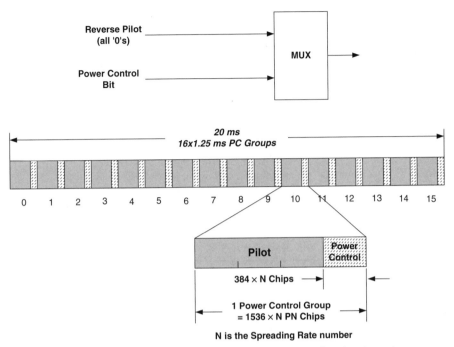

Figure 8.16   Reverse power control subchannel structure. Reproduced under written permission from TIA.

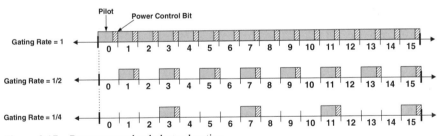

Figure 8.17   Power control subchannel gating.

mit the reverse pilot channel and the FPC bits according to the gating rate specified by the base station (see Fig. 8.17). The power control gating is typically used during the control hold state, that is, for users who are actively receiving data, and it is mainly enabled to reduce the uplink interference.

If the mobile station supports the forward supplemental channel, the 800-bps capacity of the reverse power control subchannel may be split into two lower-rate channels to allow for two independent power control loops. The primary power control subchannel would control the dedicated control or fundamental channel,

TABLE 8.2 Reverse Power Control Subchannel Configurations

FPC_MODE$_s$	Allowed in Gated Mode	Reverse Power Control Subchannel Allocations	
		Primary Reverse Power Control Subchannel	Secondary Reverse Power Control Subchannel
"000"	Yes	800 bps	Not supported
"001"	No	400 bps	400
"010"	No	200 bps	600
"011"	Yes	EIB (50 bps)	Not supported
"100"	Yes	16 bits set to QIB (50 bps)	Not supported
"101"	No	8 bits set to QIB (50 bps) (for Fundicated Channel)	8 bits set to EIB (50, 25, and 12.5 bps for 20-,
"110"	No	400 bps	40-, and 80-msec frames (for Supplemental Channel)
All other values		Reserved	Reserved

whichever is used, whereas the secondary subchannel, when used, would independently power control the supplemental channel.

The parameter FPC_MODE$_s$ defines the configuration of power control subchannels and its capacity partitioning between primary and secondary channels. Table 8.2 shows various configurations, including modes in which EIB and QIB are transmitted.

For example, when FPC_MODE$_s$ = "010," the mobile station will transmit the primary reverse power control subchannel at a 200 bps data rate and the secondary reverse power control subchannel at a 600 bps data rate.

Note that if the reverse pilot channel is in gated mode the mobile station may only transmit one reverse power control subchannel, that is, use FPC_MODE$_s$ = "000," "011," or "100." The secondary power control subchannel may only be used if the mobile supports a supplemental channel and it is not in the gated mode.

8.5.2 Reverse Traffic Channel Power Control

The basic reverse link power control process in IS2000 is very similar to IS95A. However, some changes are made to enhance the measurement part of the process, which improves the overall power control performance.

Open-Loop Reverse Power Control: An IS2000 mobile applies an open-loop power control to every uplink transmission on common or dedicated channels, including the reverse traffic channel. The total received power from the base station on the pilot channel is used as a reference to respond to forward link path loss changes. Assuming reciprocity between uplink and downlink path losses, the open-loop power control points to a higher reverse transmit power when it measures a weaker downlink power.

The mobile in RC3–RC6 sets its initial transmission power on the reverse pilot channel at a mean output level defined by:

Mean pilot channel output power(dBm) = −mean input power(dBm) + offset power
+ interference correction + ACC_CORRECTIONS + RLGAIN_ADJ$_S$,   (8.4)

where

- Mean power is referenced to the nominal CDMA channel bandwidth of 1.23 MHz for Spreading Rate 1 or 3.69 MHz for Spreading Rate 3.

- Offset power reflects the relative nominal gain of a physical channel compared with the reverse pilot channel, and it depends on the data rate and acceptable errors rate of that channel.

- Interference correction = min(max(IC_THRESH$_s$ − ECIO, 0), 7), where IC_THRESH$_s$ is a parameter set by the base station and ECIO is the E_c/I_0 (dB) per carrier of the strongest active set pilot, measured within the previous 500 ms. The E_c/I_0 (dB) must be determined by taking the ratio of the received pilot energy per chip, E_c, to the total received power spectral density (noise and signals), of at most k usable multipath components, where k is the number of demodulating elements in the mobile station receiver. In SR1 and SR2 the mobile station must determine the total received power spectral density, I_0, over 1.23 MHz and 3.69 MHz, respectively. Also, for SR3 the mobile station must calculate the total E_c/I_0 by summing the E_c from each multipath component for all three carriers before normalizing by I_0.

- The ACC_CORRECTIONS and RLGAIN_ADJ$_s$ are other correction factors based on power transmitted on the last channel before the operation of traffic channel.

In RC1 and RC2 there is no reverse pilot channel, but the above equation holds for the initial transmission power on the reverse fundamental channel. In this case, the parameter IC_THRESH$_s$ is fixed at 7 dB.

The main difference in the open-loop power calculation compared with IS95A is the interference correction factor. This factor is added to allow the mobile to distinguish between heavily loaded conditions and low-path-loss situations. In a heavily loaded situation the mobile receive power can be high because of the high forward interference floor even if the MS is far from the BS. In such cases, without the interference correction factor, the mobile may assume a low-path condition and lower its power unnecessarily.

Closed-Loop Reverse Power Control: The closed-loop power correction on the reverse traffic channel (with respect to the open-loop estimate) is achieved by the mobile's adjustments to its mean output power level in response to each valid power control bit received on the forward fundamental channel or the forward dedicated control channel. Similar to IS95A, the closed-loop power control operation consists of an outer loop and an inner loop (see Fig. 8.17).

Whereas the outer loop adjusts the BS's target E_b/N_0 set point to maintain a desired frame error rate (FER), the inner loop keeps the mobile as close to its target E_b/N_0 set point as possible.

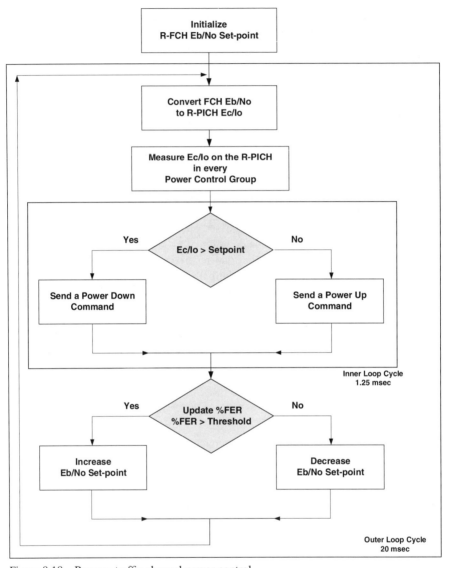

Figure 8.18 Reverse traffic channel power control.

In IS2000 the inner loop power control is similar to IS95A; however, the supporting uplink measurements are based on the reverse pilot channel instead of the traffic channel. The current target E_b/N_o for the reverse traffic channel is translated to the corresponding target reverse pilot E_c/I_o, based on the data rate of that traffic channel. The base station compares this calculated target E_c/I_o with the measured E_c/I_o of the reverse pilot channel and sets the power-up or power-down accordingly. This pilot-based power control approach allows more accurate and reliable uplink measurement, which in turn results in better power control performance.

The EIA-TIA95B/IS2000 mobiles support multiple power control step sizes defined by parameter (PWR_CNTL_STEP$_s$). The mobile applies 1 dB, 0.5 dB, or 0.25 dB power step sizes when PWR_CNTL_STEP$_s$ is set to 0, 1, or 2, respectively.

The values for initial, maximum, and minimum E_b/N_t thresholds are sent to the mobile via an extended channel assignment message (ECAM) and an extended supplemental channel assignment message (ESCAM) along with the FER thresholds. The target value of E_b/N_t threshold is reevaluated every 20 ms, and it may vary between the maximum and minimum values controlled by the outer loop power control.

The total change in the closed-loop mean output power, determined based on the accumulation of the power-up/-down steps, is applied to the total transmit power of the mobile station. The mobile station should ignore received power control bits when its transmitter is disabled. The closed-loop power control provides a dynamic range greater than ±24 dB around the open-loop estimate.

When the mobile station is unable to transmit at the requested output power level, it must reduce the data rate and/or transmission power on currently active reverse physical channels or terminate transmission on some of those channels according to their prespecified priorities. The mobile station should attempt to reduce the transmission power or the data rate or terminate transmission first on the channel with the lowest priority.

8.6 REFERENCES

1. C.S0002-A, Physical Layer Standard for cdma2000 Spread Spectrum Systems, July 2001.
2. C.S0003-A, Medium Access Control (MAC) Standard for cdma2000 Spread Spectrum Systems, July 2001.
3. C.S0005-D, Upper Layer (Layer 3) Signaling Standard for cdma2000 Spread Spectrum Systems, July 2001.

CHAPTER 9

CDMA2000 NETWORK ARCHITECTURE

9.1 INTRODUCTION

The network architectures and reference models for cdma2000 are designed by 3GPP2 to define an evolutionary and backward-compatible path from the cdmaONE legacy circuit-switched-based network to an all-IP architecture.

In most initial deployments of cdma2000 the packet data services are introduced by adding a few network elements to the existing circuit-switched-based architecture. In these cases the IP core network is separate from, but interworks with, the legacy circuit-switched core network. However, as radio access technologies and backhaul networks also evolve to IP-based frameworks the network design needs to adopt an all-IP approach.

In this chapter we present an overview of cdma2000 network architectures, starting with a review of the IS95A/B network model followed by a description of the key network elements introduced for packet data services. We then present 3GPP2 reference models for cdma2000 wireless IP and all-IP networks and discuss some of their mobility management and QoS aspects.

9.2 LEGACY CDMAONE NETWORKS

The cdmaONE network architecture is designed to support voice and low-rate circuit-switched traffic in a mobile wireless network. This architecture includes radio base stations, the circuit-switched-based transport and core networks, and other functional elements for mobility and network management.

Figure 9.1 shows the basic cdmaONE network architecture based on the TIA-EIA-TR45 model. In this figure, the blocks show network entities and circles represent network interfaces or reference points. The key network elements are briefly described in this section.

Mobile Station (MS): A MS is a mobile or fixed wireless terminal used by subscribers to access network services over a radio interface (Um).

CDMA2000® Evolution: System Concepts and Design Principles, by Kamran Etemad
ISBN: 0-471-46125-3   Copyright © 2004 John Wiley & Sons, Inc.

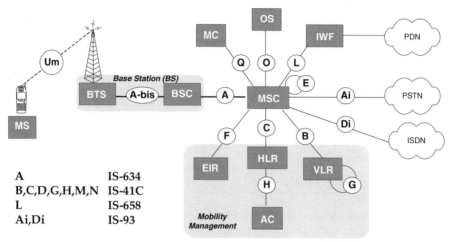

A	IS-634
B,C,D,G,H,M,N	IS-41C
L	IS-658
Ai,Di	IS-93

Figure 9.1   TIA-EIA network model for IS95A/B.

Base Station (BS): A BS is composed of a base transceiver system (BTS) and a base station controller (BSC) and provides the MS with radio access to the network services.

- The BTS is an entity, composed of radio devices, antenna, and equipment, that provides transmission capabilities across the air interface (Um).
- The BSC is an entity that provides control and management for one or more BTSs and exchanges messages with its connected BTSs and the MSC. The traffic and signaling messaging for call control, mobility management, and MS management may pass transparently through the BSC.
- The interface between a BTS and the BSC, called A-bis, is typically implementation dependent, and although it is standardized it is not widely implemented.

Mobile Switching Center (MSC): The MSC switches circuit-mode MS-originated or MS-terminated traffic and provides processing and control for calls and services. The MSC may interface to other MSCs in the same or a different network, and it may also interconnect with other public networks such as PSTN, ISDN, etc. An MSC is connected to at least one but usually multiple BSs. The interface between BS and MSC is the A interface defined in IS634.

Home Location Register (HLR): The HLR is the location register to which a user identity is assigned for record purposes such as subscriber information [e.g., electronic serial number (ESN), mobile directory number (MDN), profile Information, current location, authorization Period). The HLR is the first database that is interrogated to obtain mobile location and routing information once a MS's number is dialed by a user outside the network.

Visitor Location Register (VLR): The VLR is a location register other than the HLR used by an MSC to retrieve information for handling of calls to or from a visiting subscriber. The VLR keeps track of the mobile's location updates and works with the HLR to provide cellular mobility management. The VLR function may be integrated and collocated with an MSC. A VLR may also serve multiple MSCs.

Authentication Center (AC): The AC is an entity that manages the authentication information related to the MS. In some implementations the AC may be located within, and be indistinguishable from, an HLR. An AC may serve more than one HLR.

Equipment Identity Register (EIR): The EIR is a register to which user equipment identity may be assigned for record purposes. This record may also be used to help with the authentication process.

Message Center (MC): The MC stores and forwards short messages and may also provide supplementary services for short message service (SMS).

The integrated services digital network (ISDN) and public switched telephone network (PSTN) are defined in accordance with the appropriate ANSI T1 Standards.

Interworking Function (IWF): The IWF provides protocol conversion for one or more wireless network element. There are different IWFs to support various digital fax protocols and circuit (e.g., ISDN)-based data services.

Voice Message System (VMS): The VMS is an entity responsible for storing received voice messages and/or data, for example, E-mail, messages. The VMS also supports a method to retrieve previously stored messages, and, based on a directory number, it may also provide notifications for the presence of stored messages or a change in the number of voice and/or data messages that are waiting to be retrieved.

Operations Systems (OS): The OS, defined by the telecommunications management network (TMN), includes element management layer (EML), network management layer (NML), service management Layer (SML), and business management layer (BML) functions. These functions span across all operations systems functions such as fault management, performance management, configuration management, accounting management, and security management.

9.3 NEW NETWORK ELEMENTS IN CDMA2000

Figure 9.2 shows some of the key functional elements and interfaces defined in the cdma2000 network. Although many network elements and functionalities are similar to theEIA-TIA95A network of Figure 9.1, there are additional elements and interfaces defined to support packet data and other enhanced services. Some of the new network elements and their functions are described in this section. For more details, the reader is referred to [1].

Mobile Station (MS): The MS in cdma2000 has all the functionalities of a cdmaONE terminal plus additional features and capabilities to support new packet data services and enhanced signaling messages as it "talks" to both circuit-switched and packet-switched parts of the network.

Base Station (BS): The BS in cdma2000, in terms of basic functionalities, is similar to EIA-TIA95. However, both BTS and BSC have significant hardware and

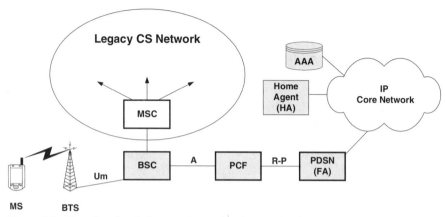

Figure 9.2   New functional elements for packet data service in cdma2000 network.

software changes to support the enhanced radio interface to provide voice, data, and multimedia traffic support. The A-bis interface has also been enhanced to facilitate more efficient transport of mixed traffic.

Packet Control Function (PCF): The PCF is an entity in the radio access network that manages the buffering and relay of packets between the BS and the PDSN. The PCF establishes, maintains, and terminates Layer 2 connection to the PDSN and maintains knowledge of radio resource status (e.g., active, dormant). The PCF also collects radio link (air interface)-related accounting information to be used by the AAA. PCF is sometimes collocated with the BSC, and in some models the combination of BS and PCF is called the radio network.

Packet Data Serving Node (PDSN): The PDSN is a new network entity with the following functions:

- Acting as a foreign agent (FA) by providing routing services according to simple IP and/or mobile IP protocols, also providing and maintaining routing tables and performing route discovery
- Managing the radio-packet (R-P) interface and PPP sessions for mobile users
- Initiating authentication, authorization, and accounting for the mobile user to the AAA server and receiving service parameters for the mobile user from the AAA server

When part of a virtual private network (VPN), the PDSN can establish a tunnel through the public packet network (PDN) using the Level 2 tunneling protocol (L2TP) to the VPN gateway, which is the access point to the private network. The private network authenticates the user and assigns him an IP address. The PDSN may optionally use the IPSec protocol to further protect the tunnel between PDSN and the VPN gateway.

Home Agent (HA): The HA is a network element within the mobile's home network that has two major functions, mobile IP registration and packet forwarding.

The HA accepts mobile IP registration requests and updates the current point of attachment of the user, that is, the current IP address to be used to transmit and receive IP packets to and from that user. The point of attachment is the last care of address (CoA) in IPv4, and/or colocated CoA in IPv6, assigned to the user by the serving FA.

The HA interacts with the AAA to receive mobile IP registration requests that have been authenticated and to return mobile IP registration responses. The HA also interacts with PDSN/FA to receive subsequent mobile IP registration requests. The HA may optionally support IPSec, to provide additional data security between HA and FA.

The HA also forwards IP packets to and from the current point of attachment through the FA where the mobile is currently registered by encapsulating IP packets for the mobile inside differently addressed packets (IP-in-IP tunneling). The HA also supports reverse link tunneling so a network operator can have both forward and reverse traffic passing through the HA to allow a more efficient use of network resources.

Authentication, Authorization and Accounting (AAA): The AAA is a network entity that provides IP-based authentication, authorization, and accounting as described in the following:

- *Authentication Function:* As part of the authentication function, the AAA provides authentication of terminal devices and subscribers. This function includes user and device identity verification for network access as well as user-based QoS requests. The authentication function also provides authentication and/or encryption keys to establish dynamic security associations between network entities.

- *Authorization Function:* To support the authorization function, the AAA decides on whether a user or device is authorized for a particular service with a specific QoS, based on user subscribers' service profile. The authorization function therefore has access to subscriber profiles, the device register, and the operator's policy repository. An entity that requests authorization from the AAA entity may request and receive a cache authorization information allowing it to make further decisions concerning services and resource allocation without a new request to the AAA. This cached authorization typically has an expiration time. The AAA also may send unsolicited messages containing policy decisions to appropriate entities.

- *Accounting Function:* This function of AAA involves collecting and storing the billing-related data concerning the offered services and their associated QoS as well as the multimedia resources requested and used by individual subscribers. The collected data includes the session details and mobility records. The session details collected from the session control manager, the core QoS manager, or other accounting servers includes information about the requesting party, requested and offered services and QoS, date and time of requests, duration of usage, and terminal used. The mobility records include administrative domain location as well as dates and times of mobiles' attach and detach.

The accounting function allows session and mobility detail records to be retrieved for further processing by OAM&P functions such as billing management.

9.4 NEW NETWORK ELEMENT INTERFACES IN CDMA2000

In addition to the new functional elements in a cdma2000 network, the interfaces between elements have also changed and been enhanced to support new signaling and user traffic.

Figure 9.3 shows some of the main reference points and their corresponding interfaces according to CDMA access network specifications [1,2].

The A reference point between BSC and MSC is implemented by the A1, A2, and A5 interfaces.

- A1: The A1 interface carries signaling information between the call control and mobility management functions of the MSC and the call control component of the BS (BSC).

- A2: The A2 interface is used to provide a path for user traffic. The A2 interface carries 64/56 kbps PCM information (for circuit-oriented voice) or 64 kbps unrestricted digital Information (UDI, for ISDN) between the switch component of the MSC and the selection/distribution Unit (SDU) function of the BS.

- A5: The A5 interface is used to provide a path for user traffic for circuit-oriented data calls between the source BS and the MSC. The A5 interface

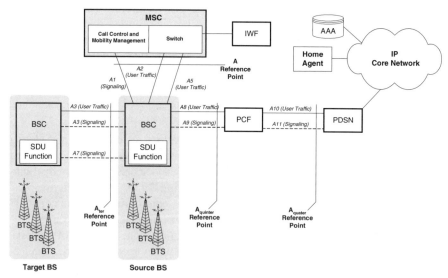

Figure 9.3   cdma2000 access network interfaces.

carries a full duplex stream of bytes between the switch component of the MSC and the SDU function of the BS.

The reference point A_{ter} is implemented by the A3 and A7 interfaces.

- A3: The A3 interface is used to transport user traffic and signaling for inter-BS soft/softer handoff when a target BS is attached to the frame selection function within the source BS. The A3 interface carries coded user information (voice/data) and signaling information between the source BS SDU function and the channel element component (BTS) of the target BS. This is a logical description of the end points of the A3 interface. The A3 interface is composed of two parts: signaling and user traffic. The signaling information is carried across a separate logical channel from the user traffic channel and controls the allocation and use of channels for transporting user traffic.
- A7: The A7 interface carries signaling information between a source BS and a target BS for inter-BS soft/softer handoff.

The $A_{quinter}$ reference point is implemented by the A8 and A9 interfaces. The A8 and A9 interfaces carry user traffic and signaling information between the BS and the PCF, respectively.

The A_{quater} reference point is implemented by the A10 and A11 interfaces. The A10 and A11 interfaces carry user traffic and signaling information between the PCF and the PSDN, respectively.

9.5 MOBILITY MANAGEMENT FOR PACKET DATA SERVICES

One of the key system aspects of cdma2000 packet data services is mobility management for packet data services including registration, authentication, and handoff procedures. In this section we briefly discuss these issues from a network perspective. For more details, the reader is referred to [3,4].

When the MS originates a call with packet data service option, a sequence of validations and resource allocations needs to be performed.

- Messages are exchanged to establish and close an R-P connection between PCF/BSC and the PDSN.
- Once the serial connection between the mobile and the PDSN is established, the PDSN begins the negotiation of a PPP session with the mobile.
- The PDSN and the mobile negotiate an authentication protocol according to a challenge handshake authentication protocol (CHAP) or a password authentication protocol (PAP).
- The PDSN sends the authentication response to the AAA server, which determines whether to authenticate the mobile.
- The PDSN constructs a network access identifier (NAI) of the form *MSID@realm*, where MSID is the mobile station identification and realm is

the Internet realm of the home network that owns the MSID. This NAI serves to identify the subscriber for the purpose of AAA server accounting.

• The mobile subscriber is identified as a valid user, and the PDSN also knows which IP service template to apply to this subscriber.

When the PPP session is established, the PDSN assigns the mobile an IP address from a pool of IP addresses that it administers. The IP address of a mobile is related to the location of the PDSN within the packet network. The mobile uses the IP address to identify itself as a client to servers in the public packet network. The routers in the packet network must be able to route any packet with this IP address to the PDSN that provides service to that mobile.

cdma2000 provides multiple levels of mobility support for packet data services through different handoff mechanisms. Figure 9.4 shows the different mobility levels and the corresponding interfaces involved in handoff.

The mobility and handoff between BTSs involves link layer soft, softer, or hard handoff procedures for active mobiles and dormant handoff procedures for inactive mobiles, as supported by the cdma2000 air interface.

MS mobility between BSCs under the same PCF is supported by A8/A9 interfaces.

Packet data handoff also occurs when a change in the location of a mobile causes the IS2000 network to divert the packet data session from one R-P interface to another.

When a handoff is between PCFs with connectivity to the same PDSN so that the serving PDSN remains the same before, during, and after handoff, it is called an intra-PDSN handoff. The A10/A11 interfaces support MS mobility between PCFs under the same PDSN.

When a handoff is between PCFs with connectivity to different PDSNs, the handoff is called an inter-PDSN handoff. In this case, the mobile IP supports MS

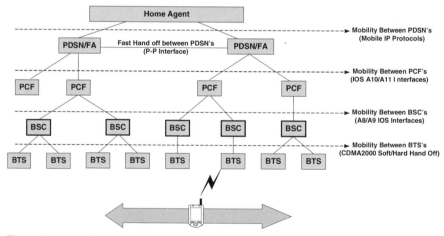

Figure 9.4   Mobility management structure in cdma2000.

mobility between PDSNs under the same home agent, while the PDSN provides the functionality of the foreign agent (FA).

The network may also support an optional P-P interface between PDSNs to support the so-called fast handoff procedures. Fast handoff with the P-P interface is used to keep the PPP session anchored when the PDSN-to-PDSN handoff is performed.

This section focuses on intra-PDSN and inter-PDSN handoff for packet data services in cdma2000.

9.5.1 Inter-PCF and Intra-PDSN Handoff

As a mobile moves from one PCF to another PCF, for every packet data service instance, a new R-P connection between the target PCF and the serving PDSN is established and the PPP session will be moved to the new R-P interface (see Fig. 9.5).

The PCF-to PCF-handoff involves:

- PDSN selection, where higher priority is given to same PDSN selection in order to maintain the existing PPP session between the PDSN and the MS. If a different PDSN is selected, an inter-PDSN handoff may be performed.
- New R-P session setup by the target PCF.
- Previous R-P session tear down. If the PDSN selected is the same serving PDSN, the PDSN triggers a release of the previous R-P session.

PCF-to-PCF handoff may occur while an MS is in the active or the dormant state. The dormant PCF handoff is supported to maintain the PPP session, whereas an MS is dormant to minimize the use of air-link resources.

During an active session handoff, the PDSN supports a low-latency PCF-to PCF-handoff by bicasting data to the target and previous PCF.

In an intra-PDSN handoff scenario the IP address will stay the same because the mobile still has an active PPP connection to the PDSN that assigned the IP

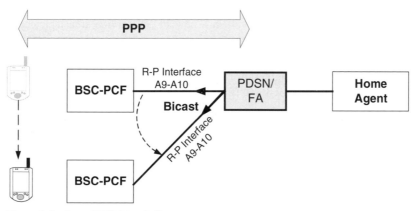

Figure 9.5   Intra-PDSN handoff.

address. The session will not be interrupted, and the handoff will be transparent to the user. The intra-PDSN handoff is also possible in the network with simple IP because it does not require mobile IP-specific features.

9.5.2   Inter-PDSN Handoff

A network based on the simple IP network does not support mobility beyond a PDSN coverage area, unless the optional fast hand off procedures are supported. Without fast handoff support the simple IP network does not allow inter-PDSN handoff and the mobility involves establishment of a new PPP session on the target PDSN and authentication by the AAA.

Thus the user must acquire a new IP address and inform the server that it has a new IP address when moving to the service area of the new PDSN, and all IP-based applications must be restarted. The old serving PDSN will continue to send packets to the old IP address, but because the packets will be undeliverable, the application sessions will time out and be torn down.

Mobile IP, however, provides the IP layer mobility management function that maintains persistent IP addresses across PDSNs. For a mobile IP-based MS to maintain a persistent IP address while moving between PDSNs, the MS reregisters with its HA.

When fast handoff is not supported, the inter-PDSN handoff in mobile IP involves the following steps:

- Establishment of a new PPP session
- Detection of a new FA via the agent advertisement message
- Authentication by AAA
- Registration with the HA

During a PDSN-to-PDSN handoff, the MS may be in an active or a dormant state. Figure 9.6 and the following sequence of events show, in more detail, the process of inter-PDSN handoff in a mobile IP-based network, when fast handoff is not supported and the MS is in an active state:

- As the MS moves to a target cell in a new PDSN service area it detects a change in packet zone ID based on the extended system parameters message received from the new base station.
- The mobile sends an origination message to the target BS requesting packet data service, and a new R-P interface to the target PDSN is established.
- The mobile starts PPP negotiation with a new PDSN.
- Once the serial connection exists between the mobile unit and the PDSN, the PDSN begins the negotiation of a PPP session with the mobile. The PDSN and the mobile negotiate an authentication protocol with visited AAA involved.
- If the visited AAA server is unable to authenticate the user the home AAA server may also be involved.

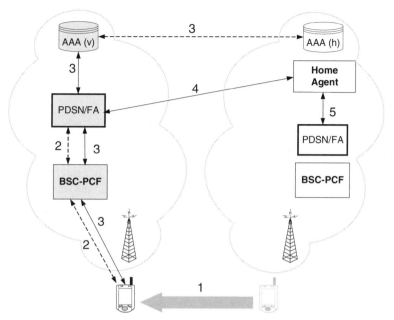

Figure 9.6   Inter-PSDN handoff process.

- When the new PPP link is established, the mobile locates PDSN/FA on the new network. In this case, the MS sends an agent solicitation and receives the agent advertisement message, and it then registers with the PDSN/FA.
- The FA assigns a temporary care-of address (COA) to the MS and notifies the HA.
- The HA updates its binding tables with the MS's new COA.
- The FA updates its visitor list with the new entry for the MS and forwards the reply to the MS.
- An IP-in-IP tunnel is established between the FA and the HA, in which the HA tunnels the traffic to the FA by encapsulating IP packets inside differently addressed IP packets.
- The FA de-capsulates the traffic before sending it to the mobile.
- The previous PPP link is dropped as the registration life timer expires.

If the MS is dormant during the transition from one cell to another, it does not use any radio resources and therefore does not need to perform radio handoff. In this case, only the PPP context information and R-P interface are maintained.

In this case, packet data handoff will not be necessary as long as the mobile moves within a packet data zone. However, if the mobile detects a change in packet data zone, it will initiate dormant hard handoff by sending an origination request with no data to the MSC, and that triggers the transfer of the dormant state session to the target PDSN. The MS then transitions to the active state and negotiates a new

PPP session with the target PDSN. The HA also updates its records with the new serving PDSN/FA.

9.5.3 Inter-PDSN Fast Handoff

If fast handoff is supported, the target PDSN initiates establishment of a P-P session with the serving PDSN. Fast handoff with the P-P interface is used to keep the PPP session anchored when the PDSN-to-PDSN handoff is performed. This allows the existing PPP session to continue, thereby reducing service interruption time and data loss.

Figure 9.7 shows the protocol reference models for mobile IP bearer data and mobile IP control, including internet key exchange protocols, during the fast handoff.

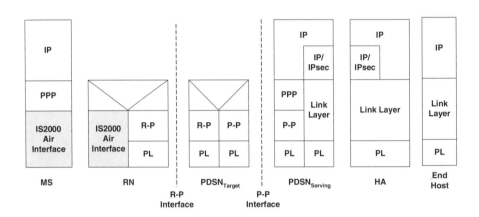

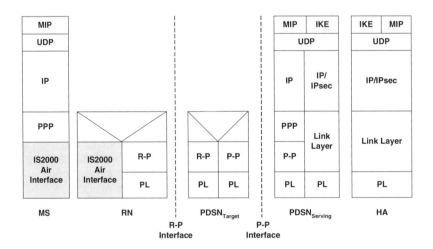

(b) Reference Model for MIP Control and IKE

Figure 9.7   Protocol reference models for MIP bearer data and control during fast handoff.

For more details on protocols reference models for mobile and simple IP, the reader is referred to [3,4].

For the forward link:

- The serving PDSN tunnels the received forward traffic through the appropriate P-P connection to the target PDSN.
- The target PDSN then forwards the traffic to the MS over the corresponding R-P connection.

On the reverse link:

- The target PDSN tunnels the received reverse traffic from the MS through the appropriate P-P connection to the serving PDSN.
- The serving PDSN then forwards the traffic to the external network.

Fast handoff implementation allows inter-PDSN handoff for an active session even in simple IP networks.

In a simple IP network with fast handoff, the PDSN-to-PDSN handoff involves the following steps:

- Establishment of a P-P connection for each associated R-P connection at the target PDSN and continuation of the current PPP session on the serving PDSN
- Establishment of a new PPP session and authentication with the AAA by the target PDSN when the MS becomes dormant or the MS renegotiates PPP
- Release of the associated P-P session while the new PPP session is being established at the target PDSN

In a mobile IP network with fast handoff, the PDSN-to-PDSN handoff involves the following steps:

- Establishment of a P-P connection for each associated R-P connection at the target PDSN and continuation of the current PPP session on the serving PDSN
- Establishment of a new PPP session by the target PDSN when the MS becomes dormant or the MS renegotiates PPP
- Release of the associated P-P connections while the new PPP session is being established at the target PDSN
- Detection of a new FA via the agent advertisement message
- Authentication by AAA
- Registration with the HA

9.6   3GPP2 CDMA2000 NETWORK MODEL

In this section we look at a more complete cdma2000 network model as defined by 3GPP2 [1]. This model, which includes all functional elements in the network, is presented in Figure 9.8 and Figure 9.9. In these figures the blocks show network entities and the circles represent network interfaces or reference points.

Figure 9.8 shows the key elements involved in data and voice traffic switching and routing as well as call processing and mobility management. Although most

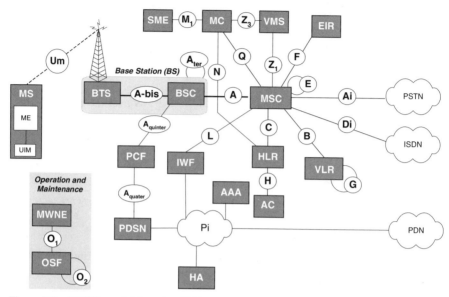

Figure 9.8   3GPP2 model for cdma2000 network (Part 1 of 2).

network elements have been introduced in previous sections, the following can be added to complete the picture.

The Mobile Station [MS; in this model the MS includes mobile equipment (ME) and a user identity module (UIM)]

ME without a UIM is only capable of accessing the network per locally defined service configuration for example, emergency services, service center.

The UIM is a part of the MS containing all the user's subscription information as well as feature subscription information. The UIM provides secure procedures in support of registration, authentication, and privacy for wireless access to network subscribers. The UIM can be integrated into any ME, or it may be a removable UIM (R-UIM). The R-UIM cards enable wireless consumers to securely transfer subscription identity to a phone utilizing different frequency or technology than used by their home network. It also allows users to roam nationally or internationally and still have access to their home services, if supported by the visited network.

The acceptance of the R-UIM standard will facilitate the rapid development and deployment of existing and next-generation features, such as global roaming and advanced services for CDMA phones worldwide by allowing CDMA users to roam in GSM and UMTS networks.

The R-UIM specifications are based on the current SIM specifications in GSM, and they include additional features and capabilities specific to cdma2000. It is intended that all upgrades to the SIM specification will also apply to the R-UIM.

The short message entity (SME): is the entity that composes and decomposes short messages. In some implementations, the SME may be located within, and be

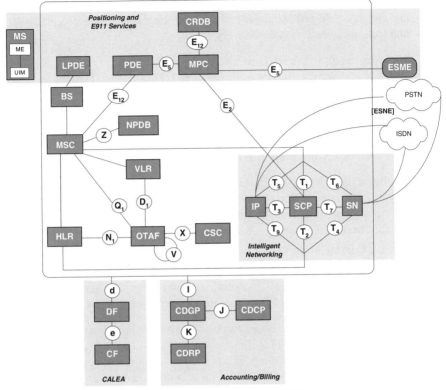

Figure 9.9   3GPP2 model for cdma2000 network (Part 2 of 2).

indistinguishable from, an HLR, MC, or MSC/VLR as defined in the EIA-TIA95 network model.

Figure 9.9 shows some of the other auxiliary network functionalities, which are described in the following:

Communication Intercept Entities: The Communications Assistance for Law Enforcement Act (CALEA) requires telecommunications carriers to ensure that their equipment, facilities, and services are able to comply with authorized electronic surveillance. In cdma2000 network architecture model the following entities are defined to support this function.

- Intercept Access Point (IAP): The IAP is an entity that provides access to the communications to or from the equipment or service that is the subject of the intercept.

- Delivery Function (DF): The DF is an entity that is responsible for duplicating and delivering intercepted communications to one or more collection functions. DF has the ability to accept and filter information over one or more data channels and to combine that information into a single data flow for each intercept subject.

- Collection Function (CF): The CF is an entity that is responsible for collecting and processing call contents and other information about intercepted communications as delivered by delivery function.

For more detail, to the reader is referred to [5].

Over-the-Air Service Provisioning Function (OTAF): The OTAF is an entity that interfaces proprietarily with CSCs to support service provisioning activities. The OTAF interfaces with the MSC to send the MS orders necessary to complete service provisioning requests [6].

Wireless Intelligence Networking (WIN) Entities: The following network entities are specifically involved in WIN:

- Intelligent Peripheral (IP) performs specialized resource functions such as playing announcements, collecting digits, performing speech-to-text or text-to-speech conversion, recording and storing voice messages, facsimile services, and data services.

- Service Control Point (SCP) acts as a real-time database and transaction-processing system that provides service control and service data functionality.

- Service Node (SN) provides service control, service data, specialized resources, and call control functions to support bearer-related services.

Number Portability Data Base (NPDB): The NPDB is an entity that provides portability information for portable directory.

Call Data Accounting Entities: The following network entities are defined to support call data generation and collection for accounting and billing purposes.

- Call Data Collection Point (CDCP) collects the *ANSI-124* format call detail information.

- Call Data Generation Point (CDGP) provides call detail information to the CDCP in *ANSI-124* format. Because the call detail information sent to the CDCP must be in *ANSI-124* format, the CDGP may need to convert the information from a proprietary format into *ANSI-124* format.

- Call Data Information Source (CDIS) is the source of call detail information that may be in a proprietary or in *ANSI-124* format.

- Call Data Rating Point (CDRP) The CDRP is the entity that takes the unrated *ANSI-124*-format call detail information and applies the applicable charge and tax-related information based on *ANSI-124* format.

Mobile Positioning and Emergency Service Entities: In the addition to the MS, the following entities are specifically involved in mobile positioning and emergency services.

- Local Position Determining Entity (LPDE) and Position Determining Entity (PDE) are entities that determine the geographic location of a wireless terminal. An LPDE or PDE may support one or more position-determining technologies. Multiple LPDEs or PDEs may serve the coverage area of a mobile position center (MPC). An LPDE typically resides at the BS.

- Mobile Position Center (MPC): The MPC selects a PDE to determine the position of a mobile station. The MPC may restrict access to position information to authorized network entities, and it may only offer this information if the MS is engaged in an emergency call.
- Coordinate Routing Data Base (CRDB) stores the position information expressed as a latitude and longitude to translate it into a string of digits.
- Emergency Service Message Entity (ESME) and Emergency Service Network Entity (ESNE): The ESME and ESNE route and process out-of-band and in-band messages related to emergency calls, respectively. These functions may be incorporated into selective routers, public safety answering ports (PSAP), and emergency response agencies.

For more details, the reader is referred to [7,8].

9.7 3GPP2 ALL-IP NETWORK MODEL

In addition to the cdma2000 reference architecture model, 3GPP2 has also defined an all-IP network model providing a common IP backbone for both CS and PS signaling and traffic [3,4]. This section describes the functional elements and architecture of the cdma2000 all-IP model as shown in Figure 9.10.

This all-IP model is based on an evolutionary strategy in which the current circuit-switched system will evolve into the "legacy MS domain" as it integrates to and operates side by side with the new packet-switched-based multimedia network. This legacy MS domain fully interworks with the legacy circuit-switched system and continues to support legacy terminals. It supports IP transport for signaling and voice except over the air interface and enables the use of legacy as well as IP-based applications provided by the service network.

One of the most important steps in the evolution of the legacy systems to the all-IP networks, adopted by both 3GPP and 3GPP2, is to split the functionality of MSC into MSC server and media gateway, as follows:

- The MSC server includes call control and mobility management for legacy voice calls and supplementary services.
- The media gateway (MGW), which acts as a gateway toward PSTNs and PLMNs, includes IOS termination for user traffic, transport interfaces, vocoders/transcoders, echo cancellers, conference call bridges, announcements machines, etc. It will also perform the necessary conversions of voice carried over IP to voice carried as circuit-switched traffic.

The introduction of MSC server and MGW implies that all *ANSI-41* operations are carried over IP retaining the same call model. Also, because in the legacy MS domain the vocoders will be located in the MGW, new control signaling and IOS standards need to be developed and used.

The IP multimedia domain has been developed for a wide range of terminal types. This step will provide support for basic voice over IP and streaming services. The multimedia services will coexist with circuit-switched services provided by

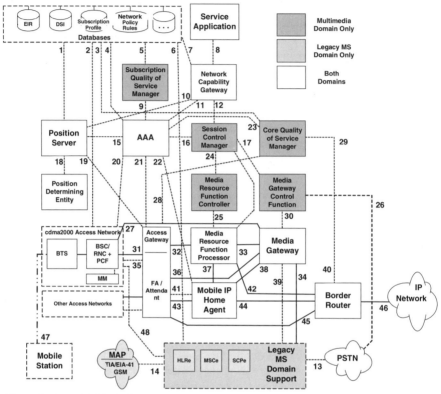

Figure 9.10   cdma2000 all-IP network model. Reproduced under written permission from TIA.

MSC server. A new radio control model is introduced that implies new signaling in IS2000. A new access-independent IP multimedia call model based on SIP is introduced.

Packet session control and mobility management are handled by the MSC in the current cdma2000 system. These functions will be refined and relocated for the IP multimedia domain in order to replace the MSC-centric model so that they can cater for all the foreseeable and unforeseeable features needed for IP multimedia services. In the network architecture model defined by 3GPP2 these functions are provided by RAN, access gateway, and Core QoS manager.

In the following we briefly describe each of the functional entities defined in the all-IP network model, where each entity represents a function that may be contained within a single physical device or be distributed over a number of physical devices. Figure 9-10 shows all the functional entities and differentiates those in the multimedia domain only, in legacy MS domain only, or in both domains.

Although some of the entities such as MS, HLR, BTS, and BSC have functionality very similar to what we described previously as part of the cdma2000 network architecture, there are new elements and functions that need to be mentioned as follows:

Legacy MS Domain Support (LMSDS): The LMSDS contains emulated functional network entities that perform the call control, mobility management, and service management functions to provide support for non-IP (i.e., legacy) MSs. The functional entities within LMSDS include the following:

- Home location register emulation (HLRe): This entity emulates the HLR function for legacy mode terminals. The HLRe, however, differs from an HLR in terms of its IP-based interfaces and some of its functionality. For legacy and dual-mode MSs the HLRe also supports roaming to legacy networks and to legacy MS domains of IP networks.
- Mobile switching center emulation (MSCe): This entity provides call processing and control as well as signaling capabilities comparable to those of a legacy MSC but has only bearer management capabilities.
- Service control point emulation (SCPe): SCPe contains the logic and processing capability required to handle wireless intelligent network (WIN)-provided service attempts. The SCPe may interact with other functional entities, for example, MSCe, other SCPes, service applications, and DBs, to access additional logic or to obtain information (service or user data) required to process a call and service logic instance. The SCPe component of LMSDS may also support multimedia domain users.

Access Gateway (AGW): In general the AGW provides the core network (CN) with access to the resources of the access network. The AGW provides the CN with a common interface to the specific capabilities, configuration, and resources of the access network, which may consist of one or a combination of different access technologies. The 3GPP2/cdma2000 AGW is one of the various types of access gateways (e.g., IS2000 RAN, UTRAN, 802.11, wire-line, etc.). The I-2000 AGW specifically consists of the PDSN and other logical functions required to interface the core network to the IS2000 RAN.

The AGW is a network component that may support both the multimedia and legacy MS domains. In the IP multimedia domain an AGW provides the following main functions:

- Foreign Agent Function: The AGW provides foreign agent (IPv4) and/or attendant (IPv6) functionality and maps network access identifiers (NAIs) into mobile station identifiers.
- Link Layer Peer Function: The AGW includes the link layer peer function for IP traffic termination from and to the mobile station such as the point-to-point protocol (PPP).
- Bearer Transport: The AGW also transports both upper layer signaling and user bearer traffic between the mobile station and the core network.
- Handoff Support: The AGW supports link layer handoffs between access networks, of the same or different technologies, when supported by the same AGW. The AGW also supports inter-AGW handoffs.
- AAA Support: The AGW also communicates with the AAA to provide the mobile with access to the network level, such as mobile IP, registration, and

authentication. It also gathers accounting information from the access network and forwards it to the AAA along with accounting information maintained by the AGW.

- QoS Support: The AGW communicates with the core Quality of service manager (CQM) for management of core QoS resources and intercepts and processes QoS requests from the mobile station. The AGW also manages the traffic to and from mobile stations and may mark a mobile's packets per the QoS profile.

In the legacy MS Domain, the AGW supports packet data services for cdma2000 legacy mobile stations.

Radio Access Network (RAN): The RAN performs mobility management functions for registering; authorizing, authenticating, and paging IP-based terminals, independent of circuit-based terminals. The RAN performs handoffs within an access network and between access networks of the same technology, and it may also support handoffs between access networks of differing technologies.

The cdma2000 access network contains the base transceiver Subsystem (BTS) and the base station controller or radio network controller (BSC/RNC), and it also includes the packet control function and mobility manager.

The BTS and BSC have functionalities similar to those described before for the cdma2000 reference model.

Mobility Manager (MM): The IP multimedia domain MM function's responsibilities include:

- Handling registration messages from the mobile station for the establishment of logical bearers through the IP multimedia domain core network

- Communicating with the AAA for access network authentication, authorization of radio link access, and accounting. Examples of access types that MM should authenticate and/or authorize with the AAA include multimedia registration, multimedia page response, inter-RAN handoff, and supporting hard handoff within and between access networks when possible.

Packet Control Function (PCF): The PCF function's responsibilities include:

- Establish, maintain, and terminate Layer 2 connection to AGW
- Interact with PDSN to support dormant handoff
- Maintain knowledge of radio resource status (e.g., active, dormant)
- Buffer packets arriving from the AGW, when radio resources are not in place or are insufficient to support the flow from the AGW
- Relay packets to and from the AGW
- Map mobile station id and connection reference to a unique Layer 2 connection identifier used to communicate with the AGW
- Collect and send radio link (air interface)-related accounting information to the AGW

In support of the legacy MS domain, the BSC exchanges messages with the BTS, other BSCs, the AGW, and the MSC server. Signaling for call control, mobility management, and MS management as well as traffic may pass transparently through the BSC. In this case, the BSC may initiate the QoS allocation request based on the address of the media gateway and QoS parameters provided by the MSC server.

In support of the IP multimedia domain, the BSC/PCF interfaces with the AGW to support bearer streams between the terminal and the IP multimedia core network, and it also communicates with BTSs and other BSCs.

Subscription Quality of Service Manager (SQM): The SQM provides management of QoS resources on a per subscription basis for users subscribed to a home network. The SQM is located in the home network and communicates with the AAA to provide authorization of resource allocations. The SQM makes policy decisions with regard to use of QoS resources for a given subscription based on applicable policy rules as well as current allocations already made with respect to that subscription.

Core Quality of Service Manager (CQM): The CQM provides management of core network QoS resources and makes related policy decisions, for example, based on service-level agreements (SLAs), within its own core network. The CQM also communicates with the AGW to provide authorization of resource allocations.

Border Router (BR): The BR connects the operator's core network with the peer networks such as Internet and other service provider networks. The BR's functions include

- Performing IP packet routing according to exterior gateway routing protocols
- Policing of incoming and outgoing traffic according to SLAs established with peer networks

Databases (DB): The core network DBs include a number of database functions including the following:

- Equipment Identity Register (EIR) is a database of equipment information and has function similar to those defined in the legacy system.
- Dynamic Subscriber Information is a database containing information about the current session registration such as the serving SCM address.
- Network Policy Rules is a database that provides network-wide policy rules specified by the operator to control the network. These policy rules are related to subscription resource usage, expected QoS, geographic service area definitions, policy rules for the applications serving a user, etc. As a result, this database is referenced by the authorization function of AAA, the subscription quality of service manager, and the core quality of service manager.
- Subscriber Profile: This database contains subscriber-specific information such as, authorized features and services, authorized service area, and creditworthiness.

Media Gateway (MGW): The MGW provides an interface between the packet environment of the core network and the circuit-switched environment of the PSTN for bearer traffic, when equipped with circuit capabilities. The MGW may provide vocoding and/or transcoding functions to the bearer traffic. The MGW may also provide modem functions to convert digital byte streams to and from audio modem tones placed on circuits and may provide the capability to terminate point-to-point protocol (PPP) connections. It also provides policy enforcement related to its activities and resources.

Media Gateway Control Function (MGCF) and Media Resource Function Processor(MRFP): The MGCF provides the ability to control a media gateway (MGW) through standardized interfaces, allowing allocation and de-allocation of MGW resources as well as modification of their usage. The MRFP, in conjunction with the controlling entity, provides multiway conference bridges, announcement playback services, tone playback services, etc.

Network Capability Gateway (NCGW) and Service Applications: The functions of these two entities are related because service applications provide value-added network-based services for wireless subscribers and the NCGW provides access to network resources needed during the execution of service application. To support the legacy MS domain, the NCGW interfaces with service application and LMSDS network entities and relays service application information between these two network entities.

Many service application types access resources (e.g., SIP server, position server, AAA) in the operator's network via NCGW for functionality required during service execution. There are also service application types that operate without interaction with the operator's network.

Position Server and Position Determining Entity (PDE): The position server provides geographic position information to requesting entities. The AGW chooses a position server and requests registration for a particular terminal with specified positioning capability. The position server selects a position determining entity (PDE) based on the positioning capabilities of the MS. The PDE determines the precise geographic position of the MS based on input data provided by the position server.

9.8 END-TO-END QOS NETWORK MODEL

The support for QoS in a network provides the ability to offer differentiated levels of service based on subscribers' application needs. The QoS model developed by 3GPP2 is based on an all-IP network model, and it clearly separates access-specific and access-independent domains. This section describes this model as it relates to end-to-end QoS support in cdma2000 networks.

The end-to-end QoS control in a network can be described based on the concept of end-to-end traffic flows. Each flow goes from one peer, for example, a mobile client, to another peer who may be a mobile or a fixed client or a server on the network. Flows are typically identified by parameters such as source/destination IP address and UDP/TCP ports along with other parameters.

The QoS requirements are associated with each flow, so in a multimedia data session each media component with a different QoS (e.g., video, audio, chat) would map to a separate flow. The purpose of providing end-to-end QoS is to ensure that the requirements for all flows are met.

Some of the key attributes of an application QoS are:

- Bandwidth is user throughput or effective data rate at the application layer.
- Latency is end-to-end or round-trip time that it takes to send a packet from a sending node to a receiving node.
- Jitter is a measure of the variation in delay between the arrival of packets at the receiver.
- Traffic (packet) loss is the rate of discarding of packets because of errors or network congestion.

On the basis of the QoS requirement associated with these attributes one can classify the applications into different groups or service classes. The following four service classes have been identified by 3GPP2 to be supported in a cdma2000 network:

- Conversational applications such as voice over IP and video conferencing, which are two way, require low delay and low packet loss rate. and are sensitive to delay variations.
- Streaming applications are the same as conversational applications, but they are one way, less sensitive to delay. and may require high bandwidth. In these applications, for example, video and audio streaming, although the time relation between information entities within a flow is preserved some fixed transfer delay may be tolerable.
- Interactive applications are two way and bursty and have variable bandwidth requirements with moderate tolerance to delay and packet loss rate. In these applications, for example, web browsing, data base retrieval, and server access, the receiving end is expecting the data within a certain time.
- Background applications are highly tolerant to delay and data loss rate and have variable bandwidth requirements. In these applications, for example, background download of E-mails, the receiving node is not expecting the data within a certain time.

Support for QoS may be guaranteed or statistical, and it may also be supported on an end-to-end or domain-by-domain basis. There are two QoS frameworks developed by the IETF and used in cdma2000 networks:

Integrated Services (IntServ) is a framework developed by the IETF, where QoS is guaranteed by reserving resources along the path of a specific IP flow. The resource reservation protocol (RSVP) is the signaling protocol typically used to reserve resources.

Differentiated Services (DiffServ) is an alternative QoS framework defined by the IETF to allow scalable service differentiation in the Internet. This is achieved by classification of traffic into aggregate classes at the edges of the "DiffServ domain" as opposed to operating on individual end-to-end flows. Unlike IntServ, in the DiffServ framework resources cannot be reserved and data service handling is only based on relative priorities of packets according to their service classes as well as specified "Per hop behaviors" for each class. QoS support across the DiffServ domains can be controlled through traffic conditioning and service level agreements (SLAs).

Although end-to-end QoS can not be guaranteed in DiffServ, by proper modeling of the traffic load and care dimensioning of the network end-to-end QoS can be reasonably controlled and ensured. The DiffServ is also believed to give a higher resource efficiency, that is, better trade-off between the service offered and the network cost compared with the IntServ approach.

The goal is to establish an end-to-end QoS Service control between a mobile terminal and a correspondent node, which may be on the same RAN or on a different access network. In principle, establishing the end-to-end service consists of three steps:

- Establishing an access bearer service, which for a RAN involves establishing and configuring the appropriate radio and physical channels according to the parameters and procedures specific to the RAN's technology
- Establishing the external bearer service across networks typically based on IETF frameworks such as DiffServ and IntServ
- Establishing the end-to-end service on top of the other two services, which involves mapping application-level requirements to resources on the access bearer service as well as mapping between the access bearer service and the external bearer service.

The cdma2000 network implementations typically use a combination of IntServ and DiffServ. Whereas the RAN provides QoS based on the IntServ framework for the access bearer service, the external bearer service of the core network uses DiffServ as a default choice. However, the 3GPP2 model in general also works with an IntServ model.

Figure 9.11 shows the end-to-end QOS network model based on the cdma2000 family of radio access technologies. In this model, the packet data serving node (PDSN) acts as an access gateway. The access gateway/router is the point at which mapping between the access bearer service and the external bearer service takes place. Other QoS-related functions of an access gateway include traffic conditioning, applying DiffServ marking, and consulting with a policy manager for core network policy decisions. In Figure 9.11 the mobile IP home agent (HA) is considered as an entity involved in the external service.

The access bearer service in cdma2000 can be divided into a radio bearer service, an IOS bearer service, and a PPP link service. The combination of these services creates logical link connections, which have one-to-one mapping to service instances.

The radio access bearer services in cdma2000 are structured based on service instances. Each mobile may have up to six active service instances identified by a reference parameter SR_ID, each with a different set of QoS attributes. Each service instance can be activated based on a service option, which describes the type of requested service. A service option may correspond to a circuit-switched or a packet-switched services instance. Also, a packet-switched service option can be offered in

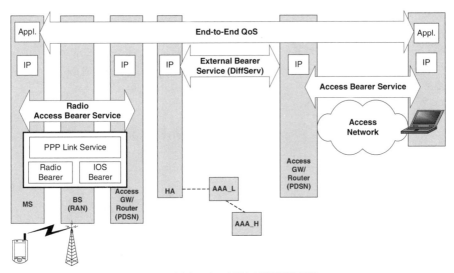

Figure 9.11   End-to-end QoS model in cdma2000 NETWORKS.

an assured or a nonassured mode. In nonassured mode the service instance is only given a relative priority compared with others, whereas in the assured mode the QoS can (to some degree) be guaranteed.

A mobile requests for new service instances through IS2000 signaling messages. The service instance may be assigned a physical channel, but it may also exist even when the assigned physical resources are released, for example, when the service in a "dormant state". The assignment/multiplexing of physical channel to service instances is controlled by the cdma2000 multiplexing function.

The interoperability specification (IOS) bearer service maps service instances to IOS A8 and A10 connections. The BSC performs the mapping of the incoming uplink/downlink user data from the radio bearer service instance onto an A8 connection and vice versa. Similarly, the PCF maps incoming uplink/downlink user data from an A8 connection onto an A10 connection and vice versa.

Establishment of packet data services may be authorized by the MSC, but the MSC does not need to authorize every packet service instance or authorize the corresponding QoS. The new service instances are established through the PDSN authorization.

PPP link service between the mobile and the PDSN provides the mapping between end-to-end IP flows and cdma2000 radio service instances, as well as the necessary support for air interface scheduling.

9.9   POSITIONING SYSTEM IN CDMA2000

Most cdma2000 positioning systems that are currently deployed around the globe are based on the Assisted Global Positioning System, or AGPS. AGPS is one of the

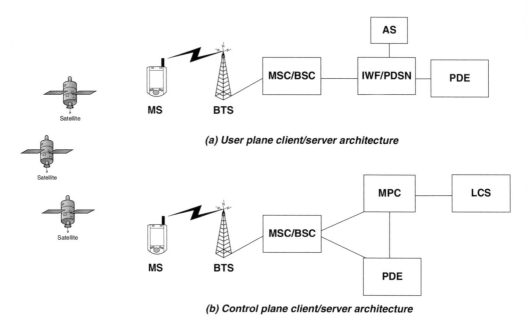

(a) User plane client/server architecture

(b) Control plane client/server architecture

Figure 9.12   AGPS signaling network structures.

most accurate and reliable positioning systems for E911 applications and, depending on the hardware and software used by the handset, its accuracy can be within a few meters.

The AGPS systems deployed in CDMA networks typically utilize their positioning signaling system either in the control plane or in the user plane part of the architecture (see Fig. 9.12).

The mobile system (MS) and base transceiver system (BTS) together with the mobile switching center/base station controller (MSC/BSC) are common to both architectures. The MS receives GPS and CDMA network measurements and communicates with the PDE via the BTS and the MSC.

- In user plane-based systems (Fig. 9.12a) the PDE is interfaced with the IWF/PDSN to communicate to the application server (AS) and the MSC/BSC. This architecture uses TCP/IP as a transport layer for IS801 messages and is well suited for quick deployment on off-the-shelf Unix platforms.

- In control plane deployments (Fig. 9.12b), the PDE communicates with the mobile positioning center (MPC) and the MSC/BTS directly. The MPC is responsible for interfacing with location services (LCS). This architecture uses network data burst messaging (DBM) as a transport mechanism of IS801 and J-STD-036 messages.

Depending on the call flow used for communication between the MS and the PDE, the PDE can provide aiding information to the GPS measurement processing of the MS. Acquisition assistance (AA) and sensitivity assistance (SA) data can be sent to the MS, and this information helps to optimize the speed and sensitivity of the search for GPS code phase measurements. In all existing deployments, the smart server or the MS-assisted mode of operation is implemented either with mobile-originated (MO) or mobile-terminated (MT) call flows. For both call flows, the navigation solution takes place at the server. This implementation simplifies the requirements for the mobile system, allowing GPS data acquisition to be performed as a background task.

Several kinds of measurements may be available to the PDE in different implementations. These can be broken into three broad categories:

- *Satellite-based measurements:* The MS acquires satellite-based measurements from global navigation satellite systems (GNSS). These are primarily GPS measurements but can also include WAAS, Glonass, and eventually Galileo measurements.

- *Terrestrial measurements*, which include those acquired by the MS on the forward link from a ground reference station such as a BTS. Terrestrial measurements can also be acquired on the reverse link, measured at the BTS.

- *Other measurements*, which include altitude aiding, based on a centralized database and/or locally available information, and system-specific timing constraints.

Regardless of their origin, all measurements that are part of a given position/velocity request are sent to the PDE for navigation calculations.

The scenario for using an AGPS consists of the following events:

- An E911 call that is initiated by a GPS-capable handset is sent to the wide area reference network (WARN) via the base station and MSC and other public networks. WARN maintains a database of all GPS satellites and their availability at different locations and times. The E911 call provides WARN with some information about the serving cell to help WARN in identifying the available GPS satellites at the user's location and time. WARN then informs the handset of the GPS satellites that it should be communicating with.

- An E911 call that is initiated by a GPS-capable handset and sent to WARN includes some information about the serving cell. WARN, which maintains information about the availability of GPS satellites in each area at a given time, informs the handset of the GPS satellites that it should be communicating with.

- When the caller's handset receives the information about the available GPS satellites, it begins searching for and inquiring from those satellites.

- The handset inquires from GPS satellites for the information it needs to calculate its own position, or it may send some supplementary data to the PDE for determining the caller's location. If the handset does not have line of sight to a sufficient number of GPS satellites, it measures its distance from the neighboring cell sites and forwards the additional information to the PDE. If the supplementary information includes distance from only one or two cell sites, the system will use the centroid of each cell site as reference points for locating the caller.

- In either case, the handset's response is processed by the PDE and the calculated position is sent to the PSAP.

- Parallel to this, a dedicated voice channel will be established between the caller and the PSAP.

Note that despite its good performance and widespread adoption, the AGPS scheme has some disadvantages as follows:

- New GPS-capable handsets are required, which have higher power consumption and potentially higher cost.

- There are additional costs of subscribing and interfacing to providers of WARN services, if outside services are required.

- In tunnels, subways, and densely urban areas with tall buildings, the handset may not have line of site with satellites or it may not see a sufficient number of satellites to calculate the location accurately.

- The GPS satellites may not be equally or sufficiently "visible" throughout the world.

When combined with network-based systems, the system becomes more complex and more expensive MPC software is required.

9.10 REFERENCES

1. 3GPP2 S.R0005-B, Network Reference Model (NRM) for cdma2000 Spread Spectrum Systems, Nov. 2000.
2. 3GPP2 A.S0011-17-A v0.3, Interoperability Specification (IOS) for cdma2000 Access Network Interfaces, July 2003.
3. 3GPP2 X.S0011-003-C cdma2000 Wireless IP Network Standard: Packet Data Mobility and Resource Management, August 2003
4. 3GPP2 P.S0001-B Version 1.0., Wireless IP Network Standard, 2002
5. J-STD-025, "TIA/EIA/J-STD-025, Lawfully Authorized Electronic Surveillance, 2000".
6. 3GPP2 C.S0016-A, Over-the-Air Service Provisioning of Mobile Stations in Wideband Spread Spectrum Systems, December 2001.
7. TIA/EIA J-STD-036-A, Wireless Enhanced Emergency Services, Phase II; Telecommunications Industry Association (TIA), March 2002
8. TIA/EIA IS-801-1, Position Determination Service Standards for Dual Mode Spread Spectrum Systems, Addendum 1, Telecommunications Industry Association (TIA), March 2001

CHAPTER 10

1XEV-DO (HRPD) AIR INTERFACE

10.1 INTRODUCTION

Throughout the standardization process for 3G systems, there was a debate on whether or not a single air interface standard should be designed and optimized to carry both high-speed packet data and voice traffic on the same RF carrier.

Some argue that because of different QoS requirements for voice and packet data applications such an integrated design approach would result in significant performance and efficiency compromises. Therefore, high-speed packet data carriers should be designed and optimized as a separate radio access technology to be deployed when needed as an overlay to the conventional voice-based cellular networks.

This argument led to the development and standardization of a "data optimized" high-rate packet data (HRPD) air interface specification (also known as 1xEV-DO) by 3GPP2. The HRPD air interface was primarily based on a proprietary high-data rate (HDR) system developed and proposed by Qualcomm Inc., which was subsequently enhanced by 3GPP2 and released as the TIA/EIA IS-856 standard [1,2]. This technology uses the same carrier size and RF parameters as cdmaONE or cdma2000 to facilitate the overlay of the data-only carriers on existing CDMA networks.

Because for typical IP-based applications the demand for capacity is highly asymmetric, with most of the traffic load and the high-rate expectations in the forward link, the design of HRPD is primarily focused on forward link enhancement.

The HRPD forward data channel employs a number of advanced physical and MAC layer techniques to maximize spectral efficiency. Some of the key design features of HRPD in the forward link are:

- A single high-speed packet "pipe" among all active users

- Adaptive and high-order modulation with peak data rate of up to 2.4 Mbps

- Physical layer hybrid ARQ using low-rate turbo codes with incremental redundancy

- Short frames and fast scheduling to achieve multiuser diversity

CDMA2000® Evolution: System Concepts and Design Principles, by Kamran Etemad
ISBN: 0-471-46125-3   Copyright © 2004 John Wiley & Sons, Inc.

- Closed-loop rate control via fast channel state feedback
- Site reselection diversity or "virtual soft handoff"
- Optional receive diversity at the mobile terminal

The original design for the reverse link of HRPD is quite different from the forward link. The reverse link design is based on lower-data rate and power-controlled traffic channels that are code division multiplexed in a way very similar to Release 0 of IS2000.

One of the key limitations of this design was the significant imbalance between the forward and reverse link throughput and latency performance. To address this limitation and provide more competitive performance to some of the other recently emerged technologies for mobile high-speed packet data applications, 3GPP2 started working on a new release of the HRPD system. These enhancements, which are incorporated in IS856 Release A, involve major changes in the reverse link to reduce latency and increase throughput. The new changes also include some improvements in the forward link for more efficient support of short packets. The key uplink PHY and MAC layer improvements are based on some of the concepts originally used in the forward link, for example, using shorter frames, higher-order modulation, and hybrid ARQ.

In the following sections we start by introducing protocol layers and channelization in HRPD (IS856) air interface, and then we focus on physical channel parameters and data transmission on the forward and reverse link. Later sections in the chapter provide an overview of recent enhancements introduced in Release A of the HRPD standard as well as some discussion of the network architecture. In many references in this chapter we will use access terminal (AT) and access network (AN) for mobile station and base station, respectively.

10.2   AIR INTERFACE PROTOCOL LAYERS

The HRPD interface is designed based on seven distinct layers with clearly defined interfaces between layers and among protocols within each layer. These layers and their protocols are shown in Figure 10.1 and briefly described in the following [2]:

- *Physical layer:* defines the air interface physical channel structure, frequency, modulation, channel coding, and spreading schemes
- *MAC layer:* provides procedures to control data rates, bandwidth allocation, and scheduling of user traffic
- *Security layer:* provides authentication and encryption services
- *Connection layer:* defines signaling procedures for establishing and maintaining the air interface connection
- *Session layer:* provides capabilities for address management, data session configuration, and management
- *Stream layer:* allows multiplexing distinct application layer streams such as signaling and user traffic over the same air interface connection
- *Application layer:* refers to applications for carrying signaling and user traffic over the air. It does not refer to user applications.

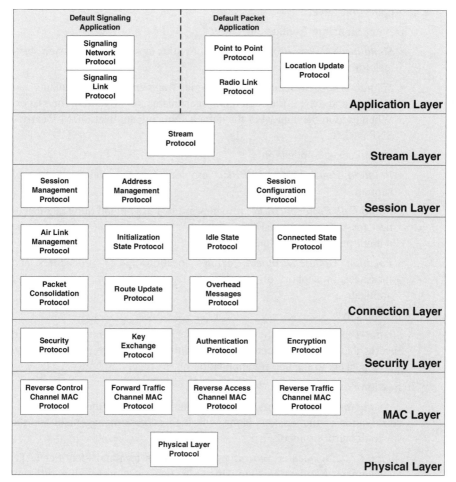

Figure 10.1   HRPD layering structure and default protocols. Reproduced under written permission from TIA.

This protocol layering does not exactly conform to the OSI seven-layer model, as the distribution of functionality among the protocol layers is not consistent with OSI definitions.

Each protocol in HRPD air interface layers is a self-contained object with predefined messages, procedures, and interfaces with other protocols, allowing independent future modifications to layers or protocols. Each protocol can be individually negotiated with its peer to provide specific functionality within that layer.

Although there may be various implementations for the same type of protocol, the HRPD standard has defined a default implementation for each protocol type. The default protocols, shown in Figure 10.1, represent the minimum set of protocols that must be supported by the AT and the AN.

The following provides an overview of the default protocols at each layer:

- In Application Layer:
  - Default Signaling Application:
    - *Signaling Network Protocol (SNP):* provides message transmission services for signaling messages.
    - *Signaling Link Protocol (SLP):* provides fragmentation mechanisms and reliable and best-effort delivery mechanisms for signaling messages. When used in the context of the default signaling application, SLP carries SNP packets.
  - Default Packet Application:
    - *Point to Point Protocol (PPP):* provides framing and multiprotocol support
    - *Radio Link Protocol (RLP):* provides retransmission and duplicate detection for an octet aligned data stream. When used in the context of the default packet application, RLP carries PPP packets.
    - *Location Update Protocol:* defines location update procedures and messages in support of mobility management for the default packet application
- In Stream Layer:
  - *Stream Protocol:* adds the stream header in the transmit direction and ensures that packets are octet aligned; removes the stream header and forwards packets to the correct application on the receiving entity
- In Session Layer:
  - *Session Management Protocol:* provides means to control the activation and deactivation of the sessions along with their address management and the session control protocols.
  - *Address Management Protocol:* provides access terminal identifier (ATI) management. AT identifiers are defined in such a way as to allow different types of user terminals.
  - *Session Configuration Protocol:* provides negotiation and configuration of the protocols used in the session
- In Connection Layer:
  - *Air Link Management Protocol* (ALMP): provides the overall state machine management that an access terminal and an access network follow during a connection. Depending on the status of connection, the ALMP triggers the appropriate protocols. Figure 10.2 provides an overview of the states and state transitions at the AT.
    The AT may be in initialization, idle, or connected states, and, depending on its state, the AT follows some of the following protocols.
  - *Initialization State Protocol:* provides the procedure that an access terminal follows to acquire an network and that an access network follows to support network acquisition

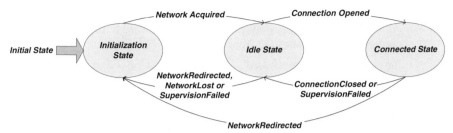

Figure 10.2   Access terminal state transitions. Reproduced under written permission from TIA.

- *Idle State Protocol*: provides the procedures that an access terminal and an access network follow when a connection is not open
- *Connected State Protocol*: provides the procedures that an access terminal and an access network follow when a connection is open
- *Route Update Protocol*: provides the means to maintain the route between the access terminal and the access network
- *Overhead Messages Protocol*: provides broadcast messages containing information that is mostly used by connection layer protocols
- *Packet Consolidation Protocol*: provides transmit prioritization and packet encapsulation for the connection layer

- In Security Layer:
  - *Key Exchange Protocol*: provides the procedures followed by the access network and the access terminal to exchange security keys for authentication and encryption
  - *Authentication Protocol*: provides the procedures followed by the access network and the access terminal for authenticating traffic
  - *Encryption Protocol*: provides the procedures followed by the access network and the access terminal for encrypting traffic
  - *Security Protocol*: provides procedures for generation of a cryptosync that can be used by the authentication protocol and encryption protocol

- In MAC Layer:

  The MAC layer of HRPD MAC consists of four protocols, each defining procedures and messages for handling of its corresponding channel.

  - *Access Channel Protocol*: defines timing, power levels, and procedures for transmission on the reverse access channel during initial system access
  - *Control Channel MAC Protocol*: defines transmission rules and procedures on the forward control channel including the scheduling of the control channel packets. The protocol also defines how the AT acquires and monitors the system using broadcast and common channel messaging on the forward control channel.

  – *Forward Traffic Channel MAC Protocol*: defines rules and procedures governing the scheduling packet transmission on forward traffic channel, based on the DRC command sent from access terminals

  – *Reverse Traffic Channel MAC Protocol*: defines rules and processes governing transmission of data, reverse rate indication (RRI), as well as soft handoff procedures

- In Physical Layer:

  – *Physical Layer Protocol*: provides channel structure, frequency, power output, and modulation specifications for the forward and reverse links

10.3 PHYSICAL AND LOGICAL CHANNELIZATION

Within a sector, the HRPD uses time division duplexing (TDM) for channelization, by allocating all power and code space resources to a single high-speed traffic channel or packet data "pipe" that is time shared by all active users. However, across cells HRPD is like a CDMA system and uses spread spectrum and interference averaging to allow a frequency reuse of 1.

In contrast to IS95A/B, where each physical channel is transmitted the entire time, using a fraction of the total sector power and single code channel, in HRPD the forward physical channels are transmitted, at full power, using the entire code space, and each occupying a certain fraction of time (see Fig. 10.3).

This section describes the overall physical channelization in the forward and the reverse link of HRPD.

10.3.1 Forward Link Channels

In the forward link of 1X-EV DO all physical channels are time multiplexed on a single composite channel and transmitted at full power at all times. Figure 10.4

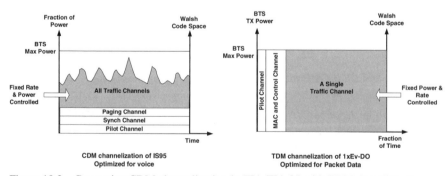

Figure 10.3 Comparing CDM channelization in TIA-TIA 95 with TDM downlink channelization in HRPD.

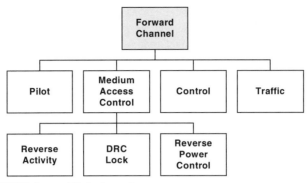

Figure 10.4   HRPD forward link channel structure.

shows these forward physical channels and the following briefly describes their function:

Pilot Channel: The main functions of the pilot channel are the same those of forward common pilot channels in cdma2000 or IS95. The pilot channel carries all "0"s, that is, no upper layer information. The power and timing of the pilot channel is used to simplify the determination of the best serving base station and to help ATs in acquiring the system timing and fast synchronization. The pilot channel also allows the AT to perform coherent demodulation in the downlink.

Medium Access Control: As suggested by its name, this channel carries MAC layer-related information to control uplink transmissions and it includes the following:

Reverse Activity Bit (RAB): This broadcast indicator (1 bit per slot) informs all the ATs in a sector about the level of activity in the reverse link for a specific sector. The network measures interference level in reverse direction, based on which it determines the activity level. The broadcast RABs are received by every connected access terminal in the sector. If a sector determines that the activity level is very high, it sets the RAB to 1; otherwise, it sets the RAB to 0. The AT uses the combination of the RAB information received from all sectors as part of its reverse rate selection process, to be discussed later.

Reverse Power Control (RPC): This indicator is specific to every active user in a sector. It provides a closed-loop power control on the active ATs, thus reducing the uplink interference.

DRCLock Channel: This is the portion of the forward MAC channel that indicates to the AT whether or not the access network can receive the DRC sent by the AT. If the AT receives a DRCLock bit set to "0" from its target sector, the AT should stop pointing its DRC at that sector.

Control Channel: This channel carries the broadcast overhead information such as the "sector parameters message" as well as user-specific common control information such as the "traffic channel assignment message." Whereas synchronous control packets are transmitted every 400 ms, the asynchronous control packets are transmitted whenever necessary.

Traffic Channel: This is a variable-rate high-speed traffic "pipe" that is time shared by many users based on a packet data scheduling algorithm. The traffic channel carries users' traffic at various rates ranging from 38.4 kbps to 2.45 Mbps depending on the choice of modulation and coding.

When a physical packet frame containing traffic or control data is sent, an all "0" preamble sequence is included at the beginning of the packet to assist the access terminal with synchronization of the following variable-rate transmission. The preamble also helps AT to identify whether the information is user traffic or control information and if it is user traffic to which active user this data packet belongs.

10.3.2   Reverse Link Channels

In the reverse link, physical channels of HRPD are code multiplexed and grouped as those used in the access mode or traffic channel mode (see Fig. 10.5). These channels are briefly introduced in the following:

Reverse Access Channel Mode: used by AT to initiate its communication with the access network or to respond to an access terminal directed message. The access channel consists of a pilot channel and a data channel.

- *Pilot Channel*: transmitted for preamble purposes as well as for time synchronization

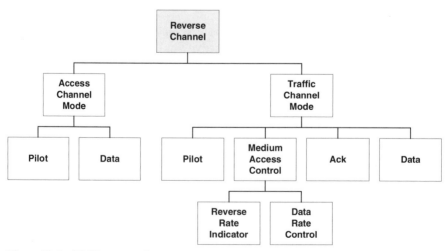

Figure 10.5   HRPD reverse link structure.

- *Data Channel*: carries any common channel control messages from AT when AT is not in *connected state* (does not have a dedicated channel). While transmitting data channel, the AT also transmits pilot channel continuously.

Reverse Traffic Channel Mode: used by the access terminal to transmit user-specific traffic or signaling information to the access network. The traffic channel mode involves transmission of the following physical channels:

- *Pilot Channel:* Similar to cdma2000, the purpose of pilot channel transmission is to provide time synchronization to the receiver and to enable coherent demodulation by the receiver.
- *Medium Access Channel:* This channel is a combination of the following subchannels.
- *Reverse Rate Indicator (RRI) Channel:* used to indicate whether or not the data channel is being transmitted on the reverse traffic channel and, if it is being transmitted, its data rate. This way, the receiver does not have to execute any rate determination algorithms.
- *Data Rate Control (DRC) Channel:* used by the access terminal to indicate to the access network the supportable forward traffic channel data rate and the best serving sector on the forward channel.
- *ACK Channel:* used by the AT to inform the AN whether or not the data packet transmitted on the forward traffic channel has been received successfully. On this channel, the AT sends ACK or NACK for each physical layer slot received. Note that; in addition to the acknowledgment at the transport layer as part of the TCP protocol, the HRPD also supports acknowledgement at the physical layer. This provides quick response from the receiver (AT) to the transmitter (AN).
- *Data Channel:* The data channel in the traffic channel mode carries both the user traffic and dedicated control messages. The data rate of this channel varies from 9.6 to 153.4 kbps.

An AT in the idle state uses the access channel mode channels, that is, the pilot and data channels, to communicate with the AN. In this mode, the AT uses open-loop power control and access probe mechanisms similar to IS95/cdma2000.

Once in the connected state, the AT uses the traffic channel mode channels and sends pilot, ACK, MAC, and data to the AN and the transmit power of AT is controlled by the AN in a closed-loop operation.

10.4 FORWARD PHYSICAL CHANNEL STRUCTURE

In HRPD system, there is only one composite forward physical channel structured as a time division multiplex of pilot, MAC, control, and traffic channels.

This physical channel is divided into 26.67-ms frames, and each physical layer frame is further divided into 16 slots. Each slot, which is 1.67 ms in duration, is also further divided into two half-slots, where each half-slot contains 1024 chips, resulting in $(1024/2)/1.67 = 1.228$ Mcps.

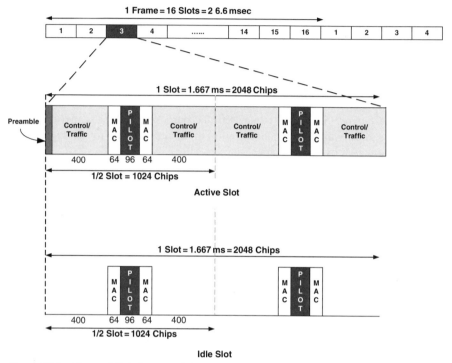

Figure 10.6   Forward link frame and slot structure.

During an active slot, the pilot, MAC, control, and traffic channels are time-multiplexed in each half-slot as shown in Figure 10.6.

When there are no traffic or control channels to be sent on the forward link, the base station sends "idle slots" that have the same time structure as an active slot but with no power transmitted during the "traffic/control" periods. All transmitted channels are sent at full power and at fixed chip positions.

From the total of 1024 chips in a half-slot; the pilot channel takes 96 chips, the MAC channel takes $2 \times 64 = 128$ chips, and the control and traffic channels have $2 \times 400 = 800$ chips. The 800 chips will carry different amounts of data depending on the modulation scheme and spreading rate. So the pilot, MAC, and traffic/control channels take about 10%, 15%, and 75% of total physical resources, respectively.

The control channel slots have a 256-slot cycle. The AN sends the broadcast and common channel messages on the control channel slots in every 256 slots = 426.67 ms. The remaining slots are used for the scheduled packet data traffic. Figure 10.7 shows the control channel cycle, where 8 slots have been allocated to control and 248 to user traffic data.

MAC Addressing: In HRPD the MAC layer assigns a MAC index (MAC_ID) to each active user as part of the traffic channel assignment. The system defines a total of 64 MAC indices. Whereas a total of 59 MAC indices are available for allocation to active users, some MAC indices are reserved for overhead operations.

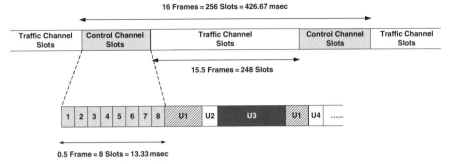

Figure 10.7 Control channel slot cycles.

TABLE 10.1 MAC_ID Usage for MAC and Preamble Channels

MAC_ID	MAC Channel Use	Preamble Use
0 and 1	Not Used	Not Used
2	Not Used	76.8-kbps Control Channel
3	Not Used	38.4-kbps Control Channel
4	RA Channel	Not Used
5–63	Available for RPC and DRCLock Channel Transmissions	Available for Forward Traffic Channel Transmissions

The MAC index has different purposes on the MAC channel and on the preamble (see Table 10.1).

- On the MAC channel: The MAC_ID is used for orthogonal channelization and code multiplexing of all active users' reverse power control channels as well as the reverse activity information bits. Whereas MAC_ID 4 is reserved for RAI, each of the MAC_ID = 5–63 are used to uniquely identify the RPC or DRCLock commands for each of the "active" users. The MAC_ID = 0–3 are not used.

- On the preamble channel: The MAC index is used for Walsh spreading to indicate whether what follows the preamble is user traffic or control information. On this channel MAC indices 2 and 3 are used to identify control channel messages at rates 3 of 8.4 and 76.8 kbps, respectively, whereas MAC indices 5–63 are used to identify data traffic of the "active" users in the sector. Other MAC_IDs are not used.

Note that each MAC index maps to either 64-ary orthogonal Walsh code when used for MAC channel channelization or 32-ary biorthogonal Walsh cover when used for control/traffic channel channelization on the preamble.

10.4.1 Forward Traffic Channel Structure

The forward traffic channel and control channel data are encoded in blocks called physical layer packets. The control channel data are sent at 38.4 or 76.8 kbps,

TABLE 10.2 Modulation Parameters for the Forward Traffic and Control Channels

Data Rate (kbps)	Number of Values per Physical Layer Packet				
	Slots	Bits	Code Rate	Modulation Type	TDM Chips (Preamble, Pilot, MAC, Data)
38.4	16	1024	1/5	QPSK	1024, 3072, 4096, 24576
76.8	8	1024	1/5	QPSK	512, 1536, 2048, 12288
153.6	4	1024	1/5	QPSK	256, 768, 1024, 6144
307.2	2	1024	1/5	QPSK	128, 384, 512, 3072
614.4	1	1024	1/3	QPSK	64, 192, 256, 1536
307.2	4	2048	1/3	QPSK	128, 768, 1024, 6272
614.4	2	2048	1/3	QPSK	64, 384, 512, 3136
1228.8	1	2048	1/3	QPSK	64, 192, 256, 1536
921.6	2	3072	1/3	8-PSK	64, 384, 512, 3136
1843.2	1	3072	1/3	8-PSK	64, 192, 256, 1536
1228.8	2	4096	1/3	16-QAM	64, 384, 512, 3136
2457.6	1	4096	1/3	16-QAM	64, 192, 256, 1536

Only 38.4- and 76.8-kbps Rates Apply to Control Channels.

whereas the user traffic can be sent at various rates ranging from 38.4 kbps to 2.45 Mbps.

Depending on the data rate chosen, the forward traffic channel and control channel physical layer packets can be transmitted in 1 to 16 slots; see Table 10.2 for details.

The traffic channel physical layer packets are turbo encoded with code rates of $R = 1/3$ or $1/5$. The turbo encoder employs two systematic recursive convolutional encoders connected in parallel, with a turbo interleaver preceding the second recursive convolutional encoder. The encoder output symbols are punctured and repeated to achieve the desired rate. Turbo encoders are discussed further in Chapter 4.

As shown in Figure 10.8, the output of the channel encoder after scrambling and interleaving is applied to a quadrature modulator with in-phase and quadrature output streams. Depending on the data rate the modulator generates QPSK, 8-PSK, or 16-QAM modulation symbols that are further repeated or punctured as necessary. The resulting sequences of rate-adjusted modulation symbols are demultiplexed to form 16 parallel pairs of quadrature (i.e., in-phase and quadrature) streams. Each of the parallel streams is then covered with a distinct 16-ary Walsh function to yield a 76.8-ksps sequence and normalized to maintain a constant total transmit power that is independent of data rate.

The scaled Walsh-covered symbols from all 16 streams are added on a chip-by-chip basis to form a single in-phase stream and a single quadrature stream at a chip rate of 1.2288 Mcps.

The resulting traffic channel chips are then time-division multiplexed with the preamble, pilot, and MAC channel chips to form the resultant sequence of chips for the quadrature spreading operation.

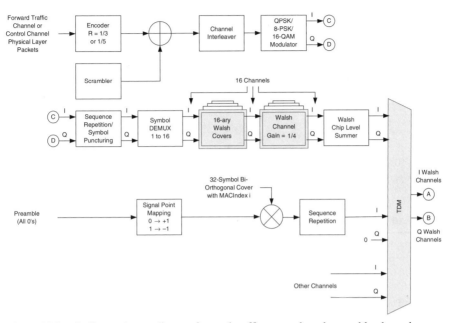

Figure 10.8   Coding and spreading on forward traffic, control, and preamble channels.

The preamble, which is transmitted on the in-phase component only, consists of all-"0" symbols spread by a 32-bit biorthogonal cover that corresponds to a unique MAC index. The biorthogonal sequence is specified in terms of the 32-ary Walsh functions and their bit-by-bit complements by

$$
\begin{array}{ll}
W_{i/2}^{32} & \text{for } i = 0, 2, \ldots, 62 \\[1em]
\overline{W_{(i-1)/2}^{32}} & \text{for } i = 1, 3, \ldots, 63
\end{array}
\tag{10.1}
$$

where $i = 0, 1, \ldots, 63$ is the MAC_ID value and $\overline{W_i^{32}}$ is the bit-by-bit complement of the 32-chip Walsh function of order i. The 32-chip biorthogonal sequence may be repeated, as needed, depending on the transmission mode and the length of the preamble.

10.4.2   Forward MAC Channel Structure

The forward MAC channel code multiplexes RAB and users' RPC/DRCLock channels with Walsh channels that are orthogonally covered and BPSK modulated. The RPC channel and the DRCLock channel are time-division multiplexed and transmitted on the same MAC channel. The DRCLock channel bits are punctured into RPC once every DRCLockPeriod slots (see Fig. 10.9).

The power control bits for the active mobiles are sent in every 1.6-ms slot, which results in 600 bps. However, as a result of DRCLock channel puncturing into

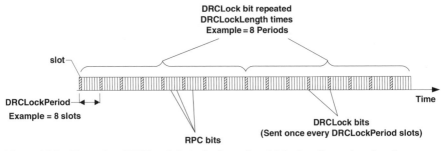

Figure 10.9   Example of RPC and drclock channel multiplexing. Reproduced under written permission from TIA.

TABLE 10.3   Modulation Parameters for the Forward MAC Channels

Parameter	RPC Channel	DRCLock Channel	RA Channel
Rate (bps)	600 × (1-1/DRCLockPeriod)	600/(DRCLockLength × DRCLockPeriod)	600/RABLength
Bit Repetition Factor	1	DRCLockLength	RABLength
Modulation (Channel)	BPSK (I or Q)	BPSK (I or Q)	BPSK (I)
Modulation Symbol Rate (sps)	2400 × (1-1/DRCLockPeriod)	2400/ DRCLockPeriod	2400
Walsh Cover Length	64	64	64
Walsh Sequence Repetition Factor	4	4	4
PN Chips/Slot	256	256	256
PN Chips/Bit	256	256 × DRCLockLength	256 × RABLength

the RPC stream the power control rate will be reduced slightly to 600/(1 − DRCLockPeriod).

Each Walsh channel is identified by a MAC_ID value that is between 0 and 63 and defines a unique 64-ary Walsh cover and a unique modulation phase.

The Walsh functions are assigned to the MAC_ID values as follows:

$$W_{i/2}^{64} \quad \text{for } i = 0, 2, \ldots, 62$$

$$W_{(i-1)/2+32}^{64} \quad \text{for } i = 1, 3, \ldots, 63 \tag{10.2}$$

Where i is the MAC_ID value. Table 10.3 captures the mapping of MAC indices to RAB and RPC channels. Figure 10.10 shows the multiplexing process for various components of the forward MAC channel as the well as the pilot channel.

MAC channels with even-numbered MAC indexes are assigned to the in-phase (I) modulation phase, whereas those with odd-numbered MAC indexes go to the quadrature (Q) modulation phase. The MAC symbol Walsh covers are transmitted four times, as 64 chips immediately preceding and following the pilot bursts, within each slot.

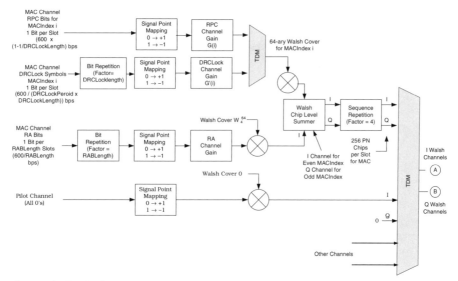

Figure 10.10   Coding and spreading of forward MAC and pilot channels. Reproduced under written permission from TIA.

The reverse pilot channel, which is an all-"0" sequence covered by Walsh zero, is time-multiplexed with other channels and assigned only to the in-phase (I) modulation phase.

10.4.3   Quadrature Spreading and Modulation

The quadrature spreading and modulation in the 1xEVDO forward channel is similar to IS2000. After orthogonal spreading, the combined time-multiplexed forward channel sequence is quadrature spread as shown in Figure 10.11. The spreading sequence is a quadrature PN sequence of length $2^{15} = 32{,}768$ chips. This PN sequence has a chip rate of 1.2288 Mcps, and its period is $32{,}768/1{,}228{,}800 = 26.667$ ms, corresponding to 75 sequence repetitions in every 2 s.

Also similar to IS2000 and cdmaONE, the pilot PN offsets allocated to sectors are identified by an offset index in the range from 0 to 511, in units of 64 chips, with respect to a zero-offset pilot PN sequence.

10.5   REVERSE PHYSICAL CHANNEL STRUCTURE

The reverse link of HRPD has a simple frame structure because most of the multiplexing is done in the code domain and there is no need for time multiplexing. The channel structure, however, is different depending on whether the AT is in the access channel mode or in the traffic channel mode.

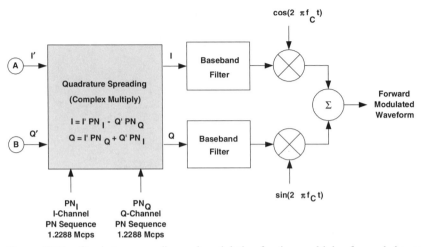

Figure 10.11   Quadrature spreading and modulation for time multiplex forward channel.

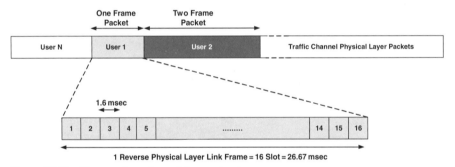

Figure 10.12   Reverse data channel frame structure.

10.5.1   Reverse Traffic Channel Structure

The reverse traffic channel is composed of pilot, data, ACK, and MAC channels as described in the following:

Data and ACK Channels: Similar to the forward link, the reverse traffic channel frames are 26.67-ms long. Each frame consists of 16 slots of 1.67-ms duration, and each slot has 2048 chips (see Fig. 10.12). The beginning of the packet transmission is aligned to the slot boundary and randomized by the frame-offset parameter. Table 10.4 shows the supported data rates and payload sizes.

The reverse link data packets have fixed length of 26.667 ms. The longer duration is chosen to provide better time diversity. The system also uses a fixed BPSK modulation scheme. Whereas the packet duration and modulation are fixed, the offered data rates can be set to 9.6, 19.2, 38.4, 76.8, and 153.6 kbps by changing the channel coding rates.

TABLE 10.4   **Reverse Data Channel Parameters**

Parameter	Data Rate (kbps)				
	9.6	19.2	38.4	76.8	153.6
Reverse Rate Index	1	2	3	4	5
Bits per Physical Layer Packet	256	512	1024	2048	4096
PHY Layer Packet Duration (ms)	26.667	26.667	26.667	26.667	26.667
Code Rate	1/4	1/4	1/4	1/4	1/2
Symbols/PHY Packet	1024	2048	4096	8192	8192
Code Symbol Rate (ksps)	38.4	76.8	153.6	307.2	307.2
Interleaved Packet Repetitions	8	4	2	1	1
Modulation Symbol Rate (ksps)	307.2	307.2	307.2	307.2	307.2
Modulation Type	BPSK	BPSK	BPSK	BPSK	BPSK

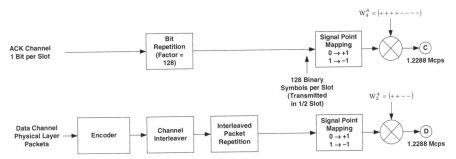

Figure 10.13   Reverse data and ACK channel structure.

The data rate is chosen by AT based on the rules and parameters set by the network. The selected rate is explicitly indicated by the RRI, to avoid the need for blind rate detection based on different hypotheses.

Figure 10.13 and Table 10.4 show the modulation and encoding structure and parameters for the reverse data and ACK channels. For example, when the user transmits at 19.2 kbps, indicated with RRI = 2, the physical layer packet contains 512 bits of user information. This packet is coded with a 1/4 rate turbo encoder, block interleaved, and repeated four times to form symbols at 307.2 ksps.

In all cases the symbols are spread with a 4-ary Walsh cover and, after BPSK modulation, are transmitted over 16 slots or 26.67 ms.

The access channel frame and reverse traffic channel frame are 26.67 ms in duration, consisting of 16 slots. Each slot is 1.67 ms in duration and contains 2048 PN chips.

Pilot Channel: For the reverse traffic channel, the encoded RRI channel symbols are time-division multiplexed with the pilot channel. This time-division-multiplexed channel is still referred to as the pilot channel. The 3-bit symbol RRI

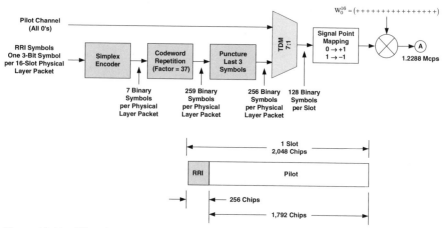

Figure 10.14   Pilot channel structure.

representing the uplink data rate is sent to the AN for every physical layer packet (at every 26.67 ms or every 16 slots). For example, RRI = "000" means a data rate of 0 bits/s or no packet transmission and RRI = "101" implies a 153.6-kbps data rate on the data channel.

This 3-bit RRI is first encoded and then repeated to generate 256 binary symbols. These symbols are then time-multiplexed with the pilot channel symbols (ratio of 1:7) and span the same time interval as the corresponding physical layer packet.

These time-division multiplexed pilot and RRI channel symbols are spread with the 16-bit Walsh cover that produces 256 chips in every slot. The RRI chips are time-division multiplexed in the first 256 chips of every slot as shown in Figure 10.14.

The RRI bits are not transmitted on the access channel because the access channel data have a fixed transmission rate of 9.6 kbps.

DRC Channel Structure: The main attributes of the DRC channel are DRC value, DRC cover, and DRC length, as described below.

When in the connected state, the AT constantly monitors the C/I of the pilot channel of all the sectors in its active set. Based on the C/I measurement, the access terminal selects the "best" serving sector from which it can receive the forward traffic channel at the highest possible data rate. The requested forward traffic channel data rate is mapped into a four-bit DRC value. This four-bit value selects rates from 38.4 kbps to 2.467 Mbps.

Also, for each sector in the active set of the access terminal, the access terminal is given a corresponding DRC cover. The DRC cover is a 3-bit number mapping to one of 8 possible Walsh codes of length 8. Once the best sector is selected, the DRC cover of the best sector is used to spread the DRC symbols. Because DRC covers are orthogonal to each other, the AN can determine the sector selected by the AT. Figure 10.15 shows the structure of DCR channel processing.

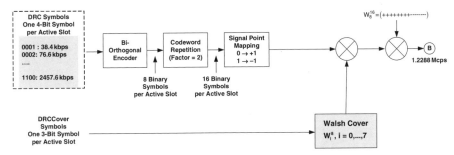

Figure 10.15   DRC channel structure.

The DRC value is mapped to a corresponding 8-bit codeword with biorthogonal encoding, and this 8-bit codeword is repeated to form a 16-bit symbol. The DRC value is sent in every slot (every 1.67 ms) and is updated periodically based on the value of DRCLength sent by the AN in the traffic channel assignment. The value of DRC and DCRCover may not be changed for DRCLength duration.

10.5.2   Reverse Access Channel Structure

In 1X-EV DO the access channel structure and protocols are very similar to those of IS95/cdma2000 Release 0. The access channel consists of a pilot channel and a data channel. The transmissions on the access channel are in the form of access probes, each consisting of a preamble followed by one or more access channel physical layer packets (see Fig. 10.16).

The data channel used during the access follows the same frame structure as traffic data channel; however, the data rate is limited to 9.6 kbps. The user may transmit one or more physical frames on the data channel, each of length 26.67 ms. In addition, because the data rate is fixed, there is no need for reverse rate indicators to be transmitted on the pilot.

During the preamble transmission, only the pilot Channel is transmitted, whereas during the access Channel physical layer packet transmission, both the pilot Channel and the data Channel are transmitted. The output power of the pilot Channel is higher during the preamble portion of an access probe than it is during the data portion of the probe by an amount such that the total output power of the preamble and data portions of the access probe are the same.

Similar to IS95 and cdma2000, the access terminal sends a series of access probes until it receives a response from the access network or the timer expires. Access probe transmission power is calculated similar to IS95/cdma2000 starting with a low initial power estimated by open-loop power control followed by power increments on every successive probe within an access attempt.

10.5.3   Quadrature Spreading and Modulation

After appropriate gains are applied, relative to the pilot channel, the Ack, DRC, and data channels are added together and the pilot. The combined channels form the I

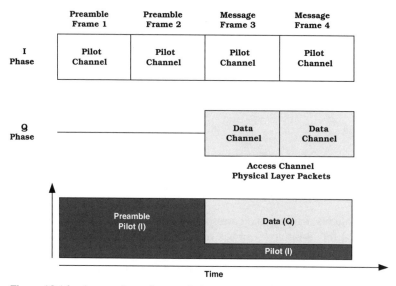

Figure 10.16   Access channel transmission structure.

and Q branches, which then go through the quadrature spreading at 1.228 Mcps with user-specific long code.

The pilot and scaled ACK channel sequences are added to form the resultant I-channel sequence, and the scaled DRC and data channel sequences are added to form the resultant Q-channel sequence. The quadrature spreading operation will be equivalent to a complex multiply operation of the resultant I-channel and resultant Q-channel sequences by the PN_I and PN_Q PN sequences, as shown in Figure 10.17.

Note that within a traffic or Access Channel the Pilot Channel, the DRC Channel, the ACK Channel, and the Data Channel are orthogonally spread by Walsh functions of length 4, 8, or 16. In addition, from user to user, each Reverse Traffic Channel is identified by a distinct user long code. Also, the Access Channel for each sector is identified by a distinct Access Channel long code.

10.6   DATA TRANSMISSION ON THE FORWARD LINK

The forward data channel is time multiplexed with the pilot and MAC channel, and it can carry dedicated traffic or control information. The data transmissions on this channel are subject to the link adaptation and scheduling procedures described in this section.

10.6.1   Forward Rate Selection

In 1X-EV DO, unlike most conventional cellular system, it is the mobile station or AT that determines the data rate at which it should receive data from the base station

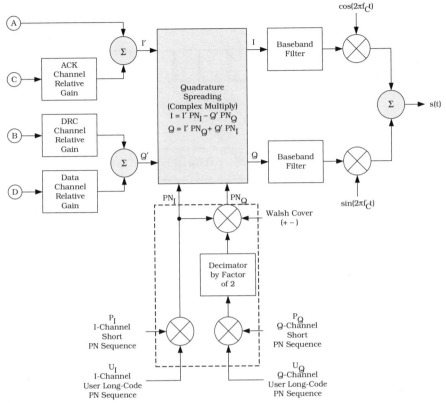

Figure 10.17 Reverse link quadrature spreading and modulation. Reproduced under written permission from TIA.

or the AN. The AN must send data to the user at the requested data rate only. The AN, however, can set the maximum allowed data rate as well as the rules and thresholds used for the AT's uplink rate selection. One can think of this rate selection as a combination of an inner loop and an outer loop process.

- In the inner loop process the AT estimates the highest supportable data rate and provides that to the access network through the DRC channel in every time slot. The AT maintains records of all the strong pilots in an active set, and for each of the pilots it measures the received signal-to interference and noise ratio (SINR) during the pilot burst (see Fig. 10.18).
- In the outer loop process the AN adjusts the minimum SINR thresholds for all data rates to achieve the target packet error rate on the forward traffic channel.

The AT uses the measured SINR of the strongest pilot and the thresholds defined by the AN to determine the highest data rate it can reliably decode as well as the identity of the corresponding best serving sector.

The AT uses the DRC channel to inform the AN of the desired data rate and the selected sector. To each pilot or sector in the active set a DRC cover is

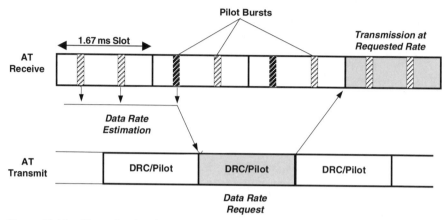

Figure 10.18   Channel estimation and data request channel timing diagram.

associated. This association is established at the connection setup time, and it is updated every time a new sector is added to the active set.

The mobile may update the information on the DRC channel in every DRC_Length of 1, 2, 4, or 8 slots. The DRC_Value transmissions during the DRC_Length slots may be repeated at relatively lower power or gated.

The AN sends DRC_Length information to the AT during connection setup, and it may be updated at any time while the AT is on the traffic channel. For example, if the DRC_Length is 4, then the AT evaluates the best rate and serving sector in every 4 slots. The AT either repeats the transmission of the same DRC information in 4 slots, at lower power, for example, about 1/4th of the nominal DRC gain, or it uses a gated transmission, in which case it sends DRC in 1 time slot at nominal gain and does not transmit on the remaining time slots. The scheduler at the AN's sector considers the DRC information sent in the next 4 slots to schedule transmission to the AT in 4 subsequent slots in the forward direction.

Note that the AN does not change the data rate once it has started the packet transmission, even if the AT changes the DRC information in subsequent slots.

10.6.2   Forward Link Packet Scheduling

In 1X-EV DO the forward link scheduler, implemented at each sector of an AN, is part of the forward traffic channel MAC protocol [3,4]. Although the details of scheduling algorithms are implementation dependent and thus not specified in the standard, their overall goal is to maximize system throughput subject to "fairness" for all active users with different amounts of data and channel conditions.

On the basis of the feedback received from the mobiles on the DRC channels, where they indicate their supportable data rates, as well as other factors, the scheduler determines which client terminal is to be served at each time slot.

Although no specific scheduling algorithm are specified in the standard the following algorithm can be mentioned as baseline:

- *Round-Robin Algorithm*: simply based on "first come first served" strategy.
- *Best-Rate or Best-Effort Algorithm*: always transmits to users with the best channel condition or data rate, indicated in their DRC.
- *Proportional Fairness*: attempts to balance the trade-off between the round-robin and best-rate methods by using a mixed criterion.

These algorithms and their comparison are discussed with more details in Chapter 4. The criterion used in proportional fairness, which is the default scheme in most implementations of HRPD, is based on the ratio of each user's requested data rate to its average received data rate. In other words, in each slot the scheduler keeps track of the moving average of each user's received data rate over N slots and normalizes the DCR rate by this moving average to identify the user with highest r_a (*i*):

$$r_a(i) = \frac{\text{DCR}_a(i)}{R_a(i)} \tag{10.3}$$

where

$$R_a(i) = \frac{(N-1) \times R_a(i-1) + \text{Current Rate}(i)}{N} \tag{10.4}$$

The current rate (*i*) is set to zero for those ATs not receiving transmission in the current slot "*i*," including ATs for which the sector has no data to send.

It should be noted that the current implementation of scheduling algorithm in 1X-EV DO does not allow support of user-specific or application-specific QoS. The evolution of 1X-EV DO, however, includes the enhanced QoS-aware scheduling options.

10.6.3 Hybrid ARQ and Multislot Data Transmission on the Forward Link

Depending on the data rate chosen, the forward traffic and control channel physical layer packets can be transmitted in 1 to 16 slots (see Table 10.2). A multislot physical layer packet is not transmitted in consecutive slots but follows a 4-slot interlacing pattern. Therefore, the multiple slots allocated to one packet are separated by three intervening slots, which are typically used for other users' packet transmissions. This timing allows the AT to process and respond with the physical layer acknowledgment (either ACK or NACK).

Furthermore, the multislot packets are encoded and transmitted based on a hybrid ARQ framework to allow for incremental redundancy and higher system throughput. Whereas the first slot or subpacket carries the systematic, that is, information bits, and some parity bits, the remaining slots may carry repeated bits and or additional coded symbols. The AT attempts to decode the packet after receiving each slot or subpacket. If the combination of all received subpackets is decoded successfully (i.e., CRC checks), the AT sends an ACK. Otherwise, the AT sends a NAK. In other words, once a multislot packet is transmitted in the forward link:

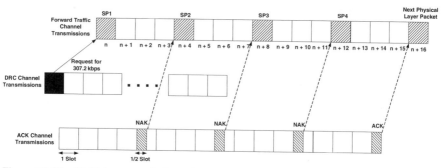

Figure 10.19   Multislot transmission with normal termination.

If the AN receives a NACK for a transmitted slot or subpacket, it continues the transmission of the next subpacket after 4 slots.

If it receives an ACK for a transmitted slot, it considers the transmission of the entire packet successful and terminates the transmission of the remaining slots of the current packet. In this scenario, the AT may schedule the freed slot to transmit a new packet, resulting in increased average throughput for the entire sector.

The following describes examples of the ACK channel timing for a normal and early termination of a four-slot transmission at a 307.2-kbps forward traffic channel. The slots from other physical layer packets may be interlaced in the intervening three-slot periods.

Figure 10.19 shows the case of a normal physical layer packet termination. In this case, the access terminal transmits NAK responses on the ACK channel after the first three slots of the physical layer packet are received, indicating that it was unable to correctly receive the packet after only one, two, or three of the nominal four slots. An ACK or NAK is also transmitted after the last slot is received, as shown. If a NAK is transmitted on the last slot, the entire packet transmission is repeated from the first subpacket.

Figure 10.20 shows the case of an early termination of the forward traffic channel physical layer packet. In this example, the access terminal transmits an ACK response after the third slot or subpacket is received, indicating that it has successfully assembled and decoded the whole physical layer packet. When the access network receives such an ACK response, it does not transmit the remaining slots of the physical layer packet. Instead, it may begin transmission of any subsequent physical layer packets. As a result of early termination, in this case the effective data rate is equal to $307.2 \times 4/3 = 409.6$ kbps.

10.7   DATA TRANSMISSION ON THE REVERSE LINK

The reverse data transmission is practically autonomous, as users' transmissions are in parallel and isolated in the code domain. The users who are permitted to transmit

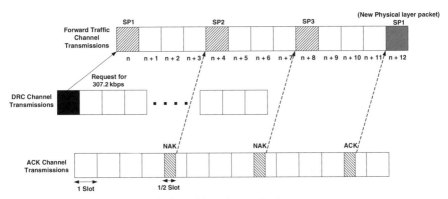

Figure 10.20   Multislot transmission with early termination.

will follow a rate selection process and transmit their data with their specific long code. Each transmission may take one or more numbers of frames, and the frames are 26.67 ms long. The transmission power also depends on the data rate chosen. In this section we describe the rate selection and power control aspects of the reverse traffic channel.

10.7.1   Data Rate Selection for Reverse Data Channel

In the reverse link direction, the AT selects the best data rate from the list of supportable rates, subject to the rules set by the AN. The rules take into account the system loading, the AT's available power, and the size of the payload. The mobile indicates the rate of transmission with the reverse rate indicators (RRI) time multiplexed with pilot. The reverse rate selection takes into account many factors and rules as described below.

ReverseRateLimit: the maximum data rate for an AT under all channel conditions. This limit is set by the access network and, although not common, can be different for different ATs.

The AN may periodically send the ReverseRateLimit message to manage usage of the reverse traffic channel. This message is used to manage overload conditions, maximize the peak bandwidth available to ATs, and maintain fairness between ATs. The AN may either broadcast the message or send it to a particular AT.

Initially, as the AT enters connected state, its rate limit is set to 9.6 kbps. The AN may change the CurrentRateLimit for one or a group of mobiles with a unicast or broadcast message, respectively. The AN sends these messages when the load on reverse carrier increases beyond implementation-defined thresholds. The mobile lowers its CurrentRateLimit parameter if the new rate indicated by the message is lower than its current limit; otherwise, the AT waits for 1 frame or 16 slots before increasing its limit to the new value. This approach is consistent with the general principle that decisions about power increases are taken conservatively.

MaxRate: the maximum rate under present channel conditions; it is also determined indicated by the access network. The MaxRate cannot exceed the CurrentRateLimit.

CurrentRate: the data rate chosen by the AT, which can vary from the lowest data rate, that is, 9.6 kbps, to the MaxRate. The AT selects the best CurrentRate based on its available transmit power and the size of the data packet.

CombinedBusyBit: the "OR" combination of RAB bits for all sectors in the active set. This combination is "1" or busy if any of the active sectors are busy with RAB = 1, and it is "0" if all active sectors are "non-busy".

Reverse Persistence Vectors: the set of upward or downward rate change probability values set by the AN at time of session setup. One of the reverse persistence vectors show the down-step transition probabilities when CombinedBusyBit = "1", whereas the other vector shows the up-step transition probabilities for the case when CombinedBusyBit = "0".

In other words, each entry contains a transition probability from one current rate to the next highest/lowest rate. For example, the entries of vector when CombinedBusyBit = "1" correspond to the probabilities of down-steps Transition019k2_009k6, Transition038k4_019k2, Transition076k8_038k4, or Transition153k6_038k4 when the current transmission rate for the access terminal is 19.2 kbps, 38.4 kbps, 76.8 kbps, or 153.6 kbps, respectively. The data rate selection process is based on the following inputs:

- Mobile's CurrentRate
- The computed CombinedBusyBit
- A random number x, uniformly distribution between 0 and 1, generated by the AT
- The two reverse persistence vectors

The mobile computes CombinedBusyBit, based on which it selects the corresponding reverse persistence vector. The mobile then generates a random number x and compares that against the entry p of the vector corresponding to transition from the AT's CurrentRate.

The AT compares the random number x against p.

- When CombinedBusyBit = "1": If $x < p$, then the access terminal decreases its rate to the next lowest rate (i.e., one step down); otherwise, it maintains its rate. In this case, if $x < p$ and the mobile was already transmitting at the minimum rate (i.e., 9.6 kbps), then the access terminal should set the MaxRate to the minimum rate (i.e., 9.6 kbps).
- When CombinedBusyBit = "0": If $x < p$, the access terminal sets MaxRate to the next highest rate relative to its current rate; otherwise, it sets MaxRate to its current rate.

For example, let us assume that the mobile has two base stations BS1 and BS2 in its active set with RAB1 = 1 and RAB2 = 0, and its CurrentRate =38.4 kbps. The CombinedBusyBit is then equal to "1". The AT then looks at the entry of persistent vectors for the "Busy" case and looks for the Transition038k4_019k2 entry, which shows a value of 0.4 that is the step-down probability. The AT then generates the random number x, which turns out to be 0.3. Because the 0.3 < 0.4 condition holds, the mobile reduces the rate to 19.2 kbps in its next transmission.

TABLE 10.5   The Required E_b/N_o and Channel Gains for Reverse Traffic Channel

Data Rate (kbps)	Data Channel Gain (dB)	Total (E_b/N_t) Required per Antenna (dB)
9.6	3.75	6.62
19.2	6.75	4.98
38.4	9.75	3.84
76.8	13.25	3.55
153.6	18.5	5.27

10.7.2   Power Control on the Reverse Traffic Channel

In HRPD the reverse link transmissions are subject to both open- and closed-loop power control. The reverse power control is directly applied to the pilot/RRI channel only, and the power levels allocated to the DRC, ACK, and data channels are adjusted by a fixed gain relative to the pilot/RRI channel. The channel gains are defined based on the coding gain, the target reliability, and the data rate for each channel to achieve the desired performance.

The required total E_b/N_t for each data rate can be calculated based on the pilot E_c/N_t, processing gain, and data, DRC, and ACK channel gains with respect to the pilot as follows:

$$
\left(\frac{E_b}{N_t}\right)_{\text{Total(dB)}} = \left(\frac{E_c}{N_t}\right)_{\text{Pilot(dB)}} + 10 \times \log\left(\frac{W}{R}\right)
$$
$$
+ 10 \times \log\left(1 + 10^{\frac{\text{DataGain}}{10}} + 10^{\frac{\text{DRCGain}}{10}} + 10^{\frac{\text{ACKGain}}{10}}\right) \quad (10.5)
$$

where the ACK gain is typically 4 dB and the DRC gain is −3 dB when DRC length is 4 slots and −1.5 dB when DRC length is 2 slots. Table 10.5 shows the default data channel gains, where the relative gain increases with the data rate so that all reverse link data rates achieve similar PER for the same value of pilot E_c/N_t.

10.7.3   Cell/Sector Reselection

In 1X EV-DO the soft handoff is only supported in the reverse link and it follows a procedure very similar to cdma2000. In the forward link, however, there is no soft handoff and the network transmits the data only on the best sector selected by the AT on the DRC channel. This is in contrast to the cdmaONE and IS2000 system in which the forward traffic information is routed through all sectors in the mobile's active set. This best-sector selection approach is taken mostly to avoid the excessive interference, overhead, and complexity associated with scheduling simultaneous transmissions of a high-rate and high-power data channel from multiple sectors. Also, because data services can tolerate larger and variable delays, the sector selection would not cause any performance problems and any marginal loss of channel reliability can be regained more efficiently through retransmissions.

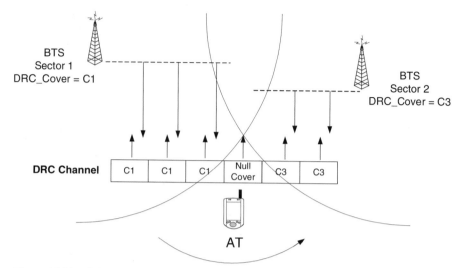

Figure 10.21   Cell reselection for forward link transmission.

On the basis of the pilot measurements provided by the AT the network iden-tifies the active set pilots corresponding to sectors with which the terminal is able to communicate. The network also instructs each sector in the active set to establish a connection with the terminal by transmitting power control bits on the RPC channel and demodulating the uplink DRC channel and the uplink data channel when needed.

Figure 10.21 shows the process of cell or sector reselection as the AT moves from sector BS1 to sector BS2:

- Initially the AT receives data transmission from sector BS1, while continu-ously measuring the C/I for remaining sectors in the active set. During this time, the AT sends its supportable data rate on the DRC channel along with DRCCover of BS1, that is, C1.

- At some point in time, the AT determines that sector BS2 provides a higher C/I and thus is the best server.

- Before immediately switching to BS2, the AT transmits a null DRCCover on one slot before it starts transmitting the DRCCover of sector BS2. In other words, the AT must transmit a null cover in one slot when it switches from one sector to another. This intermediate step notifies the BSC and prepares the network for cell reselection.

- AT then sends DRCCover for BS2, that is, C3, and then tunes to sector BS2 for receiving data on the forward channel. The BS2 will schedule transmis-sion on the forward channel based on the rate indicated by the AT once the network is ready. Note that it may take longer than one slot for the BSC and the network to switch traffic to the new "best sector" and therefore for BS2 to be ready for data transmission.

10.8   ENHANCEMENTS IN HRPD/IS856 RELEASE A

Release A of IS856 standard involves major improvements to the original HRPD system with respect to performance and features [5]. These improvements can be categorized as follows:

- Reverse link PHY and MAC layer enhancements, which result in significantly higher throughput and lower latency as well as better rate control.
- Forward link enhancements for more efficient transport of short packets that, combined with higher MAC channel capacity, allows support for many low-rate and delay-sensitive applications such as voice over IP

Enhanced mobility and QoS Support, including user and application based QoS control, broadcast and multicast services (BCMCS) and improved sector switching and hand off procedures. In this section we present an overview of some of the key features introduced in IS856-A and discuss their impact on system performance.

10.8.1   Reverse Link Enhancements

One could easily argue that the most prominent aspect of the enhancements made in Release A of HRPD is the higher packet efficiency in the reverse link. These improvements include the addition of new high-data rate physical layer packet options, with shorter framesand better link adaptation resulting in higher average throughput and lower latency. In this section we discuss some of key enhancements introduced in the reverse link of IS856 Release A.

Higher-Rate Physical Layer Packet Transmission: This enhancement is the result of the following features:

- *Higher-Order Modulation*: In addition to BPSK modulation, IS856-A adds QPSK and 8PSK modulation for the uplink data channel to provide higher-data rate options when the channel condition allows. The system would choose one of the three modulation schemes as part of the overall link adaptation.
- *Multicode Uplink Data Transmission*: This option involves the addition of a new uplink data channel D1 so that the access terminal can simultaneously transmit on two code channels. The new data channel D1 only supports QPSK, whereas the existing data channel D0 supports both BPSK and QPSK. When both data channels are used, the modulation on both channels should be QPSK. The multicode transmission is used only at higher data rates, corresponding to packet sizes of 3072 bits or larger, to reduce code rate and enhance coding gain, and it allows the AT to transmit at up to three times the data rate of D0 alone.
- *Uplink Auxiliary Pilot*: This new optional physical channel is introduced in addition to the existing reverse primary pilot channel to be used at very high data rates and specifically in conjunction with multicode transmission. Because the SNR of the primary pilot is power controlled independent of the rate of the data channel, at very high data rates the traffic-to-pilot ratio may

become so large that the primary pilot cannot provide a reliable channel estimate for data demodulation. In this case, adding an auxiliary pilot in the uplink helps the base station better estimate the channel so that the demodulation performance is not affected, without impacting the existing power control process.

Improved Link Adaptation: This improvement is the result of enhanced rate determination, smoother rate transition, and the self-correction achieved by HARQ. The key features contributing to link adaptation are described in the following:

Hybrid ARQ: Based on the same concepts used in the forward link of 1xEV-DO Release 0 and in order to improve link adaptation, channel utilization and latency performance uplink frame structure and protocols have been modified to allow fast HARQ. In this HARQ scheme each encoder packet encoder is channel encoded and the resulting symbols are grouped into 4 subpackets (SP1–SP4), with the first subpacket containing the systematic, that is, information bits. Each subpacket is transmitted over 4 slots, and up to 4 subpackets may be transmitted by the mobile with an 8-slot interval between successive transmissions corresponding to the same encoder packet.

Also, a new forward ACK channel is added by which the base station can send ACK/NAK indications for the packets. Each ACK or NAK message is transmitted over 4 slots. Figure 10.22 shows the subpacket and ACK/NAK transmission timing.

This new forward ARQ channel is time multiplexed with the reverse power control channel. The F-ARQ channel is transmitted during the MAC bursts of the ARQ slots (i.e., on RPC slots), using a Walsh cover associated with the user's MAC index and it is repeated over 3 slots. Therefore, the total MAC channel power is shared by the F-ARQ, RPC, and RAB channels.

To allow multiplexing of the ARQ channel with RPC, the power control rate must be decimated and reduced from 600 Hz to 200 Hz. The impact of lowering power control rate on the uplink performance is expected to be small to moderate, and it should be compensated by the HARQ gain, resulting in better overall link performance.

The HARQ gain can be viewed from different perspectives:

- Lower Latency: As the minimum packet duration is reduced from 16 to 4, the mean startup time is lowered from 8 slots to 2 slots. To achieve low latencies the system can force early termination after the first, second, or third subpacket by using higher transmission power.

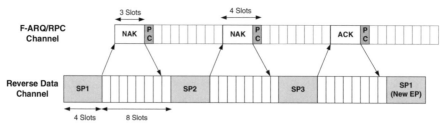

Figure 10.22   HARQ on the reverse data channel.

TABLE 10.6 Reverse Data Channel Rates for HRPD Release A

Payload (bits)	Modulation	Effective Data Rate				Code Rate			
		After 1 SP	After 2 SPs	After 3 SPs	After 4 SPs	After 1 SP	After 2 SPs	After 3 SPs	After 4 SPs
128	BPSK	19.2	9.6	6.4	4.8	1/5	1/5	1/5	1/5
256	BPSK	38.4	19.2	12.8	9.6	1/5	1/5	1/5	1/5
512	BPSK	76.8	38.4	25.6	19.2	1/4	1/5	1/5	1/5
768	BPSK	115.2	57.6	38.4	28.8	3/8	1/5	1/5	1/5
1024	BPSK	153.6	76.8	51.2	38.4	1/2	1/4	1/5	1/5
1536	QPSK D0	230.4	115.2	76.8	57.6	3/8	1/5	1/5	1/5
2048	QPSK D0	307.2	153.6	102.4	76.8	1/2	1/4	1/5	1/5
3072	QPSK D1	460.8	230.4	153.6	115.2	3/8	1/5	1/5	1/5
4096	QPSK D1	614.4	307.2	204.8	153.6	1/2	1/4	1/5	1/5
6144	QPSK D0 & D1	921.6	460.4	307.2	230.4	1/2	1/4	1/5	1/5
8192	QPSK D0 & D1	1228.8	614.4	409.6	307.2	2/3	1/3	2/9	1/5
	8PSK	1843.4							

- Better Link Margin: The link can be designed with lower power margin or $(E_b/N_o)_{min}$ as many frame errors can be corrected after multiple transmissions with incremental redundancy and thus higher FER can be tolerated.

- Better Resource Utilization: This is achieved because of support for a wider range of packet durations and smaller payloads resulting in finer "packet packing" in time.

- Inherence Rate Correction: The HARQ process can automatically adjust to conservative rate selections by early termination of subpackets, hence maximizing the effective transmission rate.

Table 10.6 shows the data rates and associated coding rates for BPSK, QPSK, and 8PSK modulation schemes based on 1–4 subpacket transmissions. This table shows the flexibility in the choice of payload size and data rate for each coding and modulation. Although longer packets provide more coding gain and time diversity and are more capacity efficient, shorter packets provide better packing efficiency (for low-rate traffic) and are more delay efficient.

Also note that in Release A, the RRI channel that was originally used to indicate the presence and rate of data packets on the reverse link is modified to show the payload size and ARQ instances on the reverse link. The enhanced RRI has six bits of information, including four bits to encode payload size and two bits to encode HARQ instance or subpacket index.

Improved Uplink Rate Control: The uplink rate control using fixed transition probabilities, as used in Release 0, lacks predictability, flexibility and the speed required by some of bursty traffic and fading channel condition. In IS856 Release

A the uplink rate control has been improved through better uplink interference control and finer rate quantization, which results in smoother and more effective rate selection.

This improved rate control IS856-Ascheme uses uplink interference rise over thermal noise (RoT) or sector loading as the fundamental resource to be managed on the CDM'ed reverse link. Each AT's contribution to the sector RoT is strongly correlated with its average traffic-to-pilot ratio (TPR or T2P), which is measurable by the AT. The AN therefore, uses TPR as a key state variable for uplink resource control. Based on the processes and parameters defined by the base station, the AT increases its TPR when the sector is unloaded and reduces the TPR as the loading increases.

Instead of random rate transitions, one step at a time, the system directly controls the TPR resource based on RAB bits. This TPR resource is puts in bucket at the AT and each packet transmission uses a portion of this resource based on the payload size and QoS.This approach effectively provides a dynamic setting of rate transition probabilities and allows aggressive rate transitions and reduced latency while maintaining the cell RoT and stability.

MAC Layer ARQ: In addition to the HARQ at the PHY layer, IS856-A also supports a MAC layer ARQ. This "outer loop" ARQ allows the base station to detect the packet erasure, that is, failure to successfully decode the encoder packet even after the 4th subpacket is received, and also to request and process retransmission of MAC layer data packets. This MAC layer ARQ provides additional time diversity to improve link layer reliability in severe channel conditions.

Enhanced Access and Control Channels: The reverse access channel in Rev. A has been improved to support higher data rates of 9.6, 19.2 and 38.4 kbps and shorter preambles. This improvement allows faster access and short data burst transmission over signaling channels. The maximum access channel data rate can be controlled by the AN and is can also be configured for each AT.

The control channel is also improved to support shorter packets of 128, 256 and 512 bits to allow shorter connection set up time for interactive peer to peer applications such as voice and gaming.

10.8.2 Forward Link Enhancements

The new forward link features in IS856-A are mainly designed to allow efficient transport of short packets with higher MAC addressing capacity in the forward link. These features can be critical for low-rate but widely used applications such as voice over IP, gaming, and instant messaging. In the following we discuss the elements of forward link enhancements in more details.

New Physical Layer Packets: Release A introduces several new physical layer packets options, shown in Table 10.7, including three short packets at low rates and one large packet at higher-rate transmissions. The two higher data rates require two new DRCs, that is, DRC = 13 for 1.536 Mbps and DRC = 14 for 3.072 Mbps, whereas the shorter packets and their rates reuse the existing low DRC values, that is, DRC = 0–5.

Table 10.8 shows the new DRC interpretations for all packet size and data rate options in IS856-A.

TABLE 10.7 **New Physical Layer Packets**

Packet Size (bits)	Nominal Duration (Slots)	Nominal Data Rate (kbps)	Code Rate	Modulation
128	6	12.8	1/5	QPSK
256	6	25.6	1/5	QPSK
512	12	25.6	1/5	QPSK
5120	2	1536	5/12	16QAM
5120	1	3072	5/12	16QAM

TABLE 10.8 **New DRC Interpretations**

DRC	Pream. Length (chips)	Packet Types (Parameters) [PktSize (bits), Tx Duration (slots), Rate (kbps)]
0	1024	[128, 6, 12.8] or [256, 6, 25.6] or [512, 12, 25.6] or [1024, 16, 38.4]
1	1024	[128, 6, 12.8] or [256, 6, 25.6] or [512, 12, 25.6] or [1024, 16, 38.4]
2	512	[128, 6, 12.8] or [256, 6, 25.6] or [512, 6, 51.2] or [1024, 8, 76.8]
3	256	[128, 4, 19.2] or [256, 4, 38.4] or [512, 4, 76.8] or [1024, 4, 153.6]
4	128	[256, 2, 76.8] or [512, 2, 153.6] or [1024, 2, 307.2]
5	128	[512, 4, 76.8] or [2048, 4, 307.2]
6	64	[1024, 1, 614.4]
7	64	[2048, 2, 614.4]
8	64	[3072, 2, 921.6]
9	64	[2048, 1, 1.2288]
10	64	[4096, 2, 1.2288]
11	64	[3072, 1, 1843.2]
12	64	[4096, 1, 2457.6]
13	64	[5120, 2, 1536]
14	64	[5120, 1, 3072]

Multiuser Packet Data Rates: In addition to support for shorter packets in Release A the system allows sharing of large packets among multiple users. A multiuser packet (MUP) is a single physical layer packet composed of one or more security layer packets addressed to different users. The base station indicates multiuser packets and their data rates with reserved MAC_IDs applied to the packet preambles. The system supports five MUPs as shown in Table 10.9.

Note that at low data rates the packing efficiency is achieved because of short frames while at high data rate improved packing is obtained with multiuser packets.

Increased Number of MAC Indices: Similar to the original IS856 standard, the MAC_IDs in Release A are used in the preamble and the MAC channel. However, to increase the MAC capacity, that is, the maximum number of supportable active users, the number of MAC-IDs is increased to 128. The MAC_IDs are represented with a 7-bit index and assigned according to Table 10.9.

The system supports up to 115 MAC addresses for active ATs in a cell. Also, to ensure backward compatibility the MAC_IDs <64 can be assigned to legacy terminals.

TABLE 10.9 New MAC-ID Interpretations for MAC Channel and Preamble

MAC Index	MAC Channel Use	Preamble Use	Preamble Length (chips)
0 and 1	Not Used	Not Used	N/A
2	Not Used	76.8 k Cont. CH.	512
3	Not Used	38.4 k Cont. CH.	1024
4	RA channel	Not Used	N/A
5	Not Used	Broadcast	Variable
6	Not Used	MUPDR: 153.6 kbps (1024 bits in 4 slots)	256
7	Not Used	MUPDR: 307.2 kbps (2048 bits in 4 slots)	128
8	Not Used	MUPDR: 921.6 kbps (3072 bits in 2 slots)	128
9	Not Used	MUPDR: 1.2 Mbps (4096 bits in 2 slots)	64
10	Not Used	MUPDR: 1.5 Mbps (5120 bits in 2 slots)	64
11–127	User-specific RPC and DRCLock Channels	Single-user Packet	DRC dependent

Note that the forward channel preambles with MAC_IDs of 6 to 10 are used to indicate the transmission of multiuser packets and their data rates. The system also reserves a single MAC_ID to indicate broadcast channel transmissions.

Broadcast and Multicast Services (BCMCS): Another aspect of forward link enhancement in IS856-A is the introduction of a new broadcast physical channel and the associated broadcast and multicast service protocols. This feature allows operators to push high rate multimedia content such as news, music and advertisements to mass audiences or large groups of users in a single transmission. The broadcast channel is used to multicast higher-layer packets addressed to specific groups of users or to broadcast information to all users who are subscribed to that particular broadcast service. The broadcast channel can also be used to more efficiently carry the forward broadcast signaling messages, which are otherwise transmitted by the existing control channel defined in IS-856.

Access to the broadcast/multicast services is controlled via link layer encryption, with keys distributed at the time of subscription.

The data on the broadcast channel are transmitted at full power at the rate of 38.4 to 614.4 kbps. To maximize the link reliability the broadcast channel uses outer Reed–Solomon channel coding and supports soft handoff. The broadcast channel operation and services do not require any major change in the reverse link except for few signaling messages.

A Multicast IP flow represents an IP content stream originating at the "Content Network" and it is identified by a multicast IP address and UDP port number. The network assigns each multicast IP flow to a "Broadcast Flow" as the flow of octets that belong to a single broadcast stream on the radio access network (RAN). Multiple broadcast flows can also be multiplexed and carried in one logical channel. Note that support for broadcast/multicast services may require some changes in the network to allow multicasting of IP among multiple access networks.

10.8.3 Improved QoS, Handoff and Call Set Up

Some of the enhancements made in IS856-A are related to the handoff performance under high mobility as well as QoS control and reduced call setup latency needed for interactive and conversational applications. Some of these enhancements are described in the following.

Multi-Flow QoS Control: One of the most important aspects of service improvement in IS856 release A is multi-flow MAC and RLP capability, which facilitates application and user class based QoS control for concurrent services. This feature provides multiple RLP flows with independent retransmission configuration and multiple MAC flows with different QoS treatments. In the reverse link multi-flow MAC is enforced using per flow resource, e.g. T2P, allocations and QoS scheduling. In the forward link the QoS control schemes mostly involve the scheduler's design, which is normally implementation specific.

Improved Cell Switching: to improve the process of sector switching in IS856-A a new physical channel called the data source control (DSC) channel is introduced. Using DSC channel the AT gives an early indication of its desire to switch the serving cell by specifying the sector from which forward link packets are requested. Therefore, as the serving sector changes and while the AN prepares redirecting RLP stream to the new sector, the AT can receive data from the old sector until the new DSC message takes effect. Once AT swiches the serving sector by changing the DRC cover, the new sector starts transmitting its data. This process avoids any interruption of traffic flow, which may be unacceptable for delay-sensitive applications.

The start of a DSC message coincides with a DRC message boundary, but the DSC message is transmitted over a longer period, typically 2 frames, to benefit from some time diversity gain resulting in higher message reliability. The DRC message contains sector information (3 bits) and data-rate/packet-type information (4 bits) and it takes effect half a slot after the end of its transmission.

The process of cell switching is also improved from the network side by multicasting the signaling from PDSN to all base stations in the user's active set and transferring the user's service states to the target cell. The data packets can also be multicast in the network to the base stations involved in the handoff to ensure the fast delivery of packets even if the connection with one base station is severed before handoff is completed. This process also assumes that the user's state and "context profile" are made available to the target base station. Note that the multicasting does not extend to the air interface because only one base station is transmitting at a time, in order to preserve air-link resources.

Flexible Sleep Time and Fast Call Setup: IS856 Release A has changed the default idle state protocol to allow faster paging by using a flexible framework for the AT's sleep period. In this framework, instead of a fixed sleep period of 5.12 s as originally defined by Release 0, the system allows a variable sleep time over a wide range of 6.7 ms to 320 s, which can gradually increase in three steps with the user's activity. After a transition from the connected to the not-connected state, the sleep time is short so that the access network and the AT can return from the sleep state to the monitor state quickly and the sleep time becomes longer over time as the AT does not show any activity. The sleep periods and the timers associated with their change can be defined for each AT according to user's class and applications and they can be renegotiated depending on the AT's remaining battery life.

This feature can significantly improve the QoS for interactive and delay sensitive applications such as gaming and push to talk, which involve frequent connection set ups and releases, with minimum impact on the AT's battery life.

10.9 HRPD NETWORK ARCHITECTURE

The network architecture in HRPD is similar to the IS2000 system with some variations in network elements, functionalities, and interface protocols. Figure 10.23 shows a simplified network diagram including the main functional elements as follows:

Access Terminal (AT): a device providing data connectivity to a user, logically the equivalent of a mobile station in IS95/cdma2000 networks. AT may be a self-contained data device such as a personal digital assistant (PDA) or a detachable module that is connected to a computing device such as a laptop personal computer.

Access Network (AN): the network equipment providing data connectivity between the packet-switched data network (typically the PDSN and Internet) and

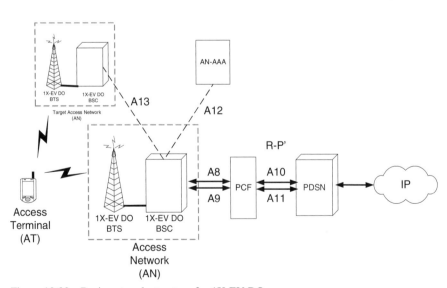

Figure 10.23   Basic network structure for 1X-EV DO.

the ATs. The AN is equivalent to the base station (BTS/BSC) in a cdma2000 network architecture. In some network specifications the BSC is referred to as the radio network controller, or RNC.

Packet Data Serving Node (PDSN): the network equipment providing data connectivity between the radio access network and a packet-switched data network. The PDSN provides connectivity to the Internet independent of the type of radio access network. If the HRPD carriers are deployed as an overlay to an IS2000 network, the same PDSN can support and be shared by both carrier types.

The air interface between the AT and the AN is defined in IS856. The interface between the AN and the PDSN is defined in the interoperability specification (IOS) in IS2001, and it is also known as the radio-packet (R-P) interface.

The basic 1X EV-DO network includes A8–A11 interfaces, which are also defined as part of the basic IOS document. The enhanced 1X-EV DO network supports enhancements in IOS interface, captured in IS878, allowing access authentication and interface between two access networks.

Like cdma2000, in HRPD systems, access authentication and packet data service authentication are defined as two distinct procedures. To support access authentication procedures the HRPD systems define and use a AAA server in the AN. This node is known as AN-AAA server and uses the remote access dial-in user service (RADIUS) protocol. The interface between the AN and the AN-AAA server, which is also based on the RADIUS protocol and is called the IOS-A12 interface, is defined in the IOS standard. Only the authentication part of the RADIUS protocol is used between the AN and the AN-AAA server.

In an enhanced 1X-EV DO system; different access networks can be connected to each other with an IOS-A13 interface defined within IOS. This interface is required to support mobility procedures when the AT moves from one AN to another. The A13 interface allows the transfer of authentication and session configuration parameters from the old AN to the new AN. The interface is based on the UDP/IP and uses messages defined in the IOS.

Figure 10.24 depicts, in more detail, the network interfaces with emphasis on the utilized IP-based protocols specified by the Internet Engineering Task Force (IETF). Some of these protocols are listed below:

- RFC-1661: logical point-to-point (PPP) protocol between AT and PDSN
- RFC 2338 and 2339: between PDSN and AAA
- RFC 2002: mobile IP protocol between PC/mobile station, PDSN, and home agent
- Higher-layer protocols such as file transfer protocol (FTP), hypertext transfer protocol (HTTP), session initiation protocol (SIP), wireless application protocol (WAP), etc., between PC/mobile station and application services

10.10   HRPD-IS2000 HYBRID NETWORKS

The HRPD system can be designed as a radio overlay on an existing IS2000 network (see Fig. 10.25). In a radio access network that supports a mix of IS2000

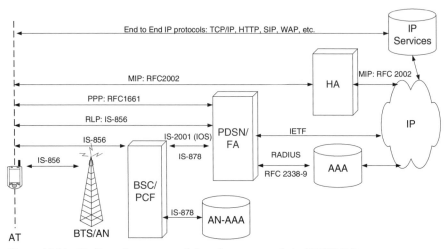

Figure 10.24   Radio and core network interfaces protocols in 1X-EV DO.

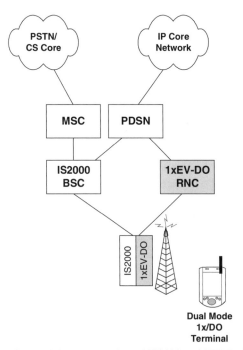

Figure 10.25   Integration of IS2000 and HRPD signaling networks.

and HRPD carriers, a hybrid or dual-mode HRPD access terminal can operate on both IS2000 and IS856 networks. These hybrid access terminals can receive voice, short messaging, and dedicated data services on IS2000 networks and high-speed packet data on HRPD networks; it is also capable of handing off between the two systems.

A hybrid terminal in the idle mode periodically monitors the IS2000 common channels (i.e., F-QPCH, FPCH, or F-CCCH/F-BCCH, whichever applicable) or the IS856 control channel. By monitoring both channels the hybrid terminal can resume high-speed packet data service on the IS856 system when needed and also receive any incoming voice and SMS on the IS2000 system.

A hybrid terminal with an active connection with the IS856 system periodically tunes to the frequency of IS2000 carriers and monitors the corresponding common channel. It then tunes back to the IS856 frequency to resume the active data session.

A hybrid terminal that is idle in the IS856 system also monitors the IS856 control channel and IS2000 common channels sequentially based on their respective slot cycles. The assigned IS856 control channel cycle should be selected such that it does not overlap with the assigned IS2000 paging slot.

Similarly, a hybrid terminal that is idle on the IS2000 system only may perform periodic off-frequency searches to discover existing IS856 systems. In this case, the system may use the preferred roaming list (PRL) to direct mobiles to search for any serving IS856 system.

In most cases a hybrid HRPD/IS2000 network uses a common PDSN for both systems. In this case, and to reduce the amount of time that a hybrid terminal performs concurrent slotted operations on the two sets of control channels, the system may transfer the IS856 data session to the IS2000 system and ask the mobile to cease monitoring the IS856 control channel.

The common PDSN sends the packets destined to the hybrid terminal to the IS2000 system, to ensure that the hybrid terminal does not miss packets that are destined for it because of not monitoring the IS856 control channel.

One of the network enhancements recently made to HRPD systems involves the integration of the signaling scheme with cdmA2000 1X to allow a dual mode IS2000/IS856 access terminal to send and receive certain signaling messages via the other network. For example a dual mode AT that tuned to EV-DO can communicate with the MSC over DO channels to receive incoming voice pages, send and receive SMS messages or performs other 1x idle mode operations (e.g., registration).

Once this "cross tunneling" feature is implemented and integrated into the network, a dual-mode MS would not be required to switch to a IS2000 carrier to look for a page message when it is "connected" to a HRPD carrier. This feature would also allow a dormant handset in an IS2000 system to maintain its IP connectivity through HRPD. In this case, the two radio access networks will share the same PDSN that usually points to 1xEV-DO RNC, as long as there is "EV-DO" coverage The MSC also views 1xEV-DO RNC as a 1x BSC and communicates with that over a standard A1 interface.

10.11 REFERENCES

1. P.Bender, P.Black, M.Grob, R.Padovani, N.Sindhusayana, A.J.Viterbi, "CDMA/HDR: a Bandwidth Efficient High Speed Wireless Data Service for Nomadic Users," *IEEE Communications Magazine*, Vol. 38, no. 7, pp70–77, July 2000
2. 3GPP2 C.S0024-0 v4.0, "cdma2000 High Rate Packet Data Air Interface Specification," 2002
3. E.Esteves, "The High Data Rate Evolution of the cdma2000 Cellular Systems," in *Multiaccess, Mobility and Teletraffic for Wireless Communications*: Volume 5, Kluwer Academic Publisher, 2000
4. *Advances in 3G Enhanced Technologies for Wireless Communications* by Jiangzhou Wang and Tung-Sang Ng (Editors), Artech House, March 2002
5. 3GPP2, C.S0024 Release A, "cdma2000 High Rate Packet Data Air Interface Specification," 2004

CHAPTER 11

CDMA2000 RELEASE D (1XEV-DV)

11.1 INTRODUCTION

Based on projected demand for mobile multimedia applications and recent technology developments, the cdma2000 standards have recently evolved to provide higher-speed and lower-latency data transport over the air interface. This backward-compatible evolution of the standard is referred to in 1xEV-DV and defined in Release C and further extended in Release D of IS2000 (see Fig. 11.1). The most important changes involve enhancements in both forward and reverse link throughput efficiencies by introduction of new radio configurations for high-speed packet data traffic channels.

During the development of the "1xEV-DV" mode the design objective was to achieve peak data rates and average throughput levels equal or higher than HRPD Rev. 0 in both forward and reverse links in a backward-compatible framework that integrates voice and data services in the single CDMA channel. More specifically, the objective was to provide a peak rate higher than 2.4 Mbps with average throughput higher than 1 Mbps in the forward link and a peak rate higher than 1.25 Mbps with average throughput higher than 600 kbps in the reverse link.

This chapter provides a detailed view of 1xEV-DV enhancements, based on specifications in IS2000 Release D [1–5], and considering Release A as a baseline.

Remember that IS2000 Release A has three main frameworks for sending data traffic:

- *Fundamental or Dedicated Control, aka Fundicated, Channels (F-FCH/F-DCCH)* are used for low-rate, circuit-switched-like data, dedicated to specific users. These channels are typically fixed rate but power controlled.

- *Supplemental Channels (F-SCH)* for higher data-rate packet and circuit-switched traffic use a time-sharing scheme with transmission times of around 80–160 ms. This mechanism allows variable spreading and data rate changes as well as power control for large packets but at a fixed modulation.

- *Broadcast Control Channels* for broadcast and multicast transmissions of low-volume SMS-like data at low rates and high latencies. Also, forward and reverse common control channels (F/R-CCCH) can be used to send extended

---

CDMA2000® Evolution: System Concepts and Design Principles, by Kamran Etemad
ISBN: 0-471-46125-3   Copyright © 2004 John Wiley & Sons, Inc.

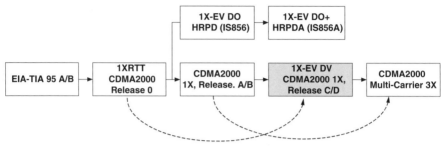

Figure 11.1   1xEV-DV in the evolution of cdma2000 networks.

SMS-type traffic or "short data burst" between individuals or groups of users in a cell at a slightly higher rates and lower latencies.

In most 2G systems, and to some extent in IS2000 Release 0/A, where the voice and low-rate circuit-switched traffic take most of the system capacity; the average utilization of system resources is relatively low. This is mostly because of the nature of voice trunking policies, which require provisioning the system for the busiest hour of the day or week to ensure the grade of service expected by voice users at all times. For example, a system designed for 1% call blocking probability has some unutilized resources 99% of the time.

One of the key design objectives in "1xEV-DV" is to maximize the utilization of air link by taking advantage of resources that are not otherwise utilized by the low-rate data and voice channels and to schedule the data packet at the highest possible rate at the most favorable channel condition. In the forward CDMA channel the main resources are Walsh code space, transmission power, and channel elements, and in the reverse link they are uplink sector loading and rise over thermal (RoT), as will be discussed later in this chapter.

Forward link packet data channel operation was introduced as part of IS2000 Release C. This mode of operation required defining a few new physical channels in the forward and reverse links and some changes in Layer 2 and 3 messaging.

This new forward packet data channel supports fast scheduling, fast link adaptation, higher-order modulation, shorter frames and hybrid ARQ, which are the key design principles used in HRPD. The main difference is that in the "1xEV-DV" the packets need to be scheduled in the presence of long-term voice and low-rate data traffic with backward compatibility with previous releases of cdma2000.

Shortly after the completion of Release C, to address the imbalance between forward and reverse link efficiency for packet data services and the need for lower-latency and higher-throughput connections in the uplink, 3GPP2 developed a new release of IS2000 (Release D) by adding a new reverse packet data channel as well as its supporting control channels and protocols. The new reverse packet data channel operation is designed based on a flexible MAC framework to allow different rate control options along with higher-order modulation with link adaptation, hybrid ARQ.

In this chapter we will study the various aspects of 1xEV-DV based on IS2000 Release D, starting with protocol layer and channelization changes and followed by more details on forward/reverse packet data channel operation and the physical layer structure for the new channels.

11.2   1X-EV DV PROTOCOL LAYERS

The protocol layers of IS2000 Release D are the same as those of IS2000 Release A, with the addition of a two new entities called forward and reverse packet data channel control function (F/R-PDCHCF) within the MAC sublayer (see Fig. 11.2).

The forward and reverse PDCHCF entities terminate all forward and reverse link physical channels associated with the high-speed packet data transmission on the forward and reverse packet data channels, respectively. The F/R-PDCHCF along with the MUX/QoS sublayer of MAC provides the mapping of the new physical channels to the appropriate logical channels in the MAC and LAC sublayers.

Some of the key PHY/MAC attributes related to F-PDCHCF, and the forward packet data mode in general, include:

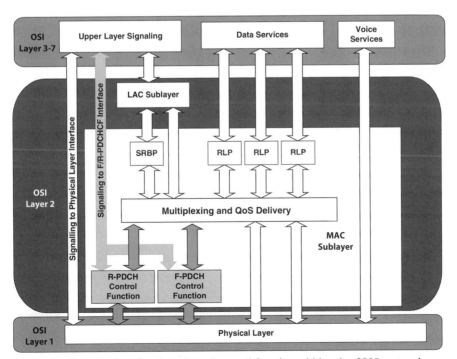

Figure 11.2   Forward packet data channel control function within cdma2000 protocol layers.

- Fast and opportunistic scheduling with short physical layer frames
- Link adaptation with high-order modulation (including QPSK, 8PSK, and 16QAM)
- Adaptive and asynchronous hybrid ARQ Type II
- Concurrent operation of four independent ARQ channels
- MAC ID matching and filtering
- Uplink channel measurement reports
- Fast sector switch as opposed to soft handoff
- Up to two simultaneous code multiplexed packet data channels
- Downlink burst rate of about 3.09 Mbps

Similarly, some of the key PHY/MAC attributes related to F-PDCHCF, and the forward packet data mode in general, include:

- Uplink scheduling with some level of QoS support
- Short physical layer frames
- Synchronous hybrid ARQ Type II
- Flexible MAC for uplink rate control
- High-order modulation (including QPSK, 8PSK and 16QAM)
- Concurrent operation of up to four ARQ channels
- Uplink peak burst rate of 1.54 Mbps

These features will be described later as part of the forward and reverse packet data channel operation.

11.3 NEW PHYSICAL CHANNELS IN 1XEV-DV (RELEASES C AND D)

This section provides an overview of the physical channels defined in cdma2000 Releases C and D to support the high-speed packet data traffic channels or "DV" mode of operation. These channels are listed in Table 11.1.

The following channels were introduced in Release C to support forward traffic channel in Radio Configuration 10 (see Fig. 11.3).

- In the forward link
  - The forward link packet data channel (F-PDCH) is the high-speed scheduled traffic channel carrying user data only.
  - The forward link packet data control channel (F-PDCCH) is a control channel that identifies the target MS to which the F-PDCH packet is transmitted and also provides the MS with the information necessary for decoding the F-PDCH. This channel is also used to broadcast other control information including Walsh space availability to all users.

TABLE 11.1   The New Physical Channels introduced in Release C and D of IS2000

			New Physical Channels Introduced in 1xEV-DV
Release D RL_RC7	Release C FL_RC10		F-PDCH Forward Packet Data Channel (0–2 Channels)
			F-PDCCH Forward Packet Data Control Channel (one per F-PDCH)
			R-ACKCH Reverse Acknowledgment Channel
			R-CQICH Reverse Channel Quality Indicator Channel
			F-ACKCH Forward Acknowledgment Channel
			F-GCH Forward Grant Channel
			F-RCCH Forward Rate Control Channel
			R-PDCH Reverse Packet Data Channel (0–1 channel)
			R-PDCCH Reverse Packet Data Control Channel (one per F-PDCH)
			R-REQCH Reverse Request Channel

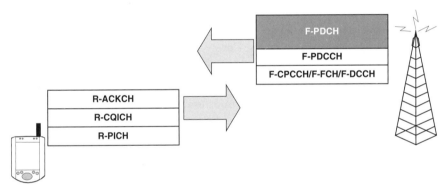

Figure 11.3   Physical channels involved in forward packet data channel operation in RC10.

    – The forward common power control channel (F-CPCCH) is the same channel defined in cdma2000 Release A that is reused to allow uplink power control when no fundicated channel is allocated to the mobile. When fundamental or dedicated channels are allocated to the MS the power control bits are sent through those channels instead of F-CPCCH.

• In the reverse link

    – The reverse link channel quality indicator channel (R-CQICH) is a control channel used to indicate to the base station the quality of the forward link pilot channel received at the mobile station and to indicate switching between base stations.

    – The reverse link acknowledgment channel (R-ACKCH) is used for the transmission of acknowledgments (ACKs/NAKs) from the mobile station to the base station in response to the data received on the F-PDCH and the F-PDCCH.

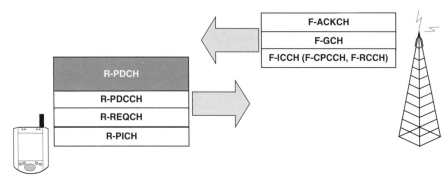

Figure 11.4   Physical channels involved in reverse packet data channel operation in RC7.

- The reverse pilot channel (PICH) is the same channel defined in Release A and serves the same purpose, namely, facilitating the coherent demodulation of uplink transmissions. However, if no forward fundicated channel is allocated to the mobile, no power control bits are punctured in this channel.

As part of the Reverse Packet Data mode introduced in cdma2000 Release D, the following new physical channels are defined to support the new packet's efficient and high-rate Radio Configuration 7 for reverse traffic channel (see Fig. 11.4).

- In the reverse link
  - The reverse packet data channel (R-PDCH) is the high-speed packet traffic channel in RC7, which carries higher-level data and control information from a MS to a base station.
  - The reverse packet data control channel (R-PDCCH) is an uplink control channel that provides the BS with information about the mobile's status and the subpacket being transmitted on the R-PDCH, including encoder packet size and transmission rate as well as subpacket identifier and QoS indicator.
  - The reverse request channel (R-REQCH) is a control channel used by the mobile station to request permission to transmit data on the R-PDCH. This channel carries the MS's data buffer and power headroom status and QoS requirement to help the BS with scheduling and QoS control.
  - The reverse secondary pilot channel is an unmodulated, direct-sequence spread spectrum signal transmitted by the MS in conjunction with certain transmissions on the reverse packet data channel. The secondary pilot channel provides additional phase reference for the R-PDCH for coherent demodulation and may provide a means for signal strength measurement.
- In the forward link
  - The forward acknowledgment channel (F-ACKCH) is a control channel used for the transmission of acknowledgments (ACKs/NAKs) from a base

station to mobile stations in response to the data received on the reverse packet data channel.

- The forward grant channel (F-GCH) is a control channel used by the base station to transmit messages to the mobile station controlling its transmission on the reverse packet data channel.

- The forward indicator control channel (F-ICCH) is a control channel used by the base station to control the power and the rate of multiple mobile stations when they are assigned the reverse packet data channel and not assigned the forward power control subchannels. The F-ICCH includes the forward common power control channel (F-CPCCH) defined in Release A as a subchannel as well as the new forward rate control subchannel (F-RCCH) that is used to control the maximum traffic-to-pilot ratio of the mobile station when operating on the R-PDCH.

More details on the usage of these channels and their physical layer structure are provided later in this chapter.

11.4 CHANNEL CONFIGURATION CAPABILITIES FOR RELEASE D

A cdma2000 Release D mobile must support certain combinations of channels to be "DV" capable, that is, to support forward and reverse packet data channel operation.

The mobile can indicate its channel configuration capability as part of its registration, origination, or page response messages to the base station. Table 11.2 shows seven channel configurations defined in Release D. Note that this table only defines the channel configurations as related to the DV mode, and in general any release C/D mobile must support all channel configurations required by Releases 0, A, and B.

For a mobile with DV capability, some combinations of channel configurations are mandatory. For example, if the mobile supports FCH it should support CC4 and CC5, and a mobile supporting DCCH must allow CC2 and CC3. In general, channel configurations are supported in one of the following sets:

TABLE 11.2 Channel Configurations for Packet Data Channel Operation in Release D

Ch. Config.	F-PDCH + R-PDCH	R-DCCH	F-DCCH	R-FCH	F-FCH
CC1	X				
CC2	X	X			
CC3	X	X	X		
CC4	X			X	
CC5	X			X	X
CC6	X	X		X	X
CC7	X	X	X	X	X

- {CC1, CC2, CC3}
- {CC1, CC4, CC5}
- {CC1, CC6, CC7}

Note that if the mobile supports R-PDCH, it has to support F-PDCH as well.

Depending on the supported physical channels, the dedicated traffic and signaling traffic are mapped to different channels. In "DV" mode, the forward/reverse packet data traffic are handled by F-PDCH/R-PDCH and supported by the respective associated control channels. The fundamental or dedicated control channels may or may not be involved in the packet data channel operation, depending on the channel configuration and the mix of traffic.

In Release C, in which the DV mode was only applicable to the forward link, the reverse traffic is carried by a fundamental or dedicated control channel, whichever is supported, along with the reverse supplemental channel for higher data rates. When FCH or DCCH is not supported, the reverse link power control commands are sent on F-CPCCH.

In Release D, the reverse packet data can be carried by R-PDCH and its associated control channels independent of any fundicated channel. Thus the CC1 configuration of Table 11.2 does not include any forward or reverse fundicated channels because the system can operate in the "DV" mode independently from any fundicated and supplemental channels. When no fundicated channel is allocated in the forward link, the MS performs the forward link supervision, that is, assessing the presence and validity of the forward link, based on the energy of the signal received on the F-CPCCH power control symbols.

11.5 FORWARD PACKET DATA CHANNEL OPERATION

The operation of F-PDCH is designed to achieve maximum resource efficiency by utilizing all the radio resources not used for voice and low-rate data channels to transmit data packets at the highest possible speeds. The design principles used in F-PDCH are very similar to those used in the HRPD forward packet data channel but subject to power and code availability imposed by voice or low-rate data traffic.

In the forward link the key limiting resources are the Walsh code space used for traffic channels and the total power allocated to them. The effective capacity of the F-PDCH is dynamically evaluated based on available power and code space, not committed to ongoing low-rate traffic and control channels, and it is time division multiplexed between different mobile stations (see Fig. 11.5).

The transmission on F-PDCH has variable data rates (from 81.6 kbps up to 3.09 Mbps) and variable subpacket durations (1.25 ms, 2.5 ms, or 5 ms). The modulation and coding on F-PDCH can be changed rapidly from frame to frame based on mobile station channel feedback in the uplink. The packet scheduling can also be performed based on channel conditions, on a framework similar to HRPD, to take advantage of multiuser diversity.

To further improve resource utilization and provide better latency control, the F-PDCH control function also supports code division Multiplexing (CDM) of up to

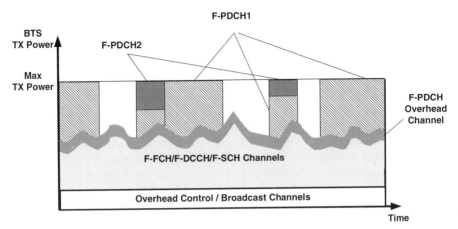

Figure 11.5   F-PDCH utilizes resources "left over" from voice and low-rate data channels.

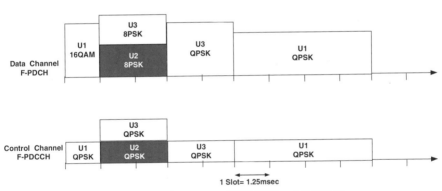

Figure 11.6   Concurrent transmissions on two F-PDCCH and F-PDCH.

two packet data channels. This option allows the base station to partition the available power and code space and transmit encoder packets to two different mobile stations at the same time. The code multiplexing is accomplished in the physical layer by using two disjoint portions of the Walsh space so that the two F-PDCHs' simultaneous transmissions are separated by the orthogonality of the Walsh codes while sharing the leftover power. Note that when two F-PDCHs are used the transmission duration of both packet data channels has to be aligned and that the two F-PDCHs cannot be dedicated to the same MS.

For each data packet channel there is one forward Packet control Channel (PDCCH) carrying associated MAC and signaling messages related to F-PDCH. A forward CDMA channel may contain up to two F-PDCCHs. There is a one-to-one correspondence between F-PDCH and its F-PDCCH, and the transmissions on the two channels must be time aligned (see Fig. 11.6). The control information for two concurrent F-PDCH transmissions, including the partitioning of the Walsh space used for orthogonal code multiplexing, is carried between the two F-PDCCHs. The base station can transmit to either one or two mobile stations during any particular transmission interval.

In 1xEV-DV, similar to HRPD, the forward packet data traffic channel transmissions are addressed through a MAC layer identifier sent concurrently on the associated packet control channel. When a F-PDCH is assigned to a mobile station, the MS is assigned an 8-bit identifier (MAC identifier, MAC_ID). This MAC_ID is used to uniquely identify the mobile station for which the data packets are scheduled.

Another key design attribute of the forward packet data channel are adaptive and asynchronous incremental redundancy (AAIR), which requires fast physical layer ARQ signaling and link adaptation. In the reverse link there are two channels defined to support AAIR on the forward packet traffic channels.

The reverse acknowledge channel (R-ACKCH) carries physical layer ACK and NAK messages from the MS to the BS after the reception and attempted decoding of each subpacket sent on the F-PDCH.

The reverse channel quality indication channel (R-CQICH) carries feedback information from the mobile station about the received signal quality to be used for link adaptation on the forward packet data channel. This information can be used by the base station to determine the transmission power, data rate, and possible need for packet data channel handoff. This information may also be used by the base station in deciding when to schedule a particular mobile station on the forward packet data channel as part of an opportunistic scheduling algorithm. In addition, the mobile station uses the R-CQICH to indicate its selection of serving sector via the Walsh cover applied to the R-CQICH signal.

Figure 11.7 shows a simplified massage flow for F-PDCH channel setup. After an origination or a page response message the BS sends an extended channel assignment message whereby it assigns the F-PDCH to the mobile station. It also starts sending power control bits to the mobile on one of the F-CPCCH subchannels. The MS then acquires the forward link channel and indicates that to the BS by sending a preamble on a reverse dedicated traffic channel, which depending on the channel configuration may be mapped to different physical channels. The mobile station also starts sending the channel quality indicator report. The BS then sends an ACK order message to MS to ensure a healthy ACK loop, and once it receives the ACK it starts the service and session establishment. This session setup involves RLP and PPP synchronization before user traffic frames are exchanged.

The following sections provide more details of F-PDCCH processing, hybrid ARQ, and uplink measurement report and cell/sector switching.

11.5.1 F-PDCCH Processing

The F-PDCCH0 and F-PDCCH1 physical layer channels carry control information to direct the mobile stations that are assigned to the packet data channel. Once the mobile station is assigned a MAC_ID, it must monitor the two F-PDCCH channels according to the procedures specified in this section.

Because the transmissions on F-PDCCH can be of variable length (1, 2, or 4 slots long), the mobile station must attempt to decode messages of any of the

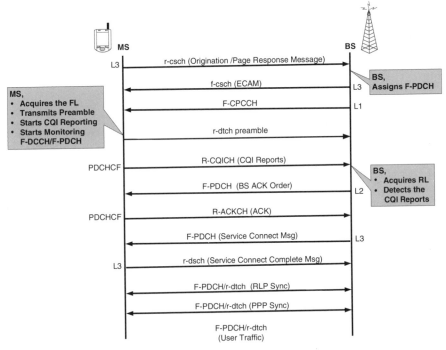

Figure 11.7   Message flow for F-PDCH call setup.

supported slot lengths beginning on any 1.25-ms slot boundary. When the physical layer successfully decodes a F-PDCCH0 message, that is, when the CRC is valid, the mobile station evaluates the fields in the F-PDCCH message to determine the appropriate action based on the following scenarios shown in Figure 11.8.

- If the F-PDCCH0 message contains a MAC_ID = "00" the remainder of this message contains a broadcast Walsh mask bitmap rather than a packet data channel assignment for a specific mobile station. In this case, in addition to the 8-bit MAC_ID, there are 13 bits in the message that comprise a bitmap corresponding to specific Walsh indices that are to be omitted from the Walsh space that is used to decode the F-PDCH transmission. All mobile stations that receive this message save this bitmap and apply it to subsequent F-PDCCH assignments including any assignment on F-PDCCH1 during the same time interval. This is used by the base station to exclude portions of the Walsh code space (e.g., when the Walsh codes are in use for F-FCHs, F-DCCHs, or F-SCHs assigned to certain mobile stations).

- If the F-PDCCH0 message contains the MAC_ID for this mobile station, the mobile starts decoding the subpacket on the F-PDCH0 based on the information contained in the message. In this case, all of the information needed by the physical layer and the PDCH control function to demodulate and to

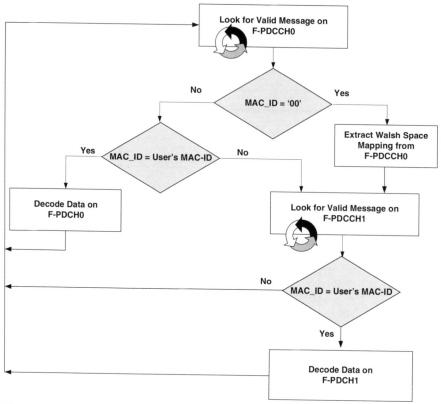

Figure 11.8   Monitoring and processing F-PDCCH0 and F-PDCCH1.

(attempt to) decode the encoder subpacket is included in the F-PDCCH0 message. More specifically, the F-PDCCH message would include the following information:

– User's MAC_ID (8-bits) showing which user is to demodulate the data packet on F-PDCH

– ARQ channel identifier (ACID, 2 bits) to enable up to 4 parallel H-ARQ channels

– Subpacket ID (SPID, 2 bits), identifying subpacket numbers for retransmissions

– Encoder packet size (EP_SIZE, 3 bits)

– Last Walsh code index (LWCI, 5 bits), identifying the last Walsh code, among the 28, used by F-PDCH

– ARQ identifier sequence number (AI-SN, 1 bit), toggles between 1 and 0 to indicate the start of a new packet transmission

These parameters are further discussed as part of the H-ARQ description later in this chapter. Note that in this case the MS uses Walsh code set [0, . . . ,LWCI] to decode the subpacket on the F-PDCH0. The mobile then continues monitoring the F-PDCCH0 channel.

- If the F-PDCCH0 message does not contain the MAC_ID for this mobile station, then the mobile attempts to decode the message on the F-PDCCH1, received during the same time interval as the F-PDCCH0.

- If the F-PDCCH1 message contains the MAC_ID for this mobile station, then the mobile attempts to decode the second code multiplexed F-PDCH1. The structure and content of F-PDCCH1 are the same as described in the previous scenario, providing all the information needed to process received subpacket. In this case, however, the Walsh space information for this code division multiplexed transmission is determined from both F-PDCCHs.

- If the F-PDCCH0 contains the broadcast Walsh code mask message, the MS applies the mask to capture the set of Walsh codes not to be used in the F-PDCH decode attempt.

- Otherwise, that is, when F-PDCCH0 was targeted at another user, the MS determines the applicable codes by excluding the portion of the Walsh space indicated by the LWCI saved from the F-PDCCH0 message. In other words. the MS decodes the F-PDCH1 by using the Walsh code set [LWCI0+1, . . . , LWCI1] where LWCI0 and LWCI1 are the last Walsh code index values indicated on F-PDCCH0 and F-PDCCH1, respectively (see Fig. 11.8).

Note that the values of MAC_ID between "00000001" and "00111111" inclusive are reserved and do not currently represent valid MAC_ID address assignments.

If the mobile is in the control hold state, it should monitor the two F-PDCCH channels without demodulating the F-PDCH channel.

11.5.2 Hybrid ARQ on F-PDCH

The hybrid ARQ in the forward packet data channel provides time diversity and incremental redundancy gains associated with soft combining of multiple subpackets. This ARQ scheme is both adaptive and asynchronous.

The adaptive feature allows modulation and coding rate changes on a per subpacket basis, whereas the asynchronous feature facilitates better opportunistic scheduling. This fast physical layer ARQ also works as a self-correction mechanism for link adaptation, allowing the system to use more aggressive rate selection and operate at a higher average FER.

One ARQ instance is associated with each encoder packet, and up to four concurrent ARQ instances can be defined, which avoids the blocking of a new encoder packet transmission while a previous one is being completed. Each of the four ARQ instances is identified by a 2-bit ARQ channel identifier (ACID) included in the corresponding F-PDCCH message.

Each encoder packet of F-PDCH is turbo encoded, and the coded symbols are grouped and selected into four subpackets. The four subpackets are distinguished by their subpacket identifiers (SPID = "00","01","10" or "11"). The symbol selection

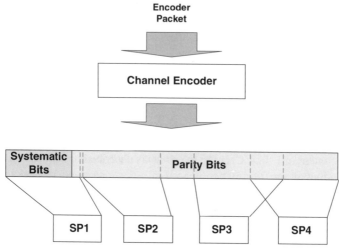

Figure 11.9   Flexible symbol selection at output of the channel encoder.

is very flexible, allowing some symbols to be transmitted in multiple subpackets or not be transmitted in any of them (see Fig. 11.9). The base station always starts with the first subpacket SPID = 00, which contains the systematic part of the code or the information bits.

The base station can send any number of subpackets associated with an encoder packet in any order and at any time (as long as the mobile station is still assigned to the packet data channel and the total number of subpackets does not exceed 8).

The ARQ is based on a stop and wait protocol, and it is ACK based. When the decoding operation is successful (i.e., the CRC on the encoder packet is valid), the mobile station sends an ACK on the R-ACKCH channel and the base station can discontinue sending subpackets for this encoder packet. If the decoding operation is unsuccessful (i.e., the CRC on the encoder packet is invalid) the mobile station sends a NAK on the R-ACKCH channel; the base station can use this feedback to transmit additional subpackets. Note that the ACK/NAK indications are in reference to the encoder packet, not the individual subpackets.

All packet data channel physical channels operate on a 1.25-ms slot basis. Transmissions on the F-PDCH and the F-PDCCHs occur over 1-, 2-, or 4-slot durations (i.e., 1.25-ms, 2.5-ms, or 5-ms frames). Transmissions on the R-ACKCH and R-CQICH occur on 1.25-ms slot intervals.

The retransmission and decoding process is based on adaptive and asynchronous incremental redundancy. When the mobile station receives a subpacket, it attempts to decode the encoder packet based on all of the subpackets that have been received associated with this encoder packet (i.e., in this transmission and from previous transmissions).

Subpackets that are part of the same encoder packet can be transmitted within 1.25-, 2.5-, or-5ms frames, each using different modulation schemes (see Fig. 11.10).

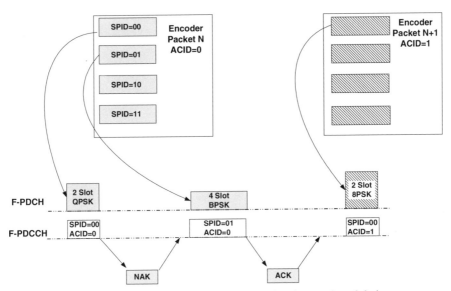

Figure 11.10   Subpackets of an EP can have different durations and modulations.

In fact, it is possible for the base station to retransmit the same subpacket of the same encoder packet more than one time using different transmission lengths and/or modulation. Therefore, the data rate for subpacket transmission can be adaptively changed from one subpacket to the next based on the mobile's channel condition feedback provided on the R-CQICH.

The subpacket retransmission times are also flexible or asynchronous, so that the base station can send a subpacket whenever the radio channel condition for a mobile is favorable and the required radio resources are available.

The combination of incremental redundancy with link adaptation and asynchronous retransmission provides the system with a powerful and flexible ARQ mechanism, which increases channel utilization and throughput efficiency. Comparing this scheme with HARQ in the HRPD, one can see that IS856 does not support modulation changes or variable subpacket frame sizes and it is synchronous.

As the mobile station receiver is combining subpackets with previously received subpackets of the same encoder packet, it needs to determine when a su-packet within the same ARQ channel (i.e., same ACID) represents the start of a new encoder packet transmission. Thus, in addition to SP_ID and ACID, the F-PDCCH message also supplies the receiver with a single-bit ARQ instance sequence number (AI_SN) that toggles between "0" and "1" on each alternate encoder packet being transmitted within the same ARQ channel. Every time the AI_SN toggles, the mobile station treats the subpacket as a subpacket from a new encoder packet. Thus the mobile station discards any previously saved subpacket for that ARQ instance, to avoid combining subpackets belonging to two different encoder packets.

Also note that despite the flexible timing of subpacket transmissions in the forward link there are specific timing requirements for the acknowledgments and

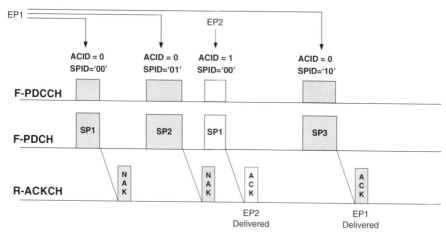

Figure 11.11   Asynchronous transmission of subpackets in two ARQ channels with synchronous ACKs/NAKs.

negative acknowledgments that are sent on the R-ACKCH. Because the R-ACKCH carries only a binary ACK/NAK message to the base station, the explicit timing of the ACK or NAK transmission on the R-ACKCH is needed by the base station to determine which specific subpacket transmission is being acknowledged or negatively acknowledged (see Fig. 11.11).

The ACK/NAK messages are transmitted on the R-ACKCH at ACK_DELAY slots after the complete reception of a F-PDCH frame. Depending on the mobile station's capability, the ACK_DELAY can be 1 or 2 slots. Also, ACK/NAK can be repeated 1, 2, or 4 times as defined by the BS.

11.6   CHANNEL QUALITY MEASUREMENT AND CELL SWITCH

Once the F-PDCH is assigned to a mobile station, the mobile station starts sending the channel quality feedback information on the reverse channel Quality indicator channel (R-CQICH) every 1.25 ms. The channel quality indicator is estimated based on the signal-to-noise ratio of the base station's pilot. Each R-CQICH transmission is directed (by a distinct Walsh cover) to one particular pilot/sector from which the mobile station desires to receive packet data channel transmissions. The mobile station determines the pilot from the packet data channel active set, based on the relative signal strengths received from the pilots in the packet data channel active set. This active set is typically a reduced subset of the FCH active set.

The information provided by the mobile on R-CQICH can be used by the base station to estimate the channel condition to be used to perform opportunistic scheduling by giving priority to users with better channel condition. The report also helps the BS to determine appropriate modulation and coding, that is, data rate and its transmission power levels to the mobile station as well as timing of switching trans-mission on the F-PDCH from one pilot (e.g., sector or BTS) to another sector or BTS.

Because of the asynchronous and channel opportunistic nature of scheduling on F-PDCH this channel does not go into soft handoff. Instead, fast cell switching is used to ensure that the MS receives data from the best possible serving sector from F-PDCH's reduced active set. The mobile station decides to switch based on pilot strength from the serving sector and the target sector, the switching hysteresis parameter, well as the cost of switching (in terms of delay). The procedures for pro-cessing the R-CQICH transmissions and determining the target pilot that is selected for packet data transmissions are not specified.

During the switching process, the MS indicates the target sector as part of the CQI channel reports. The serving sector may complete or interrupt its pending packet transmission, while the target sector detects the switching indications and prepares for F-PSCH transmission. The serving BS may end the switching process early and order the MS to switch to the target sector immediately by sending a message on F-PDCCH.

After the completion or early termination of the switching process the MS resumes the regular R-CQICH reports with the new sector.

The CQI report patterns, which are repeated every 16 slots or 20 ms, are deter-mined and can be altered by the serving base station. Figure 11.12 shows a sample CQI report in a 20-ms frame. The patterns may differ in different aspects:

- Full vs. differential report modes
- Cell switching
- Pilot gating mode for control hold mode
- Repetition of CQI reports

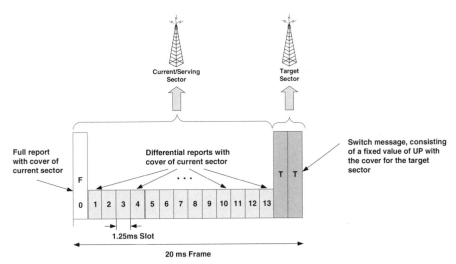

Figure 11.12   Reverse CQI channel report structure.

Full/Differential Report: There are two modes of operation of the reverse channel quality indicator channel: full C/I feedback mode and differential C/I feedback mode. Differential CQI values are a positive or negative increment to the most recently transmitted full CQI. The best current CQI estimate at the base station is the most recently received full CQI value plus the sum of all differential CQI values that were subsequently transmitted. In the full C/I feedback mode, only full C/I measurement reports are sent, whereas in the differential C/I feedback mode, a pattern of full and differential C/I reports is sent. To avoid transmission of full CQI reports at exactly the same 1.25-ms time slots, a time offset parameter set by the base station is applied by each mobile.

Cell Switching: When the mobile station selects a new member of the packet data channel active set to be the next serving sector, it initiates a switching procedure by transmitting a distinctive switching pattern on the R-CQICH for some number of 20-ms periods. During the switching period, the reverse channel quality indicator channel transmissions are modified to use the Walsh cover of the target base station in certain 1.25-ms frames. Sector switching slots carry all up bits to facilitate reliable and fast cell switching detection, and they are positioned at the end of each frame. There can be 2, 4, or 7 switching slots in a frame as configured through L3 messaging. The length of the switching period depends on whether the source and target pilots in the packet data channel active set are within the same BTS or are in different cells.

Pilot Gating and Control Hold Mode: The report pattern is also subject to the current reverse link pilot gating rate. In other words, when the reverse pilot channel gating is enabled, the mobile station does not transmit on the R-CQICH during the 1.25-ms slots in which the reverse pilot channel is gated. The gating can be 1, 1/2, or 1/4. The gating is particularly performed during the F-PDCH control hold state, that is, when the mobile is not receiving data on the packet channel, to reduce the uplink interference. The MS's transition to the control hold state for the packet channel is based on lower layer triggers as opposed to L3 messaging used in IS2000A traffic channels.

Repetition of CQI Reports: The CQI reports may be repeated 1, 2, or 4 times as defined by the base station to provide higher reliability of received reports. Figure 11.13 shows examples of CQI reports with and without cell switching, gating. or repetitions.

11.7 REVERSE PACKET DATA CHANNEL OPERATION

IS2000 Release D significantly improves the reverse link packet data performance by defining a reverse packet data channel (R-PDCH) with a new MAC and PHY design. The PHY improvements are primarily due to higher order modulation and fast Hybrid ARQ and the MAC enhancements are based on adaptive scheduling and flexible rate control.

In the following we focus on some of the key features of R-PDCH channel operation.

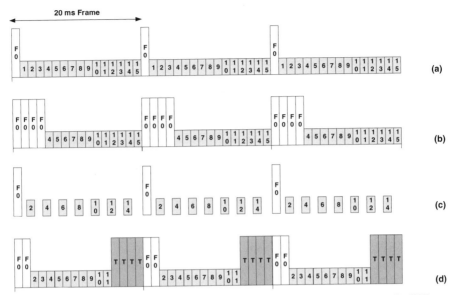

Figure 11.13   CQI report patterns: (a) simple pattern without sector switching, (b) CQI pattern with repetition, (c) CQI pattern with gating in control hold, (d) CQI reports with sector switching and 2 times repetition.

Flexible Rate Control: Designing a rate control protocol for a cdma2000 uplink requires some trade-off considerations. Typically, the tighter the BS's control on the rate, the higher the required signaling overhead. Also, whereas the delay-sensitive traffic requires a fast ramp up in data rate, a graceful ramp up is more desirable to avoid bursty uplink interference to the ongoing CDM channels.

In Release D, the flexible MAC design for R-PDCH provides means of meeting different QoS types per user and allowing different levels of signaling overhead. This flexible design supports three different modes of rate control:

- Autonomous transmission, where uplink transmission rates can be selected autonomously by the mobile up to a maximum rate defined by the base station. This mode provides some level rate control with a minimum amount of MAC/signaling overhead.
- Differentiated rate control, which allows fast and closed-loop rate control of mobile stations. The forward rate control channel may be used by the BS to transmit Up, Down, or Hold rate control information to control the variable data rates on the reverse packet data channels from one or more MSs, in a dedicated or common rate control mode:

- In the dedicated rate control (DRC) mode one indicator control subchannel is assigned to a single reverse packet data channel.
- In the common rate control (CRC) mode multiple reverse packet data channels are assigned to the same indicator control subchannel.

• Absolute rate grant or scheduled mode, where the transmission rate to be used by the mobile is indicated explicitly in a message by the base station.

The two physical channels involved in rate control are the forward grant channel (F-GCH), which provides messaging needed for the scheduled mode, and the forward rate control channel (F-RCCH), which accommodates both dedicated rate control (DRC) and/or Common rate control (CRC) commands. The F-RCCH carries an Up, Down, or Hold indication to the target MSs to increase, decrease, or maintain its authorized traffic-to pilot-power (ATPR) ratio, respectively.

Although the F-GCH does not support soft handoff, the F-RCCH can be used in SHO and L3 messaging determines whether the MS should listen to RC bits from the serving sector only or all BSs in the active set.

Depending on the application and state of the packet data session, one or a combination of these rate control mechanisms can be used. For example, for high-rate applications with sensitive QoS requirements the absolute rate grant may be used. The autonomous rate transmission, on the other hand, may be better suited for low and variable non-real-time application to reduce the overhead. Also, the rate may initially be allocated through absolute grant to provide a "quick start" mechanism and subsequently adjusted through dedicated or common rate control messaging. Note that a rate grant message takes precedence over rate control.

Hybrid ARQ: To further improve the uplink PHY performance the R-PDCH is designed to operate based on a synchronous hybrid ARQ (HARQ) scheme. The HARQ lowers the required E_b/N_o by providing time diversity and soft combining of multiple subpacket transmissions per encoder packet. This physical layer ARQ also lowers the likelihood of using higher-layer ARQ schemes, which requires more overhead cost.

Similar to the forward link the subpackets are derived by a predefined symbol selection process at the output of a systematic 1/5-rate turbo encoder. The HARQ in the reverse link, however, is somewhat simpler than in the forward link. The reverse HARQ is synchronous; implying that after the first subpacket transmission the timing of subsequent subpackets is predefined and fixed by the system (see Fig. 11.14).

Also, each subpacket is transmitted in a fixed size frame of 10-ms duration. The ACK/NAK indications on the forward acknowledgment channel are time multiplexed and sent in every 10-ms frame.

The mobile station can send the next subpacket 30 ms after the first transmission ends. Depending on the ACK or NAK response received from the base station, a new subpacket for the current or the next encoder packet will be sent. This HARQ scheme allows a maximum of two retransmissions per subpacket.

Similar to the forward link, the R-PDCH supports up to four parallel ARQ channels to provide higher channel utilization while the mobile is waiting for ACK/NAK messages.

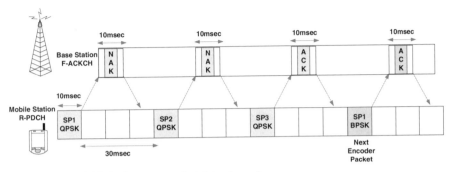

Figure 11.14   HARQ of reverse packet data channel.

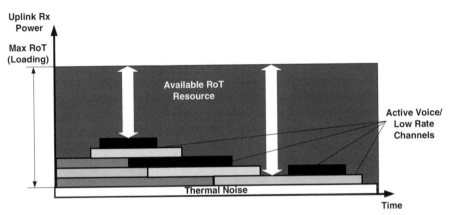

Figure 11.15   The uplink RoT is the key resource to be shared by different users or applications in the reverse link.

The R-PDCH also supports multiple modulation and coding rates, but the modulation and coding cannot be changed for different subpackets of the same encoder packet. Instead, the transmission power can be changed on a per subpacket basis, allowing for a trade-off between packet delay and power efficiency.

QoS Control with Scheduling and Rate Determination: During the operation of the R-PDCH the mobile station reports its transmission buffer status as well as its power headroom to provide BS with information needed for rate determination, scheduling, and QoS control. The new rate selection methodology relies on the fact that the fundamental resource on the CDM reverse link is loading or rise over thermal (RoT) noise level.

The base station considers the current uplink (RoT) noise level as well as the expected MS's contribution to RoT as part of the scheduling and rate allocation processes (see Fig. 11.15). The RoT is highly correlated with the MS's average traffic-to-pilot ratio (TPR), which is typically considered as the key state variable to be used for opportunistic scheduling subject to fairness constraints. Note that for the same channel condition the higher the RoT allowance to a MS, the higher its achievable uplink data rate. The resource control may allow higher levels of TPR in the uplink when the sector is unloaded and reduce the TPR as the loading increases to maintain the overall fairness goals.

11.8 PHYSICAL LAYER STRUCTURE FOR THE NEW CHANNELS

This section describes the structure and usage of all new physical channels introduced in 1xEV-DV to efficiently support high-speed packet data in the forward and the reverse link.

11.8.1 Forward Packet Data Control Channel Structure

The forward packet data control channel is defined in RC10 and used by the base station to send control information for the associated F-PDCH or a Walsh mask to the mobile station.

A forward CDMA channel may contain up to two forward packet data control channels identified by a channel identifier (PDCCH_ID). If the base station supports two F-PDCHs, the base station should also support two F-PDCCHs. All forward packet data control channels and forward packet data Channels transmitted simultaneously should have the same start time and the same duration.

The data rates and frames sizes used on the F-PDCCH are shown in Table 11.3. The frame duration is specified by n = NUM_SLOTS (= 1, 2, or 4) times 1.25-ms slots, corresponding to 1.25-, 1.5-, and 5-ms frames. Depending on the frame duration the data rates may be at 29,600, 14,800, and 7400 bps. Figure 11.16 shows the channel coding process for the forward packet data control channel. The frame consists of the 13-bit (SDU[8 . . . 20]) control information, scrambled by bit-by-bit modulo-2 adding a 13-bit scrambler sequence plus the 8-bit frame quality indicator-covered SDU[0 . . . 7], the 8-bit inner frame quality indicator (CRC), and

TABLE 11.3 Forward Packet Data Control Channel Parameters

Parameter	Data Rate (bps)		
	29,600	14,800	7,400
Frame Duration (msec)	1.25	2.5	5
PN Chip Rate (Mcps)	1.2288	1.2288	1.2288
Code Rate	1/2	1/4	1/4
Modulation Symbol Rate (sps)	38,400	38,400	38,400
QPSK Symbol Rate (sps)	19,200	19,200	19,200
Walsh Length (PN Chips)	64	64	64
Processing Gain (Pn Chips/Bit)	41.51	83.03	166.05

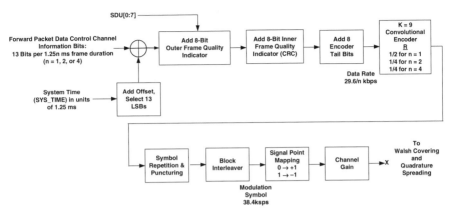

Figure 11.16   Coding structure for forward packet data control channel.

the 8 encoder tail bits. Depending on the number of slots n, the convolutional coding rate may be 1/2 or 1/4. The encoded symbols are punctured/repeated as needed and block interleaved before Walsh channelization spreading with W_n^{64}, where $1 \leq n \leq 63$ as specified by the base station.

11.8.2   Forward Packet Data Channel Structure

The information bits on the F-PDCH are divided into forward physical data channel packets corresponding to payload sizes of 386, 770, 1538, 2304, 3074, or 3842 bits. Each encoder packet consists of the information payload plus 16 CRC bits and 6 encoder tail bits, which can be transmitted in 1, 2, or 4 slots or 1.25, 2.5, or 5 ms, respectively (see Fig. 11.17).

For example, the smallest payload size of (408 = 386 bits + 16 CRC + 6 Tail) transmitted in the longest frame of 5 ms results in the lowest data rate of 408 bits/5 ms = 81,600 bps. The highest data rate is achieved by sending a 3842-bit payload over 1.25 ms, resulting in (3842 + 16 + 6) bits/1.25 ms = 3.091 Mbps. Table 11.4 shows all supported data rates in radio configuration 10 for the packet data channel.

All encoder packets are initially encoded with a 1/5 rate turbo encoder. The coded symbols then go through the channel interleaving process, which involves symbol separation, subblock interleaving, and symbol grouping as shown in Figure 11.17.

The channel interleaver's output sequence, for each input encoder packet, is scrambled and is subsequently divided into subpackets. Each encoder packet can be transmitted as 1 to 4 subpackets, where the symbols in each subpacket are formed by selecting specific sequences of symbols from the scrambler's output. Each subpacket may be transmitted in 1, 2, or 4 time slots or in 1.25, 2.5, or 5 ms, respectively.

The packet data channel supports QPSK, 8PSK, and 16QAM, allowing for adaptive selection of the modulation scheme based on radio channel conditions. The

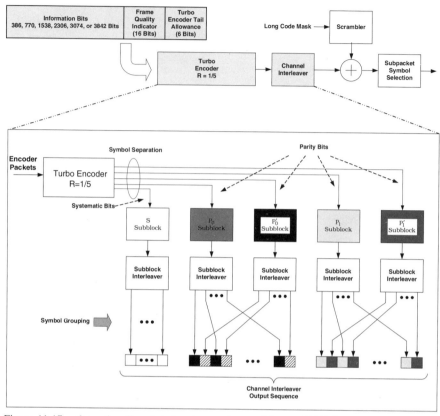

Figure 11.17   channel coding, interleaving, and scrambling on F-PDCH.

modulation order is not explicitly signaled, but it can be determined based on certain parameters available to the mobile. The modulated I and Q symbol for each sub-packet are de-multiplexed into N_{Codes} streams, each covered with a different Walsh cover of length 32 (see Fig. 11.18).

The number of symbols in each subpacket is related to the number of Walsh channels N_{Codes}, the number of slots N_{Slots}, and the chosen modulation scheme. For each subpacket a modulation order coder rate product factor (μ) that reflects the combination of coding and modulation rate can be defined as:

$$\mu = \frac{N_{\text{bits/subpacket}}}{48 \times N_{\text{Codes}} \times N_{\text{Slots/subpacket}}} \tag{11.1}$$

where the factor 48 is calculated based on the chip rate, 1.228 Mcps, multiplied by 1.25-ms slot size and divided by the Walsh size of 32 chips.

Given the information needed in this equation, the base station and mobile, that is, the transmitter and receiver end, can calculate μ and thus select the appro-

TABLE 11.4 Code Rates and Modulation Orders on F-PDCH

N_{bits}/Encoder Packet	No. of Walsh Channels	Subpacket Data Rate	N_{slots}/Subpacket	Modulation Order	Effective Subpacket Code Rate
408	19	326.4	1	2	0.2237
408	8	326.4	1	2	0.5313
792	11	158.4	4	2	0.1875
792	15	633.6	1	2	0.55
792	8	633.6	1	4	0.5156
1560	20	624	2	2	0.4063
1560	8	624	2	4	0.5078
1560	13	1248.00	1	4	0.625
2328	8	465.6	4	3	0.5052
2328	4	465.6	4	4	0.7578
2328	8	931.2	2	4	0.7578
2328	28	1862.40	1	3	0.5774
3096	16	619.2	4	2	0.5039
3096	21	2476.80	1	4	0.7679
3864	16	772.8	4	2	0.6289
3864	12	772.8	4	3	0.559
3864	9	772.8	4	4	0.559

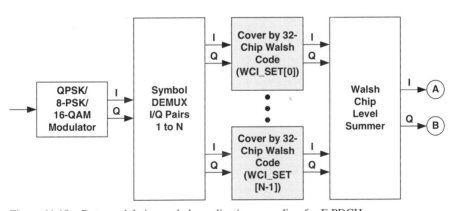

Figure 11.18   Data modulation and channelization spreading for F-PDCH.

priate modulation order factor. Avoiding the higher end of coding rates for each modulation m, one can suggest the ranges shown in Table 11.5.

Table 11.5 also shows examples of code rate and modulation orders for different encoder packet sizes. In this table, the subpacket data rate is the number of encoder packet bits divided by the subpacket duration.

The effective subpacket code rate is defined as the number of bits in the encoder packet divided by the number of subpacket binary code symbols in all of the slots and Walsh channels of the forward packet data channel subpacket. This is

TABLE 11.5 Ranges of Modulation Order Factor μ for F-PDCH

Modulation	Modulation Order (m)	Coding Rate* (R)	Range for m
QPSK	2	$0 < R < 2/3$	$\mu < m/R = 1.5$
8PSK	3	$1/2 < R < 2/3$	$1.5 < \mu < m/R = 2$
16QAM	4	$1/2 < R < 4/5$	$\mu > m/R = 3.2$

* Higher rates may be possible but not used.

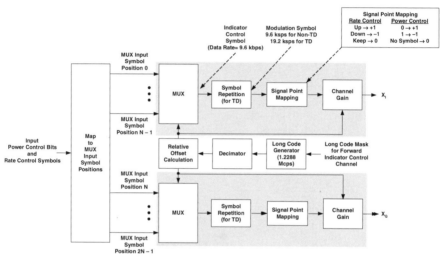

Figure 11.19   Forward indicator control structure.

the effective subpacket code rate of the kth subpacket of an encoder packet and is equal to N_{EP}/L_k.

$$R_{\text{Sub-Packet}} = \frac{N_{\text{bits/Encoder Packet}}}{48 \times N_{\text{Codes}} \times N_{\text{Slots/subpacket}} \times \text{Mod. Order}} \quad (11.2)$$

Forward Indicator Control Channel: In SR1 this channel consists of indicator control subchannels that are used for the common power control channel and the forward rate control channel. For SR3 the forward indicator control channel only consists of the forward common power control channel.

Each indicator control subchannel consists of 2, 4, or 8 indicator control symbols, with a total of 192 indicator control symbols per 10-ms frame and a transmission rate of 9.6 ksps on each of the I and Q arms (see Fig. 11.19).

The BS uses the common power control channel to transmit power control information to one or more mobile stations, by assigning one indicator control subchannel for each MS. Each common power control subchannel uses 8, 4, or 2 of the

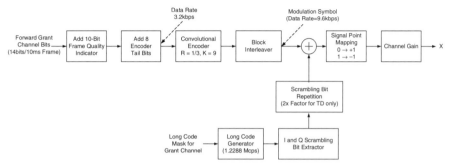

Figure 11.20   Forward grant channel structure.

192 indicator control symbols per 10-ms frame, resulting in rates of 800, 400, and 200 updates/s, respectively.

The common power control channel in the forward link is used for the power control of all reverse physical channels expect for fundicated channels, which have their own associated power control subchannels.

The forward rate control channel may be used by the BS to transmit rate control information to control the data rates on the reverse packet data channels from one or a group of MSs, in dedicated or common rate control mode. The rate control information on each indicator control subchannel consists of rate control symbols at a fixed rate of 100 updates/s.

Forward Grant Channel: The base station uses the forward grant channel to grant mobile stations operating with Spreading Rate 1 permission to transmit on the reverse packet data channel. The grant channel gives permission to the mobile station to transmit for one or more encoder packets.

Figure 11.20 shows the structure of the forward grant channel. The base station transmits information on the forward grant channel at a fixed data rate of 3200 bps and in 10-ms frames. The F-GCH supports discontinuous transmission, and the decision to enable or disable this channel is made by the base station on a frame-by-frame basis.

Forward Acknowledgment Channel: The forward acknowledgment channel consists of 192 forward acknowledgment subchannels, each carrying ACK/NAK symbols to a mobile station's transmission on the R-PDCH. A "+1" or "0" bit corresponds to an ACK or NAK response, respectively. Also, the forward acknowledgment subchannel is gated off when a NAK is transmitted.

In each 10-ms frame, the forward acknowledgment subchannels, numbered from 0 through 191, are multiplexed into separate data streams on the I and Q arms of each forward acknowledgment channel, as shown in Figure 11.21. The data rate on each of the 192 forward acknowledgment subchannels is 100 bps, resulting in a total rate of 19.2 kbps for the F-ACKCH.

Table 11.6 summarizes the data rates, frame sizes, and Walsh codes used for channelization spreading of the new forward physical channels introduced in 1xEV-DV.

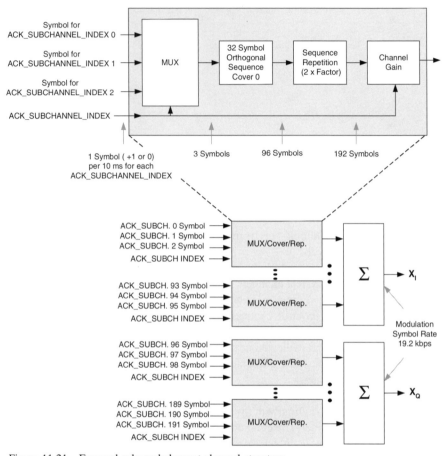

Figure 11.21   Forward acknowledgment channel structure.

After the coding, data modulation, and Walsh spreading, the I and Q channel symbols for all new forward physical channels go through the same quadrature spreading and modulation as used in IS2000 Release A and shown in Figure 11.22.

11.8.3   Reverse Packet Data and Control Channels

This section provides details of the physical layer structure for the new reverse physical channels introduced in 1xEV-DV.

Table 11.7 shows the Walsh channelization codes, frame sizes, and data rates for these channels, and their specific coding, spreading, and modulation attributes are described in the following:

The Reverse Acknowledgment Channel (R-ACKCH) is used by the mobile station to send ACK or NAK responses to forward packet data control channel messages.

TABLE 11.6   Data Rates of New Forward Channels Introduced in 1xEV-DV

Channel Type	Walsh Functions	Data Rates (bps)
Forward Indicator Control Channel	W_n^{128}, where $1 \le n \le 127$	19,200 (9,600 sps per I and Q arm)
Forward Grant Channel	W_n^{256}, where $1 \le n \le 255$	3,200 (10-ms frames)
Forward Acknowledgment Channel	W_n^{64}, where $1 \le n \le 63$	19,200 (10-ms frames)
Forward Packet Data Control Channel	W_n^{64}, where $1 \le n \le 63$	29,600 (1.25-ms frames) 14,800 (2.5-ms frames) 7,400 (5-ms frames)
Forward Packet Data Channel	1 to 28 code channels W_n^{32}, where $1 \le n \le 31$	3,091,200, 2,476,800, 1,862,400, 1,248,000, 633,600, or 326,400 (1.25-ms frames) 1,545,600, 1,238,400, 931,200, 624,000, 316,800, or 163,200 (2.5-ms frames) 772,800, 619,200, 465,600, 312,000, 158,400, or 81,600 (5-ms frames)

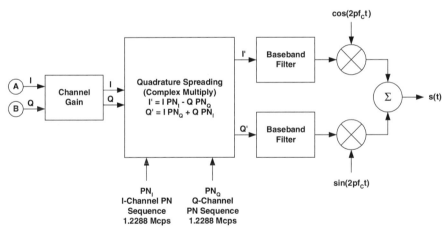

Figure 11.22   Quadrature spreading and modulation in the forward link.

In each 1.25-ms frame the mobile sends a "0" or a "1" bit for an ACK or NAK response, respectively, and when no feedback is needed the R-ACK channel is gated off. The ACK/NAK bits are repeated for some error projection and subsequently covered by Walsh W_{16}^{64} (see Fig. 11.23).

The Reverse Channel Quality Indicator Channel is used by the mobile station to indicate the channel quality measurements of the member of the packet data channel active set from which the mobile station has selected to receive F-PDCH transmissions.

TABLE 11.7 Data Rates and Walsh Code Allocations for New Reverse Pysical Channels Introduced in 1xEV-DV

Channel Type	Walsh Function	Data Rates*
Reverse Secondary Pilot Channel	W_{32}^{64}	0
Reverse Packet Data Control Channel	W_{48}^{64}	700 bps (10-ms frames)
Reverse Request Channel	W_8^{16}	3200 bps (10-ms frames)
Reverse Acknowledgment Channel	W_{16}^{64}	800 bps
Reverse Channel Quality Indicator Channel	W_{12}^{16}	3200 or 800 bps
Reverse Packet Data Channel	W_1^2 and/or W_2^4	6.4 kbps to 1.5384 Mbps (10-ms frames)

* See Table 11.8 for all the rates.

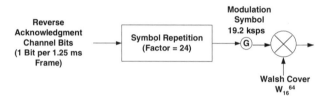

(a) Reverse Acknowledgement Channel (R_ACKCH) Structure

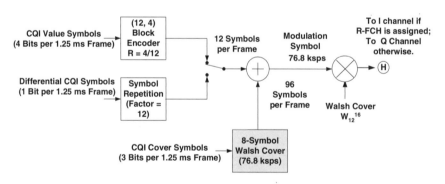

(b) Reverse Channel Quality Indicator Channel (R-CQICH) Structure

Figure 11.23   Reverse ACK and reverse quality indicator channel structure.

The reverse channel quality indicator channel structure is shown in Figure 11.23. The transmissions on R-CQICH in each 1.25-ms frame are in the form of full or differential quality reports. When a full report is sent, a 4-bit value is mapped into 12 symbols using a (12, 4) block code. When differential reports are sent, a single quality indicator channel bit is used that is repeated 12 times to form 12 symbols per 1.25-ms frame.

An 8-ary Walsh function specified by the MAC is used to spread the R-CQICH transmission to indicate the selected base station identity. The resulting symbols at a rate of 76.8 ksps are further spread for orthogonal channelization by Walsh W_{12}^{16}.

The Reverse Packet Data Control Channel (R-PDCCH) is used by the mobile station to send the control information associated with a reverse packet data channel. This control information includes the data rate on R-PDCH, the subpacket identifier, and the QoS indicator. The R-PDCCH also carries a single mobile status indicator bit (MSIB) that is used by the mobile station to indicate whether it has enough power and data to increase the transmission rate on the reverse packet data Channel.

The mobile station transmits information on the R-PDCCH at a fixed data rate of 700 bps and in 10-ms frames. The MAC layer instructs the physical layer whether to transmit on the reverse packet data control channel on a frame-by-frame basis.

Transmission on the reverse packet data control channel should be aligned with the transmission of data on the corresponding reverse packet data channel.

The R-PDCCH information in each 10-ms frame consists of 6 bits and the Mobile status indicator bit that is used at the output of the encoder packet. The 6-bit control information is coded with a (64, 6) orthogonal Walsh block code. This code involves mapping every 6 input bits to a corresponding 64-bit Walsh code at the output code. This mapping is similar to the orthogonal Walsh modulation used in the reverse link of IS95 or RC1/RC2 of cdma2000. The output of the (64, 6) orthogonal block code is bit-by-bit modulo-2 added to the mobile station information bit to form a (64, 7) biorthogonal block code (see Fig. 11.24). The encoded bits are then repeated, resulting in a modulation symbol rate of 19.2 ksps before orthogonal spreading with W_{48}^{64}.

The reverse packet data channel is the portion of the traffic channel of Radio Configuration 7 that carries the higher-layer data. The structure of the reverse packet data channel is shown in Figure 11.25. The channel coding applied to R-PDCH is very similar to turbo coding and subpacket formation in the F-PDCH as shown in Figure 11.17. This coding includes a 1/5 rate turbo encoding, block interleaving, and

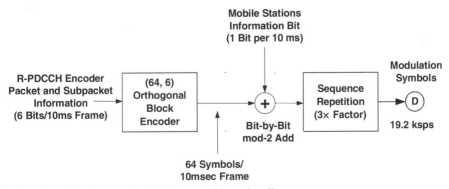

Figure 11.24   Reverse packet data control channel coding.

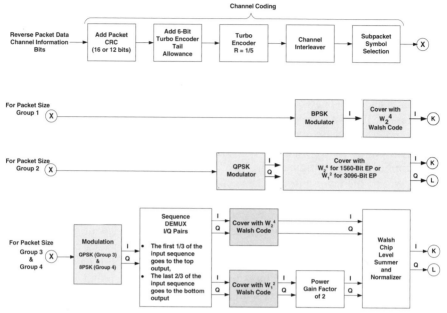

Figure 11.25   Channel coding and data modulation on R-PDCH.

subpacket symbol selections, for up to three subpackets. The main difference is that the frame size of R-PDCH is the fixed 10-ms frame size.

The type of data modulation and Walsh covering used for this channel depend on the size of encoder packet (EP) size. Figure 11.25 shows the details of the modulation and spreading process for four packet size groups defined in Table 11.8.

The mobile station uses BPSK modulation for EP sizes of 192, 408, and 792 bits, QPSK modulation for EP sizes of 1560, 3096, 4632, 6168, 9240, 12,312, and 15,384 bits and 8PSK modulation for EP size of 18456 bit. Also, for encoder packets of size 4632 bits and higher the QPSK or 8PSK modulator output is de-multiplexed into two sequences, each covered by a different Walsh code. Note that after the Walsh covering, the symbols covered with W_1^2 are amplified by a power gain factor of 2 with respect to the symbols covered by W_2^4, and the resulting sequences are summed and normalized to obtain a single sequence.

Because the encoder packet may be transmitted after several subpackets, the data rate shown in Table 11.8 is calculated as the number of encoder packet bits divided by the duration of all of the transmitted subpackets of the encoder packet. Also, the effective code rate in this table is defined as the number of bits in the encoder packet divided by the number of binary code symbols in all of the corresponding transmitted subpackets.

Figure 11.26 summarizes the Walsh code allocations to physical channels and their mapping to I and Q branches in the reverse link.

TABLE 11.8 Effective Coding and Data Rates on R-DPCH

Packet Size Group & Modulation	Walsh Codes	No. Bits/EP	Trans. Subpacket ID	Data Rate (kbps)	No. Binary Code Symbols Trans.	Effective Code Rate
Group 1 BPSK (on I)	W_2^4 ++--	192	2	6.4	9,216	0.0208
			1	9.6	6,144	0.0313
			0	19.2	3,072	0.0625
		408	2	13.6	9,216	0.0443
			1	20.4	6,144	0.0664
			0	40.8	3,072	0.1328
		792	2	26.4	9,216	0.0859
			1	39.6	6,144	0.1289
			0	79.2	3,072	0.2578
Group 2 (QPSK)	W_2^4 ++--	1,560	2	52.0	18,432	0.0846
			1	78.0	12,288	0.1270
			0	156.0	6,144	0.2539
	W_1^2 +-	3,096	2	103.2	36,864	0.0840
			1	154.8	24,576	0.1260
			0	309.6	12,288	0.2520
Group 3 (QPSK)	W_2^4 ++-- & W_1^2 +-	4,632	2	154.4	55,296	0.0838
			1	231.6	36,864	0.1257
			0	463.2	18,432	0.2513
		6,168	2	205.6	55,296	0.1115
			1	308.4	36,864	0.1673
			0	616.8	18,432	0.3346
		9,240	2	308.0	55,296	0.1671
			1	462.0	36,864	0.2507
			0	924.0	18,432	0.5013
		12,312	2	410.4	55,296	0.2227
			1	615.6	36,864	0.3340
			0	1,231.2	18,432	0.6680
		15,384	2	512.8	55,296	0.2782
			1	769.2	36,864	0.4173
			0	1,538.4	18,432	0.8346
Group 4 (8PSK)	W_2^4 ++-- & W_1^2 +-	18,456	2	615.2	82,944	0.2225
			1	922.8	55,296	0.3338
			0	1.845	27,648	0.6675

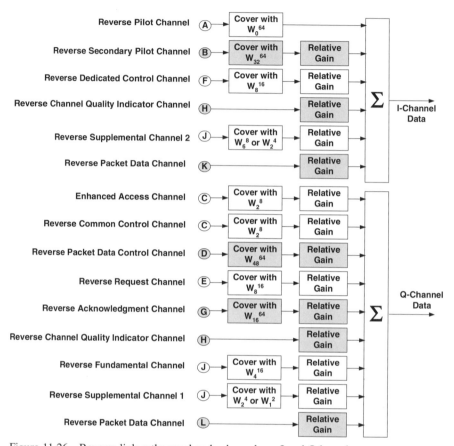

Figure 11.26 Reverse link orthogonal code channels on I and Q branches.

11.9 OTHER ENHANCEMENTS IN 1XEV-DV

In addition to the forward and reverse packet data channel operation in 1xEV-DV, there are a number of other system enhancements made as part of standards evolution.

This section briefly describes some specific enhancements, namely, broadcast/multicast service and fast call setup and MEID support. For more details on these features and other new enhancements, the reader is referred to the IS2000 standard specification documents.

11.9.1 Broadcast/Multicast Service (BCMCS)

The IP-based broadcast/multicast service provides the ability to transmit the same information stream to multiple users simultaneously to achieve the most efficient use of air interface and network resources. BCMCS are delivered via the most effi-

TABLE 11.9   Broadcast/Multicast Services Modes

MS State	BCMCS Mode	Attributes
MSs in mobile station idle state, with no active uplink channel	Idle	Multiple MSs are assigned the same FL Walsh code on a designated F-BSCH All MSs use the same public long code mask (PLCM)
MSs in mobile station control on the traffic channel state with active uplink channels	Dedicated	Multiple MSs are assigned same forward link Walsh codes Traffic on F-SCH Signaling on F-DCCH Individual MS are addressed via MS long code mask
	Shared	Multiple MSs assigned same FL Walsh code Traffic on F-SCH Signaling on F-FCH All MSs use the same public long code mask (PLCM)

cient transmission technique based on the information (type of media) being transmitted and available radio resources.

The type of information transmitted is not limited to text but can be any type of data, including multimedia (e.g., voice) and streaming media. Also, the information may be received by all users or may be restricted to a subset of users via encryption.

Retransmission and acknowledgment in BCMCS are not required, because the type of transmission is "one to many."

cdma2000 Release D defines a new framework to efficiently support broadcast and multicast traffic to be used for group messaging and subscription-based broadcast data services, in which the same traffic frames can be sent once from the base station and received and used by all or a select group of mobile stations.

Depending on the state of mobiles receiving the BCMCS messages and the desired type of message delivery, different BCMCS modes are defined. These modes and their respective attributes in Release D are summarized in Table 11.9.

Note that BCMCS to mobiles in the idle mode typically does not require any active channels in the reverse link and it is suited for subscription downstream programs such as news and entertainment programs. The dedicated and shared modes, however, require mobiles to maintain an uplink channel for signaling purposes including possible acknowledgments. These modes are more suited for group messaging for a smaller number of mobile stations, such as fleet calls and group chats.

Release D defines a new forward broadcast traffic channel as a point-to-point logical channel that carries broadcast/multicast data. This logical channel is typically mapped to a specific supplemental channel designated as the forward broadcast supplemental channel (F-BSCH) using Radio Configuration 5 only.

The Broadcast/Multicast (BMAC) multiplex sublayer is used by broadcast services to transmit data to the mobile station via the broadcast traffic logical channels. The broadcast logical channels are mapped to broadcast supplemental channels. Each broadcast supplemental channel has an identifier (FSCH_ID) associated with it. The mobile station Layer 3 commands the BMAC multiplex sublayer to start or stop receiving a particular broadcast logical channel carried in a particular broadcast supplemental channel.

To support BCMCS on F-BSCH, the supplemental channel in Radio Configuration 5 supports a new outer coding, beside the default Turbo encoding, to provide additional coding gain for better coverage when the F-SCH is used to carry broadcast/multicast service traffic across the cell area. When outer coding is used, the outer code rate of F-SCH is specified by the base station. In this case, the information bits are stored in the outer coding buffer and then outer encoded with Reed–Solomon codes. The outer coded symbols of the outer coding buffer are transmitted in 64 consecutive 20-ms forward supplemental channel frames with Radio Configuration 5 and a data rate of 115.2 kbps.

11.9.2 Fast Call Setup

One of the important new features in Release D is fast call setup. This capability allows the network to reduce call setup latency for all or some mobiles based on specific application types or mobile classes. The fast call setup is achieved by a combination of different features including the following:

Reduced Cycle Index: To reduce the time needed for waking up a mobile station and thus the overall call setup time, IS2000 Release D defines a set of reduced paging slot cycles. In Release A the supported slot cycle indices are SCI = 0, 1, 2, ..., and 7 corresponding to 16, 32, 48, ... and 128 paging slots, respectively. In Release D negative SCI are defined as SCI = −1 (8 slots), −2 (4 slots), −3 (2 slots), and −4 (1 slot). Note that the smaller the SCI, the lower the paging cycle and the smaller the call setup latency, which is critical to some applications. However, the smaller SCI requires the mobile to check paging channels more frequently, causing more battery usage and a shorter MS battery life. MS can register with reduced SCI, where the maximum and minimum SCI allowed are defined by the base station and the network. To reduce the impact of reduced SCI on the system, the MS and the BS may only operate temporarily in reduced SCI mode after a call release. Also, this option may be used only for specific applications such as PTT and in preparation for quick traffic channel setup shortly after a channel release.

11.9.3 Other Signaling Enhancements

There are a number of other enhancements made or planned for IS2000 Release D or future releases that are beyond the scope of this book. In the following we briefly describe some of these features and list others for the reader's reference.

Direct Channel Assignment and Mobile Tracking: With this option enabled, the BS can send an enhanced channel assignment Message (ECAM) directly without sending paging and/or receiving a page response. Also, BS can optionally request a page response in unassured mode. This capability is also helpful when the system needs to set up a new traffic channel shortly after a call release, for example, in a push to talk scenario.

Another related feature is capability to provide mobile tracking reports after a call release for a specified period of time to help waking up the mobile quickly.

Mobile Equipment Identifier (MEID) Support: The MEID is defined based on harmonization between 3G technologies as a universal mobile equipment identifier to facilitate tracking mobiles under global roaming conditions. MEID is a 56-bit new identifier for the MS to replace ESN. The fields in the MEID are coded with hexadecimal coding, with an addressing space that is so large that exhaustion issues are not expected.

The Global Equipment Identity, or GEID (i.e., IMEI and MEID) provides not only the manufacturer identity of the mobile equipment but also information such as type allocation code or serial number.

By means of and with the assistance of a manufacturer's database lookup, MEID may help service providers identify the ME to the levels of model, manufacturer factory, and lot number. The information can be used for corrective or preventive actions to improve service quality and to maintain a list of MEs that have been stolen or denied service.

Staring in Release D, the IS2000 standard supports MEID as a mandatory feature for both MS and BS. The MS uses 32-bit pseudo-ESN when talking to BSs with older releases.

The following list includes some of the other enhancements expected in IS2000 Revision D and future releases.

- Access control based on call type (ACCT)
- CDMA card application toolkit (CCAT, in support of R-UIM)
- Enhanced call recovery
- Enhanced OTASP procedures
- Enhanced packet data air interface security
- Internet over-the-air handset configuration management (IOTA)
- Link layer-assisted robust header compression (LLA ROHC) for VoIP
- Multimedia message service (MMS) and multimedia streaming
- Prepaid service support for HRPD
- Presence service
- Priority service
- Realm configured Packet data session inactivity timer
- Rescue channel
- Support for common channel-only capable devices
- Video conferencing service

- Wideband speech codec for cdma2000 systems
- Wireless applications management

For more details on these enhancements, the reader is referred to [1–5].

11.10 REFERENCES

1. C.S0001-D, Introduction to cdma2000 Standards for Spread Spectrum Systems, February 2004.
2. C.S0002-D, Physical Layer Standard for cdma2000 Spread Spectrum Systems, February 2004.
3. C.S0003-D, Medium Access Control (MAC) Standard for cdma2000 Spread Spectrum Systems, February 2004.
4. C.S0004-D, Signaling Link Access Control (LAC) Standard for cdma2000 Spread Spectrum Systems, February 2004.
5. C.S0005-D, Upper Layer (Layer 3) Signaling Standard for cdma2000 Spread Spectrum Systems, February 2004.

CHAPTER 12

RADIO PERFORMANCE AND NETWORK PLANNING

12.1 INTRODUCTION

There are many ways to evaluate the performance of an access network technology with respect to its resource efficiency and quality of services offered. For radio access technologies there are two key performance figures that have direct impact on planning and dimensioning of the networks and therefore on the cost and complexity associated with their deployment. These two performance figures or sets of figures, in general, reflect the coverage efficiency and spectrum efficiency of the system.

In this chapter we discuss the coverage and spectrum efficiency of cdma2000 radio access networks. We also provide some practical guidelines for planning and dimensioning these networks based on RF coverage and capacity considerations.

An exact coverage or capacity analysis for cdma2000, supporting a mixture of voice and data services, requires extensive radio channel and traffic modeling as well as computer-based simulations, which are beyond the scope of this book. However, by analyzing the system under some simplified traffic and radio propagation assumptions one can arrive at reasonably close performance estimates. These calculations and the corresponding estimated performance numbers are very helpful in understanding the important design trade-off and network planning considerations.

In the following sections we study the coverage efficiency of a cdma2000 system based on the appropriate link budget analysis (LBA) for voice and data services. We will also look at the system capacity for voice, data, and mixed-traffic scenarios. Using the same methodology we will look at HRPD system performance for data applications.

After performance figures are discussed, we will then present a simple, but not necessarily most accurate, method for dimensioning a cdma2000-based network. Using this method one can arrive at the approximate number of network elements needed to provide coverage, capacity, and quality of service for all of the subscribers in the network.

CDMA2000® Evolution: System Concepts and Design Principles, by Kamran Etemad
ISBN: 0-471-46125-3 Copyright © 2004 John Wiley & Sons, Inc.

12.2 CDMA2000 COVERAGE AND LINK BUDGET ANALYSIS

One of the classic approaches to evaluating the coverage efficiency of a radio access technology is the link budget analysis (LBA). The link budget or power budget analysis summarizes the main gains and losses between a transmitter node and the receiving node in a path loss equation and, under some typical assumptions, estimates the maximum allowable link path loss. This analysis can be performed for the forward and reverse links, and on the basis of the results, which are typically different for the two links, some adjustments may be made in RF parameters to balance the links.

In addition to path balancing, the results of LBA can also be used to estimate the average cell size in a given propagation environment. By estimating the average cell sizes for different parts of a network, one can estimate the number of cells and sites needed from coverage perspectives, which is part of the overall network dimensioning process.

From a high level, the maximum path loss is the difference between the maximum radiated power from the transmitter and the minimum acceptable signal level for reliable communications at the receiving end. However, a detailed calculation of the link budget takes into account certain margins allocated for fading, interference, as well as the gains associated with handoff and other diversity schemes.

In this section we provide a brief overview of key parameters involved in LBA for voice and data services in cdma2000, followed by some discussions on LBA issues in 1xEV-DO systems.

In conventional voice-based systems the links are analyzed mostly in the reverse link, which is the limiting link in most deployment conditions and the link that is easier to analyze. Therefore, we first present the reverse link budget for basic voice services in CDMA systems, including IS95 and cdma2000. This simple analysis provides insight into the key concepts and factors that affect the power budget not only for voice but also for data services at different rates. We will then apply similar concepts and procedures to forward LBA of cdma2000.

12.2.1 IS2000 Reverse Link Budget Analysis

Figure 12.1 shows major gains and losses in the reverse link radio path, which are to be captured in the reverse link budget calculation.

On the transmitter side, we start by estimating the mobile station's effective isotropic radiated Power (EiRP) by adding the signal power at the output of the MS's power amplifier (PA) to the antenna gain and applying the losses associated with cabling if any.

$$EiRP_{MS} = P_{Amp\_MS} - L_{C\_MS} + G_{MS} \qquad (12.1)$$

For a portable terminal ,the cabling loss L_C is 0 and the antenna gain G in (dBi) is a small positive or negative number.

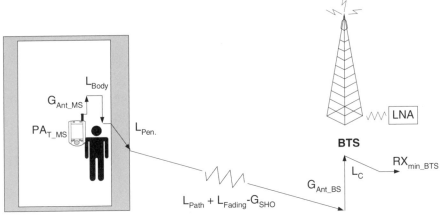

Figure 12.1 Gains and losses in reverse link. Reproduced under written permission from Telecommunications Industry Association.

The mobile station's radiated power goes through radio channel degradation as a result of static distance-based path loss as well as shadow fading, which are accounted for by L_{Path} and L_{Fading}, respectively.

The parameter L_{Fading} is considered as a safety margin against shadowing in the path of the radio signal causing fading effects on the average received signal. This long-term shadow-fading effect, when represented in decibel scale, usually follows a normal distribution, and therefore it is also referred to as log normal fading. The fade margin is calculated for a given propagation environment and for a target coverage reliability or outage probability. The fade margin depends on the standard deviation of log normal shadow fading, the path loss slope, and the coverage reliability. L_{Fading} is higher for terrains with a higher degree of variation, for example, urban regions, and is lower for flat terrains. Also, for higher level of coverage reliability a larger fade margin needs to be considered. For details on how fade margin is estimated, see [1].

A CDMA system supporting soft handoff at the edges of the cells benefits from a macrospace diversity gain, which effectively reduces the fade margin needed to ensure certain link reliability. Therefore, the level of soft handoff gain depends on the variance and correlation of shadowing effects, the number of base stations in handoff, and the target coverage reliability. Assuming independent shadowing effects from two base stations covering the handoff areas, one can estimate the soft handoff gain to be in the range of 2–4 dB on average.

When the user terminal is used indoors, there is an additional in-building penetration loss L_{Bldg} that needs to be considered to account for signal attenuation as it penetrates through the building structure and walls. Obviously, L_{Bldg} depends on the building structure and the position of the MS within the building, and as a result an average but conservative number is usually assumed.

In many cases, and specifically for voice applications, the user terminal is used very close to the user's body, which acts as a conductor. In these cases, the antenna pattern is affected by the user's body and the received power is lower. This effect is captured as a body loss (L_{Body}) factor in the calculations.

As a result of all the above gains and losses in the path, the received signal power at the base station can be estimated as

$$RX_{BTS} = EiRP_{MS} - L_{Body} - L_{Bldg} - L_{Path} - L_{Fade} + G_{SHO} + G_{BTS} - L_{C\_BTS} \quad (12.2)$$

This equation can be rearranged to show path loss as function of other parameters:

$$L_{Path} = EiRP_{MS} - L_{Body} - L_{Bldg} - RX_{BTS} - L_{Fade} + G_{SHO} + G_{BTS} - L_{C\_BTS} \quad (12.3)$$

Based on this equation the maximum allowable path loss can be calculated when the mobile's transmit power is at its maximum and the received signal at the BTS is at its minimum acceptable level.

The minimum acceptable signal level for a given data rate and FER is called receiver sensitivity, and it is calculated based on the following equation:

$$RX_{min\_BTS} = R \times (E_b/N_o)_{min} \times kT \times Nf \quad (12.4)$$

or in decibel scale:

$$RX_{min\_BTS(dB)} = 10 \times \log(R) + (E_b/N_o)_{min,dB} + 10 \times \log(kT) + NF_{dB} \quad (12.5)$$

The parameters in this equation are described as follows:

- $(E_b/N_o)_{min}$, which is the required traffic channel energy per bit-to-total noise power spectral density ratio, is the key receiver performance figure. The value of $(E_b/N_o)_{min}$ depends on coding, spreading, and modulation used for each data rate and also on the average mobile speed and the required FER. In CDMA systems, the required $(E_b/N_o)_{min}$ typically includes a two-branch receive diversity gain and power control gain.

- Noise figure, which is the increased effective input noise level as result of thermal noise added by the receiver's RF front end. The noise factor Nf is the linear scale factor, whereas the noise figure NF is the decibel scale representation.

- Thermal noise power density, which is modeled as $No = kT$, where $k = 1.38 \times 10^{-28}$ J/K is the Boltzmann constant and T is the temperature in Kelvins. Typically in most LBAs the temperature is assumed at about $T = 290$ Kelvin resulting in $No = -174$ dBm/Hz.

Note that in this receiver sensitivity calculation only the thermal noise is considered and the effect of uplink cochannel interference is not incorporated. Therefore, the maximum allowable path loss (MAPL) without any loading consideration is estimated as:

$$MAPL_{No\_Load} = EiRP_{max\_MS} - L_{Body} - L_{Bldg} - RX_{min\_BTS} - L_{Fade} + G_{SHO} + G_{BTS} - L_{C\_BTS} \quad (12.6)$$

The effect of loading is the increase of interference rise above noise. The uplink cochannel interference increases the background "noise floor" at the BS's receiver and thus degrades the receiver sensitivity and decreases the maximum allowable path loss. The amount of degradation is directly proportional to the ratio of total noise and interference to noise spectral density, which, as discussed in Chapter, 2 is equal to $1/(1 - L)$, where L is capacity loading factor. Therefore, in decibel scale the MAPL is reduced by the so-called interference margin as shown in the following equation:

$$\text{MAPL} = \text{MAPL}_{\text{No\_Load}} - 10 \log\left(\frac{1}{1 - L}\right) \qquad (12.7)$$

Table 12.1 shows a summary of the reverse link LBA calculation described above using some typical assumptions for voice in IS95 and in cdma2000.

The reverse link budget for voice service in cdma2000 is very similar to IS95A with these minor differences:

- In the reverse link of cdma2000 a portion of the mobile transmitted power has to be allocated to reverse pilot channel power, whereas in IS95 all of the MS's

TABLE 12.1   Reverse Link Budget Analysis Example for 2G and 3G Voice

Reverse Link Budget Analysis (Voice)			
Input Parameters	Units	IS95 Voice	CDMA2000 Voice
Vocoder Rate	kbps	9.6	9.6
Receiver Parameters			
Target E_b/N_t	dB	7.00	4.20
Thermal Noise Spectral Density No = $10 \log(KT)$	dBm/Hz	−174.0	−174.0
BTS Noise Figure	dB	4.0	4.0
Abolute Rx. Sensitivity	dBm	−123.18	−125.98
BTS Cabling Loss (L_C)	dB	3.00	3.00
BTS Antenna Gain (G_BTS)	dB	16.00	16.00
Additional RX Diversity Gain	dB	0.00	0.00
Transmitter Parameters			
MS PA Power	dBm	23.00	23.00
MS Antenna Gain (G_MS)	dBi	0.0	0.0
MS EIRP	dBm	23.0	23.0
Channel Parameters			
Probability of Service at Cell edge	%	90%	90%
Composite Slow Fading Std. Deviation	dB	8.0	8.0
Fade Margin (L_Fading)	dB	10.3	10.3
Building Penetration Loss (L_Pen)	dB	15.0	15.0
User Body Loss (L_Body)	dB	3	3
SHO Gain (G_SHO)	dB	4	4
(Total Margin-Soft Handoff Gain)	dB	24.25	24.25
Max. Allowable Path Loss (No Load)	dB	134.92	137.72
Sector Loading	%	50%	50%
Interference Margin (Loading Factor)	dB	3.0	3.0
Max. Allowable Path Loss (With Loading)	dB	131.9	134.7

transmitted power can be used for traffic channel. However, the portion of power allocated to the pilot is small, and it does not significantly affect the link calculations described above.

The required traffic channel E_b/N_o requirements for IS95, and the total traffic plus pilot E_b/N_o requirement for cdma2000, to achieve 1% target FER is 4–4.5 dB for cdma2000 vs. 7 dB for IS95. This difference is the result of coherent demodulation of uplink traffic channel in cdma2000, which improves the receiver sensitivity by about 2.5–3 dB.

As a result, the cdma2000 carriers, for the same vocoder, have a about 3-dB of link advantage over IS95 carriers, allowing simple one-to-one overlay of 3G carriers with marginally improved link or voice quality.

One can apply the same set of calculations to estimate MAPL for different uplink data bearer services in cdma2000. These estimates are captured in Table 12.2, where the 3G-voice column from the previous table is repeated for comparison.

Comparing the LBA calculations for data and voice, one can see the following differences:

- $(E_b/N_o)_{min}$ is different for each combination of coding and spreading, and therefore it varies with data rate. Also the target FER for the packet data application can be higher than 1%, resulting in more relaxed $(E_b/N_o)_{min}$ requirement. In addition, the higher data rates the lower the $(E_b/N_o)_{min}$ required because of higher coding gains.

- Despite the lower $(E_b/N_o)_{min}$ for higher data rates, the resulting receiver sensitivity level increases with data rate as reflected in the $10 \times Log(R)$ factor in the RX_{min} equation.

- Another important factor is the body loss, which is expected to be smaller for data applications than for voice. This is because of the fact that for data usage the MS device is typically positioned farther from the user's body.

Given the assumed input parameters and the above MAPL calculations one can compare the uplink coverage performance for different voice and data.

For each case, one can also translate the MAPL to a range or cell radius by using an appropriate path loss model. Although various path loss models have been proposed and used for different propagation environments, most models can be simplified for our purpose to the following form:

$$L_{dB} = A + B \times \log(d) \tag{12.8}$$

where A and B are the intercept and slope associated with the path loss function. For more details on various path loss models, the reader is referred to [1,2]. Using this generic model as a basis one can compare the uplink radii for different bearer services by comparing them to the cell radius for 2G voice service.

$$
\begin{aligned}
MAPL_S - MAPL_{2G\_Voice} &= (A + B \times \log(d_S)) - (A + B \times \log(d_{2G\_Voice})) \\
&= B \times \log(d_S / d_{2G\_Voice}) \\
d_S / d_{2G\_Voice} &= 10^{\frac{MAPL_S - MAPL_{2G\_Voice}}{B}}
\end{aligned}
\tag{12.9}
$$

TABLE 12.2   Reverse LBA for Different Bearer Data Rates in cdma2000

cdma2000 Reverse Link Budget Analysis (Voice and Data)

Input Parameters	Units	Voice	Data				
Bearer Data Rate	kbps	9.6	9.6	19.2	38.4	76.8	153.6
Receiver Parameters							
Target E_b/N_t	dB	4.20	3.90	3.60	3.20	2.20	2.20
No = $10\log(KT)$	dBm/Hz	−174.0	−174.0	−174.0	−174.0	−174.0	−174.0
BTS Noise Figure	dB	4.0	4.0	4.0	4.0	4.0	4.0
Abolute Rx. Sensitivity	dBm	−125.98	−126.28	−123.57	−120.96	−118.95	−115.94
BTS Cabling Loss	dB	3.00	3.00	3.00	3.00	3.00	3.00
BTS Antenna	dB	16.00	16.00	16.00	16.00	16.00	16.00
Additional RX Diversity Gain	dB	0.00	0.00	0.00	0.00	0.00	0.00
Transmitter Parameters							
MS PA Power	dBm	23.00	23.00	23.00	23.00	23.00	23.00
MS Antenna Gain	dBi	0.0	0.0	0.0	0.0	0.0	0.0
MS EIRP	dBm	23.0	23.0	23.0	23.0	23.0	23.0
Channel Parameters							
Cell Edge	%	90%	90%	90%	90%	90%	90%
Composite Slow Fading Std. Deviation	dB	8.0	8.0	8.0	8.0	8.0	8.0
Fade Margin	dB	10.3	10.3	10.3	10.3	10.3	10.3
Building Loss	dB	15.0	15.0	15.0	15.0	15.0	15.0
User Body Loss	dB	3	1	1	1	1	1
SHO Gain	dB	4	4	4	4	4	4
(Total Margin- Soft Handoff Gain)	dB	24.25	22.25	22.25	22.25	22.25	22.25
Max. Allowable Path Loss (No Load)	dB	137.72	140.02	137.31	134.70	132.69	129.68
Sector Loading	%	50%	50%	50%	50%	50%	50%
Interference Margin	dB	3.0	3.0	3.0	3.0	3.0	3.0
Max. Allowable Path Loss (With Loading)	dB	134.7	137.0	134.3	131.7	129.7	126.7

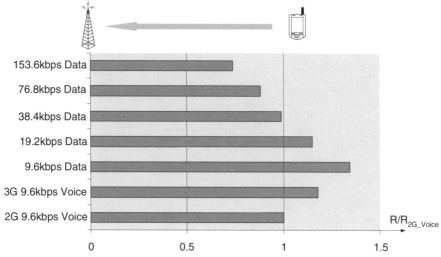

Figure 12.2 Uplink range comparison for voice and data services in cdma2000.

This comparison is particularly appropriate for cases in which the cdma2000 system is to be deployed as a one-to-one overlay on existing 2G/IS95 sites. Figure 12.2 shows this comparison assuming an average path loss slope of $B = 40\,\text{dB/dec}$.

Based on all the assumptions we have made so far, one can see from the results that the cdma2000 system can offer at least 19.2- and in some cases 38.4-kbps data rate in a cell design that is optimized for 2G voice service. Also, the higher data rates can be offered closer to the base station.

Note that these comparisons are made in terms of distance or range and to convert them to areas one needs to look at each number squared. For example, to provide 76.8-kbps minimum data rates the cell radius should be considered to be 0.88 times the 2G-Voice radius and cell areas will be $0.78 = (0.88)^2$ times the 2G-Voice based cells.

12.2.2 IS2000 Forward Link Budget Analysis

In principle, the forward link budget analysis is similar to that of the reverse link. However, there are some differences due to partitioning of BTS power among multiple dedicated traffic and broadcast/common control channels and path balancing.

Figure 12.3 shows the gains and losses in the forward path.

$$RX_{MS} = 10 \times \log(\alpha_T) + EiRP_{BTS} - L_{Path} - L_{Fade} + G'_{SHO} + G_{MS} - L_{Bldg} - L_{Body}$$

$$L_{Path} = 10 \times \log(\alpha_T) + EiRP_{BTS} - RX_{MS} - L_{Fade} + G'_{SHO} + G_{MS} - L_{Bldg} - L_{Body} \quad (12.10)$$

where α_T is the fraction of total power allocated to a traffic channel. Also, the mobile station receiver sensitivity RX_{MS} and the base station's total effective isotropic radiated power $EiRP_{BTS}$ are calculated based on equations similar to those used for the reverse link.

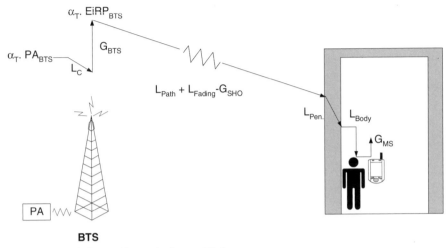

Figure 12.3   Gains and losses in forward link.

Table 12.3 provides the results of LBA calculations for the forward CDMA traffic channels of different data rates. In this table it is assumed that

- A fixed percentage of total power is allocated to each traffic channel regardless of its data rates. This approach is taken merely to show the differences in the radio links and is not necessarily a practical assumption.

- The effect of same-cell and other-cell interference is captured in an interference margin with a typical value of 3 dB. This margin, which is not to be confused with the loading factor of the reverse link, is the result of total forward link interference rise above noise at the mobile station. The interference includes both other-cell and same-cell interference.

- The other-cell interference is caused by signal power received from all base stations in the area that are not in the traffic channel's active set.

- The interference from the same cell is caused by other traffic channels as well as common and broadcast control channels transmitted from the serving base station. In the absence of multipath this portion of the received signal should be orthogonal to the desired traffic channel signal because of orthogonality of Walsh channelization codes. However, the multipath version Walsh codes are no longer orthogonal to each other, and therefore part of the orthogonality is lost at the receiver's end, resulting in nonzero same-cell interference. The total interference is a summation of I_{oc} and $I_{sc} = \varphi \cdot P_{RX}$, where φ is the orthogonality factor, which is assumed to be 0.1–0.25.

- The handoff in the forward link also provides some macrospace diversity gain that is indicated in the Table 12.3. However, because the supplemental channel uses a reduced active set and typically performs sector selection as opposed to maximum ratio combining, the effective soft handoff gain for high-rate channels is slightly lower than for voice channels.

TABLE 12.3 cdma2000 Forward Link Budget Analysis Using the 10% Maximum Power Allocation for All Traffic Channels

CDMA2000 Forward Link Budget Analysis (Voice and Data)

Input Parameters	Units	Voice	Data				
Bearer Data Rate	kbps	9.6	9.6	19.2	38.4	76.8	153.6
Receiver Parameters							
Target E_b/N_t	dB	4.40	4.40	4.20	4.20	3.40	3.00
No = $10\log(KT)$	dBm/Hz	−174.0	−174.0	−174.0	−174.0	−174.0	−174.0
MS Noise Figure	dB	8.5	8.5	8.5	8.5	8.5	8.5
Abolute Rx. Sensitivity	dBm	−121.28	−121.28	−118.47	−115.46	−113.25	−110.64
MS Cabling Loss	dB	0.00	0.00	0.00	0.00	0.00	0.00
MS Antenna Gain	dB	0.00	0.00	0.00	0.00	0.00	0.00
Transmitter Parameters							
BTS Max PA Power	W	20.00	20.00	20.00	20.00	20.00	20.00
BTS Max PA Power in dBm	dBm	43.01	43.01	43.01	43.01	43.01	43.01
BTS Cabling Loss	dB	3.00	3.00	3.00	3.00	3.00	3.00
BTS Antenna Gain	dB	16.00	16.00	16.00	16.00	16.00	16.00
BTS Max EIRP	dBm	56.0	56.0	56.0	56.0	56.0	56.0
Max %PA Power Allocation/ Traffic Channel	%	10.00	10.00	10.00	10.00	10.00	10.00
Max EIRP per Traffic Channel	dBm	46.01	46.01	46.01	46.01	46.01	46.01
Channel Parameters							
Probability of Service at Cell edge	%	90%	90%	90%	90%	90%	90%
Composite Slow Fading Std. Deviation	dB	8.0	8.0	8.0	8.0	8.0	8.0
Fade Margin	dB	10.3	10.3	10.3	10.3	10.3	10.3
Building Penetration Loss	dB	15.0	15.0	15.0	15.0	15.0	15.0
User Body Loss	dB	3	3	3	3	3	3
SHO Gain	dB	4	4	3	3	3	3
TX Diversity Gain	dB	0	0	0	0	0	0
(Total Margin- Soft Handoff Gain)	dB	24.25	24.25	25.25	25.25	25.25	25.25
Max. Allowable Path Loss (No Load)	dB	143.04	143.04	139.22	136.21	134.00	131.39
Interference Margin	dB	3.0	3.0	3.0	3.0	3.0	3.0
Max. Allowable Path Loss (With Loading)	dB	140.0	140.0	136.2	133.2	131.0	128.4

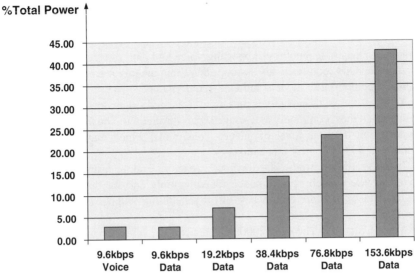

Figure 12.4 Comparing forward power allocation, required for different data rates, to meet reverse link coverage in cdma2000.

- Transmit diversity is assumed to be zero because this optional feature is not commonly used in early deployments.
- Based on all these assumptions and other typical input parameters the results of Table 12.3 can be used to compare system gain differences for different bearer services in the forward link.

An alternative and more practical way of looking at the forward link budget is to set the maximum path loss to be the same as the reverse link path loss for the lowest target data rate and calculate the differences in required power allocation for forward traffic channels.

Figure 12.4 shows the results of such calculations using the 134.7-dB maximum path loss that was derived previously for cdma2000 reverse link voice service. Note that although the actual numbers represent are valid with our specific set of assumptions, the overall trends are applicable in general.

Note that the link budget calculations presented in this section are applicable to all releases of cdma2000. The addition of higher-rate radio configurations in Releases C and D typically does not affect the cell coverage, as the coverage is determined mostly by the voice and low-rate traffic channels. The major impact of high-rate traffic channels and other enhancements in release C and D is on system capacity and spectrum efficiency.

12.3   CDMA2000 VOICE AND DATA CAPACITY PERFORMANCE

For any radio access technology supporting voice and data services, characterizing the system capacity requires many assumptions about the mixture of services as well as their respective traffic models and QoS requirements.

Whereas the voice capacity is measured in carried Erlang load, the data capacity is measured most commonly as average sector throughput. The throughput capacity is also tied to some level of latency constraints, which may vary based on scheduling and QoS control mechanisms.

In this section we study the cdma2000 capacity by taking a simplistic approach and viewing the capacity performance for voice-only and for data-only services. Also, for data we assume only a best-effort scheduling without any specific QoS requirement.

Voice Capacity: The cdma2000 cell capacity for voice services is higher than the IS95 capacity, mainly because of coherent uplink demodulation and fast forward power control. Using the same capacity equation that we used in Chapter 2 for an IS95 example we can estimate the cdma2000 cell capacity:

$$N_{rev} = \frac{W/R}{(E_b/N_o)_{min} \times \overline{v}(1+f)} \times L \tag{12.11}$$

where Table 12.4 shows the key assumptions used in this equation. The main difference with the IS95 case is the lower $(E_b/N_o)_{min}$ requirement for cdma2000, i.e. 4.5 dB vs. 7 dB, which results in about 78% increase in the number of voice channels and about 100% increase in Erlang capacity.

Table 12.5 provides the estimated cell capacity for a single-carrier system for different vocoder type or encoding modes, supported by cdma2000.

Note that in the backward-compatible mode, and when serving legacy mobiles, the cdma2000 system does not benefit from the new link enhancements and thus offers the same 2G-voice capacity as IS95, 13.1 Erlangs in our example.

As we discussed in Chapter 4 the SMV vocoders, depending on the selected modes of operation, provide different levels of capacity improvement over conven-

TABLE 12.4   Assumptions Used for cdma2000 Voice Capacity Estimation

Parameter/Assumption	Value
W = Channel bandwidth	1.25 MHz
R = Data Rate	9.6 kbps
$(E_b/N_o)_{min}$	4.5 dB
v = Voice Activity	40%
F = Other Cell to Same Cell Ratio	0.6
L = Loading Factor	50%
EVRC Capacity (No. Voice Channels)	35.4
Erlangs @ 2% Blocking Probability	26.5 Erlangs

TABLE 12.5 cdma2000 Voice Capacity for Different
Vocoders

Vocoder Type/Mode	Erlang Capacity per Carrier
EVRC (Legacy MS)	13.1 Erlang
EVRC	26.5 Erlang
SMV (Mode 0)	26.5
SMV (Mode 1)	$26.5 \times 1.34 = 35.5$
SMV (Mode 2)	$26.5 \times 1.61 = 42.7$
SMV (Mode 3)	$26.5 \times 1.75 = 46.4$

tional EVRC vocoders. Thus for SMV the effective Erlang capacity is equal to or higher than EVRC depending on the vocoder mode.

In a cdma2000 network supporting different generations of mobile station with different protocol revisions (P_REVs), the effective voice capacity depends on the mix of vocoders and mobile station types.

Let $\alpha_{\text{EVRC-2G}}$, $\alpha_{\text{EVRC-3G}}$, and α_{SMV} be the percentage of mobile stations using 2G-EVRC, 3G-EVRC, and SMV, respectively. Also assume that α_0, α_1, α_2, and α_3 represent the percentage SMV users operating in Modes 0, 1, 2, and 3, respectively. Then one can approximately estimate the average cell capacity by taking a weighted average of all Erlang capacities as follows:

$$C_{\text{AVG}} = C_{\text{EVRC-2G}} \times \alpha_{\text{EVRC-2G}} + C_{\text{EVRC-3G}} \times \alpha_{\text{EVRC-3G}} + C_{\text{SMV}} \times \alpha_{\text{SMV}}$$

$$\text{Where } C_{\text{SMV}} = C_{\text{Mode0}} \times \alpha_0 + C_{\text{Mode1}} \times \alpha_1 + C_{\text{Mode2}} \times \alpha_2 + C_{\text{Mode3}} \times \alpha_3 \quad (12.12)$$

Example: Using the Erlang capacity examples in Table 12.5 and assuming 20% legacy mobiles, 50% cdma2000 mobiles with EVRC, and 30% with SMV, from which half are in Mode 0 and other half in Mode 1, we can estimate the cell capacity to be:

$$C_{\text{SMV}} = 26.5 \times 50\% + 35.5 \times 50\% = 31$$

$$C_{\text{AVG}} = 13.1 \times 20\% + 26.5 \times 50\% + C_{\text{SMV}} \times 30\% \approx 25 \quad (12.13)$$

Obviously, this is only an example, and it is valid as long as the underlying assumptions apply. The average capacity per cell will be higher as the percentage of legacy mobile decreases over time and more and more mobile stations use lower rates of SMV.

Almost all cdma2000 systems have been deployed with multisector base station and multicarrier sector configurations. For these networks, the voice capacity offered by each base station is approximately equal to the per carrier capacities mentioned above times the number of carriers and multiplied by the sectorization gain. The sectorization gain is ideally equal to the number of sectors, for example, equal to 3 for three-sector sites, but in practice it is slightly smaller, for example, 2.5–2.8, because of overlap and interference between "co-site" sectors. In the dimensioning procedures in this section, we assume perfect sectorization, that is, 3 time capacity for three sectors.

Data Capacity: There are a number of values reported for cdma2000 data throughput, each with a different set of radio configurations, number of active users, application types, interference levels, and user mobility models.

One of the most common radio configurations used in cdma2000 deployments is the combination of RC4 in the forward link and RC3 in the reverse link. For this combination the average sector throughputs for best-effort packet data service in the forward and reverse links are about 300 kbps and 150 kbps, respectively. These numbers are average values without any voice users present, and they depend on the number of active users and their distribution around the base station. Assuming a uniform user distribution in the cell, the sector throughput typically increases as the number of active users increases from 1 to 8 or 10 and then decreases for higher numbers of users.

For 1xEV-DV or cdma2000 Release D with the new radio configurations, namely, RC10 in the downlink and RC7 in the uplink, the throughput values are higher because of link adaptation, multiuser diversity, and enhanced MAC procedures. Preliminary reports on the sector throughput performance figures show an average of 1 Mbps in the downlink and 600 kbps in the uplink without any voice services. Note that these numbers are examples, and the actual throughput values would depend on many previously mentioned factors, especially the number of users and their mobility.

Because of the asymmetry of packet data traffic with the majority of demand on the downlink, the forward link throughput is the main limitation that defines cdma2000 data capacity per sector.

In the case of mixed voice and data services one can only use simulation data to assess the system performance; however, following our simplified approach so far we can approximate the capacity under mixed traffic using a linear interpolation between all voice and all data scenarios (see Fig. 12.5).

12.4 1XEV-DO COVERAGE AND CAPACITY PERFORMANCE

In this section, we focus on the coverage and capacity performance of the HRPD system by looking at the link budget and throughput.

The coverage of a HRPD cell is typically determined by its reverse link; therefore, we will focus on the reverse link budget, which is very similar to cdma2000 as presented in the previous sections. The main differences are the following [3,4]:

- The different sets of required E_b/N_o for each data rate as captured in Table 12.6. Note that these E_b/N_o requirements are per antenna.

- Lower soft handoff gain of about 2 dB as the reverse link employs selection diversity on a frame-by-frame basis, instead of the cdma2000 fully soft handoff.

- Slightly lower power available for reverse data channel because of the uplink MAC-related DRC and ACK channel.

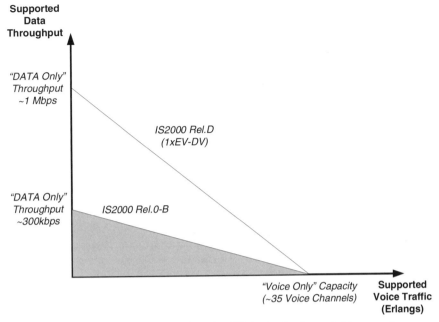

Figure 12.5   An approximate model for cdma2000 mixed voice and data capacity.

TABLE 12.6   Required E_b/N_o per Antenna for HRPD(IS856) Reverse Link

Data Rate (kbps)	Total (E_b/N_t) Required per Antenna (dB)
9.6	6.62
19.2	4.98
38.4	3.84
76.8	3.55
153.6	5.27

Accounting for all these differences, the maximum uplink path loss in HRPD for a 9.6-kbps channel can be calculated and shown to be:

- About 0.5–1 dB higher than IS95 voice
- About 1.5 dB lower than cdma2000 voice
- About 3.5 dB lower than 9.6-kbps data channel

As a result, the HRPD system that is typically deployed as an overlay on a cdmaONE or cdma2000 network does not require any additional sites for coverage. Also, as these overlay radios are deployed mostly in high-traffic areas, the underlying

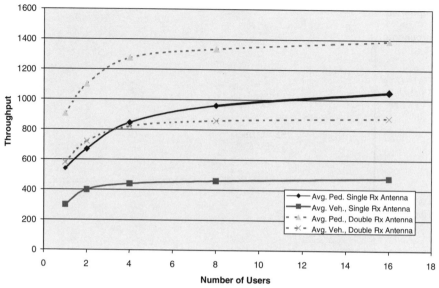

Figure 12.6   Average downlink sector throughput for HRPD system.

cdma2000 cells are typically placed based on capacity limitations, and these have smaller size than what the link budget suggests.

The key issue in dimensioning HRPD radios in most cases is the throughput and not necessarily the coverage.

Figure 12.6 shows the single-carrier forward link sector throughput for the 1xEV-DO (IS856) system. This throughput performance is shown for average pedestrian as well as average vehicular channels using single or dual antenna receivers for purposes of comparison. The results show the following:

- The achievable throughput highly depends on users' mobility and the number of active users in the cell. At lower mobility, the closed-loop rate control mechanism of 1xEV-DO provides more reliable channel state information to the scheduler, resulting in higher throughput for pedestrian users.

- The performance can be significantly improved by using dual RX antenna at the access terminal.

- The increase in throughput with the number of users is the result of the adaptive scheduling and multiuser diversity gain, which were conceptually discussed in Chapter 4.

- As expected, the multiuser diversity gain is less significant for the dual Rx antenna case, as some of the channel dynamics is removed, and benefited from, as a result of diversity combining by the receiver.

- The actual sector throughput in a real network would depend on the percentage of access terminals with dual-receive antenna.

- The reverse link throughput for 1xEV-DO is significantly lower than the forward link, and it is subject to different limitations. The average sector throughput is approximately 170 kbps per carrier for up to eight active users, but it decreases as the number of users increases. This can be explained by the fact that as the number of active users in a sector increases, a larger number of pilot and DRC and possibly ACK channels are transmitted, increasing the uplink sector loading without proportionally benefiting the throughput performance [4,5]. As a result, although the uplink transmissions do not benefit from scheduling/multiuser diversity, a higher number of active users leaves a proportionally smaller rise over thermal (RoT) resource for data traffic and therefore lowers the overall sector throughput.

12.5 RADIO NETWORK DIMENSIONING AND PLANNING ISSUES

Radio network dimensioning is the process of estimating the required number of radio network elements to meet both coverage and capacity objectives for all target services in a network. Although the exact planning of the network requires propagation and traffic simulation with realistic user distributions, site locations, and terrain profiles, an approximate but objective dimensioning process can be very helpful in predicting network size and therefore cost.

In the following we provide examples of such procedures based on heuristic and practical guidelines mostly for educational purposes and to show the main trade-off considerations.

The cdma2000 system may be built as new networks in a Greenfield spectrum or more commonly as a 3G overlay in the migration of the cdmaONE network. In the following we first discuss the Greenfield deployments scenario and then present a variation of the process based on an existing 2G/2.5G network.

12.5.1 Greenfield Deployment

In this case, no existing sites or carriers are deployed and there is no existing traffic to be considered for dimensioning process. Therefore, the process starts with estimation of the minimum number of sites needed to provide the desired level of coverage, followed by traffic and capacity considerations.

Estimating Number of Sites for Coverage: Assuming a uniform propagation environment captured in a tuned path loss model, one can use the maximum allowable path loss obtained from the link budget to estimate the average cell size and cell count. This simple process is shown in Figure 12.7.

The path loss equation presents the path loss L as a function F of distance D, for example $L = F(D)$. By setting L equal to maximum path loss from the link budget for the lowest rate or voice, one can obtain the maximum range or cell radius. Also, assuming circular or hexagonal cells, one can estimate the average cell area size from the radius. This area may need to be reduced by a small percentage because of overlap between adjacent cells.

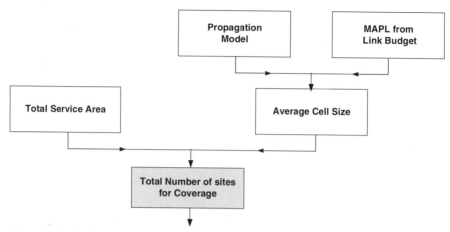

Figure 12.7 Estimating the required number of sites for coverage.

We can then estimate the number of cells by dividing the total market area by the average cell area. The following example helps in further clarifying the steps.

Example: A cdma2000 network is to be deployed in "Market" to provide minimum data rate of 19.2 kbps in the uplink within a service area that is about 1000 km^2. For this market, a verified path loss model is available and can be simplified to $L = A + B\log(D)$, where $A = 130$ and $B = 45$ are the slope and intercept of path loss L as a function of distance D. The objective is to find the number of sites needed assuming a 10% area overlap between cells.

We start by setting L in the path loss equation equal to maximum path loss for the lowest rate service, which in the case of 19.2 kbps is 134.3 dB. Thus we can write $L = 134.3 = 128 + 45 \times \log(D)$, which when solved for D results in d = 1.38 km as our estimate for the radius. Based on this radius and assuming circular cells we can estimate the average cell area as $A_{\text{Cell}} = \pi D^2 = 6$ km^2.

Accounting for $\theta = 10\%$ area overlap between adjacent cells, we can we can estimate the required number of sites as follows:

$$N_{\text{Sites}} = \frac{A_{\text{Service Area}}}{A_{\text{Cell}} \times (1 - \theta)} \qquad (12.14)$$

In this example, the number of sites will be $1000/[6 \times (1 - 0.1)] = 185$ for the target service area.

Traffic Load Projection: For voice and data traffic projection ,we use a simple method based on parameters most commonly used by operators, namely, the average monthly minutes of usage (MoU) for voice and megabytes of data per subscriber for all data services (see Fig. 12.8). Note that more detailed procedures differentiating various application types and their QoS can be devised, but they are outside the scope of our discussion.

We start with voice traffic and expected monthly MoU per subscriber, which is projected based on marketing surveys and current usage pattern in similar net-

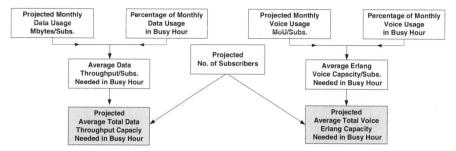

Figure 12.8 Projecting the total voice and data demand in the network.

works. This number may be the projected usage for few quarters after the service is launched. The monthly usage is scaled down to obtain minutes of calls during the busy hour.

The scaling factor should also be obtained from historic data reflecting real traffic patterns in similar networks. If this is not possible, this factor can be approximately estimated based on other information, such as expected percentage of daily calls made during the busy hour, number of days in the month, and number of hours in a day.

We can then calculate the Erlang capacity demand during the busy hour by dividing the projected minutes of use in the busy hour by 60 (minutes/hour). The result is our estimate of voice demand in terms of busy hour Erlangs per subscriber. This usage can also be scaled for future time frames proportionally with MoU growth.

$$E_{\text{Subs.}} = \text{MoU} \times S_{\text{BH\_Voice}} / 60_{\text{min/BH}} \qquad (12.15)$$

Example: Assuming 500 minutes of use per average subscriber and 10% of daily calls made during the busy hour we need to calculate the Erlang usage.

In this example, the busy hour scaling factor is $(1/30 \text{ days}) \times 10\% = 0.0033$ and per subscriber Erlang usage is equal to $500 \times 0.0033/60 = 27.7\,\text{mElangs/Subs./BH}$.

For data we take a similar approach and start by scaling down the monthly data usage to arrive at the usage during the busy hour. We then translate the results to bits of data to be transferred in the busy hour by taking into account the forward/reverse link asymmetry. In most cases the forward link is the limiting link, and that will be our focus in this procedure. The result is the number of downloaded bytes/BH, which can be translated to the average downlink throughput required per subscriber as follows:

$$TP_{\text{Subs.}} = \{(\text{N}_{\text{Mbytes/Month}}) \times S_{\text{BH}} \times S_{\text{Asym}} \times (8_{\text{bits/byte}}) / (3600_{\text{sec/BH}})\} \times S_{\text{Burstiness}} \quad (12.16)$$

where:

- S_{BH} is the busy hour scaling factor or the percentage of monthly usage expected during the busy hour.

- $S_{Asym} < 1$ reflects the asymmetry of traffic load in terms of the percentage of data traffic for the link of interest, that is, the forward or reverse link.
- $S_{Burstiness} > 1$ accounts for nonuniform demand for throughput during the busy hour.

Example: Assuming that an average subscriber uses 5 Mbytes of data per month with 80% of data transferred in the downlink, we need to calculate the required forward link throughput per user. In this calculation, we assume that about 1% of monthly data usage is needed during the busy hour and we also account for 50% higher throughput margin demand due to traffic burstiness. The result is

$$TP_{Subs.} = 5_{Mbytes/Month} \times 1\% \times 80\% \times 8_{bits}/3600 \times (1 + 50\%) \approx 133\,bps \qquad (12.17)$$

The total required Erlang and throughput for all subscribers can be derived simply by multiplying the "per subscriber" values calculated above by the projected total number of subscribers. For this example, assuming a total of 500,000 subscribers in the service area ,the total Erlang and throughput will be $500,000 \times 27.7\,mE = 13,850$ Erlangs and $500,000 \times 133 = 66.6\,Mbps$, respectively.

The next step is to estimate the required capacity per cell and by comparing that with the capacity available determine whether additional carrier or sites are needed. Figure 12.9 shows a simplified process to achieve this.

We begin with voice and divide the total voice demand by the number of sites previously calculated based on the link budget, to obtain the required voice capacity per cell site. We then calculate the required number of carriers per site needed for voice, dividing the capacity demand per cell by the cell site capacity per carrier.

If the number of required carriers in higher than a maximum allowed based on spectrum or hardware configuration limitation, we add a few sites and recalculate the traffic load per site, Otherwise, we proceed with the data capacity consideration.

Example: Following our previous examples and assuming that all terminals use EVRC vocoders and all sites have three sectors, we calculate the number of carriers needed per sector in the following steps:

- For three-sector sites with a typical sectorization gain of 3, the capacity is about $3 \times 26.5 = 79.5$ Erlangs per carrier.
- Following our previous example, with the total demand of 13,850 Erlangs and 185 sites the Erlang per site is about $13,850/185 = 75$ Erlangs.
- To support 75 Erlangs and with 79.5 Erlangs per site we only need to deploy a single carrier per sector, which is assumed to be feasible.

After capacity planning for voice is performed and the number of carrier and sites needed for voice is determined, we start data throughput analysis by estimating the "left over" data capacity in the system as follows:

Any circuit-switched voice system that is designed to ensure a small probability of blocking has some inherent leftover capacity for packet data traffic. This capacity is proportional to the difference between the number of voice channels supported and the carried Erlang traffic. Using the simplified linear model for mixed voice and data capacity shown in Figure 12.5 one can arrive at the following:

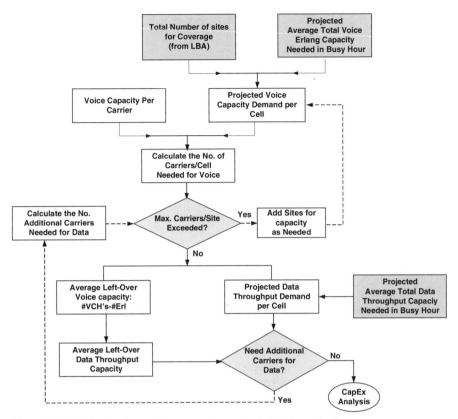

Figure 12.9   Mixed voice and data capacity analysis for dimensioning.

$$C_{\text{Data}} = \frac{N_{\text{VCHs}} - E_{\text{Voice}}}{N_{\text{VCHs}}} \times (N_{\text{Carriers}} \times TP_{\text{Carrier}}) \times N_{\text{Sector}} \times \eta_{\text{Data}} \qquad (12.18)$$

where N_{VCH} is the total number of voice channels per sector for all carriers, E_{Voice} is the average Erlang voice capacity demand per sector, N_{Carrier} is the number of carriers per sector, TP_{carrier} is the maximum data throughput per carrier, and N_{sector} is the number of sectors per site. The factor $\eta_{\text{Data}} < 1$ is a utilization factor that accounts for the inherent burstiness of packet data demand.

This capacity needs to be compared with throughput demand per site, which is the projected total throughput demand divided by the number of sites.

If capacity is higher than demand we are done with the data analysis and can move on to any cost analysis if needed; otherwise, we will need to add carriers to address data demand. The number of additional carriers needed per sector depends on the difference between demand and capacity.

Example: Continuing our capacity analysis in the chain of examples so far we can calculate the following:

- N_{VCH} = 1 carrier × 35 = 35 channels/sector based on the all-EVRC vocoder scenario
- E_{Voice} = (75 Erlangs per site)/3 = 25 Erlang/sector based on previous demand calculations
- $TP_{carrier}$ = 300 kbps, from Figure 12.5 and with cdma2000 Release A
- Assuming 75% capacity utilization, the available data throughput can be estimated as:

$$C_{Data} = [(35 - 25)/35] \times (1 \times 300\,\text{kbps}) \times 3\ \text{sectors} \times 0.75 = 193\,\text{kbps/site}$$

- The demand, on the other hand, is 66.6 Mbps/185 sites = 360 kbps/site.
- Because the demand is higher than capacity by 360 − 193 = 167 kbps, we need to add the second carrier to meet the data traffic.

12.5.2 Migration or Expansion of an Existing Network

In this scenario dimensioning is performed on a cdma2000 system that already exists or is being deployed as an overlay, typically one on one, to an exiting cdmaONE network. The process in this case is very similar to the one we presented for the Greenfield case with the following differences:

- Rather than estimating the required number of sites based on LBA and the path loss model we will normally use the existing site count as a basis. However, if the minimum bearer data rate to be offered as part of the migration is large and requires a higher number of sites than the number of existing ones, in this case we need to reevaluate the cell count based on LBA as we did for the Greenfield case.
- The current voice and data traffic usage and distribution can be used as a reference for projecting future voice and data capacity demand and its distribution in the service area.
- The traffic mix in this case may include a number of 2G-voice users, which have an impact on the cell capacity. In this case, the expected cell capacity should be proportionally lowered or the effective traffic load should be increased to account for lower capacity efficiency for 2G voice services. This can be achieved by doubling the estimated 2G Erlang voice before adding that to the cdma2000 voice traffic load.

In the following example we go through all the steps in the dimensioning process for the expansion of an existing network.

Example: A cdma2000 network currently has 250 base stations, each with three sectors and two carriers per sector. The network has 800,000 current subscribers and covers about 2000 km² of service area. The operator, who has 5 MHz of paired spectrum, plans to expand the network in preparation for 25% growth in subscriber number and new packet data services. The projected subscriber usage is 600 minutes of use and 5 Mbytes of data for cdma2000 mobiles. About 10% of subscribers are expected to use legacy mobile stations with voice-only capability.

The operator also intends to offer a minimum uplink data rate of 19.2 kbps with a link budget assumption consistent with Table 12.1, with the exception of in-building loss, which is assumed to be 12 dB. We will use $L = 128 + 40x\log(D)$ as our path loss model.

- Cell Count: For 19.2 kbps the adjusted path loss for 12 dB in-building loss instead of 15 dB will be $134.3 + 15 - 12 = 137.3$. Assuming 10% cell area overlap, the cell count is calculated as follows:

$$137.3 = 128 + 40x\log(D) \rightarrow D = 1.7 \, \text{km} \rightarrow A = \pi D^2 = 9.2 \, \text{km}^2$$

Number of cells = $2000 \, \text{km}^2 / [9.2 \, \text{km}^2 \times (1 - 0.1)] = 242 < 250$ existing sites

Because the 242 is lower than the existing number of sites, we will proceed with 250 as our reference for the cell count.

For capacity analysis we make the following design assumptions:

- The voice and data usage in the busy hour are 0.25% and 3% of monthly usage, respectively. Also, for data service we assume 90% traffic in the downlink, a burstiness factor of 1.4, and a maximum of 50% capacity utilization.

Projected Subscriber Base:

First note that the projected number of subscribers, based on 25% growth, is $800 \text{K} \times 1.5 = 1$ million, of which 900,000 subscribers have cdma2000 terminals with data capability and EVRC; the remaining 100,000 use EVRC but with legacy mobiles for voice-only services.

Voice Demand:

Erlang/Subscriber = $(600 \times 0.25\%)/60 = 50$ mErlangs

total Erlangs = $50 \times 100,000 \times 2 + 50 \times 900,000 = 55,000$ Erlangs

Erlangs/site = $55,000/250 = 220$

Dimensioning for Voice Capacity:

required number of carriers/sector = 220 Erlang/$(26.5 \times 3) = 2.7 \rightarrow 3$, so we need 3 carriers on each of the 3 sectors, which is feasible with 10-MHz spectrum.

Data Demand:

throughput/Subs. = $(5 \text{ Mbytes} \times 3\% \times 90\% \times 8/3600 \text{s}) \times 1.4 = 0.42 \, \text{kbps}$

total throughput = $900,000 \times 0.42 \, \text{kbps} = 378 \, \text{Mbps}$

throughput per cell = $378/250 = 1.5 \, \text{Mbps/cell}$ or $500 \, \text{kbps/sector}$

Dimensioning for Data Capacity:

$N_{\text{VCH}} = 3$ carriers $\times 35 = 105$ channels/sector based on all-EVRC vocoder scenario. Note that 2G voice is accounted for by increasing the effective traffic load.

$E_{\text{Voice}} = (220 \text{ Erlangs/site})/3 = 73$ Erlangs/sector

$TP_{\text{carrier}} = 300 \, \text{kbps}$, from Figure 12.5 and with cdma2000 Release A

With 70% capacity utilization, the available data throughput is:

$$C_{\text{Data}} = [(105 - 73)/105] \times (3 \times 300 \, \text{kbps}) \times 3 \, \text{sectors} \times 0.70 = 571 \, \text{kbps/site}$$

Because the data capacity of 571 kbps is higher than the 500 kbps demand, no additional carrier is needed for data. Therefore, our dimensioning process for this scenario suggests that adding a third carrier to all sectors in the network would be sufficient to meet both voice and data capacity objectives.

12.6 REFERENCES

1. *Microwave Mobile Communications* by William C. Jakes, John Wiley & Sons, 1975.
2. Peter J. Black and Qiang Wu, "Link Budget of cdma2000 1xEV-DO Wireless Internet Access System," PIMRC 2002.
3. *Advances in 3G Enhanced Technologies for Wireless Communications* by Jiangzhou Wang and Tung-Sang Ng (Editors), Artech House, March 2002.
4. P. Black and M.Gurelli, "Capacity Simulation of cdma2000 1xEV Wireless Internet Access System," IEEE MWCN 2001, Aug 2001.
5. *Wireless Communications, Principles and Practice* by T. Rappaport, 2nd Edition, IEEE Press, 2002.

CHAPTER 13

OVERVIEW OF OTHER IMT2000 STANDARDS

13.1 INTRODUCTION

The IMT2000 family of standards includes cdma2000, UMTS-FDD (WCDMA), and UMTS-TDD [TD-(S)CDMA] air interfaces commonly referred to as 3G systems. In this chapter we provide a high-level but technical overview of WCDMA and TD-CDMA radio access technologies, focusing on their key design features to help in their comparison with cdma2000. Detailed descriptions of these technologies are beyond the scope of this book and can be found in [1–3].

13.2 IMT2000-DIRECT SPREAD, UMTS-FDD (WCDMA)

The frequency division dDuplex (FDD) mode of UMTS terrestrial radio access (UTRAN) is based on wideband code division multiple access (WCDMA) technology. This technology has also been adopted as a standard by the ITU under the name "IMT-2000 Direct Spread." Table 13.1 captures some of the key attributes of WCDMA physical layer design.

Using direct sequence spreading over wide, for example, 5 MHz, carriers, the WCDMA system can support mobile voice and data applications with up to 2 Mbps within a local area access or 384 kbps within a wide area access. WCDMA is mainly used in Europe and in the context of migration from GSM to UMTS.

Starting with Release 99, the WCDMA system employs many concepts and techniques used in cdma2000: direct sequence spreading with variable orthogonal length spreading codes for channelization, turbo encoding, soft handoff, coherent demodulation, and fast power control on both forward and reverse links. This system also supports beam forming and transmit diversity but with slightly different schemes than those used in cdma2000.

The spreading code structure, cell search, and synchronization process in WCDMA are designed such that the system can operate with asynchronous base stations. The system can also operate with GPS-based synchronous base stations when possible.

CDMA2000® Evolution: System Concepts and Design Principles, by Kamran Etemad
ISBN: 0-471-46125-3 Copyright © 2004 John Wiley & Sons, Inc.

TABLE 13.1 Some of the Key Physical Layer Features of WCDMA

RF Channelization	
Frequency Band and Duplexing	IMT2000 FDD Spectrum
	1920 MHz–1980 MHz UL and 2110 MHz–2170 MHz DL
Minimum frequency band required	~2 × 5 MHz
Frequency reuse	1
Channel raster	200 kHz
Carrier spacing	4.4 MHz–5.2 MHz
Coding, Modulation, and Spreading	
Voice coding	AMR codecs (4.75 kHz–12.2 kHz, GSM EFR = 12.2 kHz and SID (1.8 kHz)
Channel coding	1/2 and 1/3 rate convolutional coding, turbo code for high-rate data
Channel bit rate	5.76 Mbps
Modulation	QPSK w/o HSPDA
	QPSK and 16QAM with HSDPA
Pulse shaping	Root raised cosine, rolloff = 0.22
Chip rate	3.84 Mcps
Physical layer spreading factors	4-256 UL, 4-512 DL
Frame Structure	
Frame length	10 ms (38400 chips)
Number of slots/frame	15
Number of chips/slot	2560 chips
Handoff and Power Control	
Handovers	Soft, Softer, (interfrequency: Hard HO)
Power control Rate	1500-Hz rate
Power control step size	0.5, 1, 1.5, and 2 dB (variable)

Figure 13.1 shows the general frame structure for dedicated channels in WCDMA. Within each 0.625-ms slot the Layer 2 user data and Layer 1 control information are time multiplexed in the downlink but code multiplexed in the uplink.

Similar to cdma2000, there are many reported estimates for capacity performance of a WCDMA system. Table 13.2 provides some examples of simulation-based capacity numbers.

13.3 HIGH-SPEED DOWNLINK PACKET ACCESS (HSDPA)

Starting with Release 4 specifications, the WCDMA provides efficient IP support enabling provision of services through an all-IP core network, and Release 5 specifications focus on a new high-speed downlink packet access mode (HSDPA) to provide data rates up to approximately 10 Mbps for packet-based multimedia services.

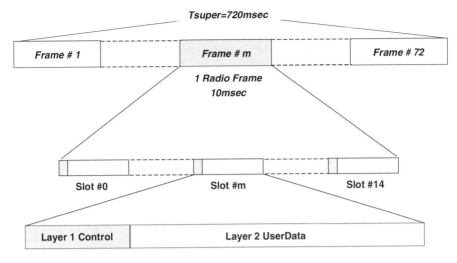

Figure 13.1   Frame structure of WCDMA.

TABLE 13.2   The Data and Voice Capacity Performance of WCDMA

	Capacity Performance
Data type	Packet and Circuit Switched
Maximum number of voice channels in 2×5 MHz	~196 (spreading factor 256 UL, AMR 7.95 kbps) ~98 (spreading factor 128 UL, AMR 12.2 kbps)
Maximum user data rate (physical channel)	~2.3 Mbps (spreading factor 4, parallel codes (3 DL/6 UL), 1/2 rate coding), but interference limited.
Maximum user data rate (offered)	384 kbps (year 2002), higher rates (~2 Mbps) HSPDA will offer data speeds up to 8–10 Mbps (and 20 Mbps for MIMO systems)
Carrier aggregate throughput	~1 Mbps DL, ~1 Mbps UL

In HSDPA the spreading factor remains fixed, but the coding rate can vary between 1/4 and 3/4. The HSDPA specification supports the use of 5, 10, or 15 multicodes. The higher rates are achieved by

- Adaptive modulation and coding
- A fast channel-based scheduling function, which is controlled in the base station (BTS), rather than by the radio network controller (RNC)
- Fast hybrid ARQ and packet retransmissions with shorter frames, soft combining, and incremental redundancy

One can see that the concepts used in HSDPA are very similar to those adopted for high-rate forward packet data channel in RC10 of IS2000 Release C or the 1x-EVDV

TABLE 13.3 HSPDA and CDMA2000 1xEV-DV Comparison

Features	HSDPA	1xEV-DV (RC10 in IS2000-C)
Downlink frame size	2 ms TTI (3 slots)	1.25-, 2.5-, 5-, 10-ms variable frame size (1.25-ms slot size)
Channel quality feedback	Reported in every 2 ms (i.e., @ 500 Hz)	Reported in every 1.25 ms (i.e., @ 800 Hz)
User data traffic multiplexing	TDM/CDM	TDM/CDM
Adaptive modulation	QPSK & 16-QAM	QPSK, 8-PSK & 16-QAM
Physical layer Hybrid-ARQ	Chase or incremental redundancy (IR)	Async. incremental redundancy (IR)
Spreading factor	Fixed spreading; Combination of 5, 10, or 15 orthogonal OVSF (Walsh) codes with SF = 16	Fixed spreading; Combination of up 28 orthogonal Walsh channelization code of length 32
Control channel approach	Dedicated channel pointing to shared channel	Common control channel

mode. Table 13.3 compares some of the features on HSDPA and 1xEV-DV (IS2000 Release C). Note that the HSPDA in Release 5 mainly improves the forward link throughput and latency and using similar techniques for the uplink is being studied for future releases.

Another important development to further enhance the HSPDA spectrum efficiency is the so-called multiple-input multiple-output (MIMO) structure included in Release 6, which is expected to provide higher data transmission rates up to 20 Mbps. In MIMO multiple antennas are implemented at both base station and mobile terminals. At the transmitter, the information bits are divided into several bit streams and transmitted through different antennas. The transmitted information are recovered from the received signals at multiple receive antennas by using an advanced receiver.

13.4 IMT2000-TDD (TD-CDMA)

The IMT2000 TDD standard is also based on the UTRAN framework and uses a combination of TDMA and CDMA for its channelization and multiple access. The TD-CDMA has the same physical layer features as those shown in Table 13.1.

Specifically, the system uses QPSK modulation and convolution encoders as well as turbo code and has the same chip rate and frame size as WCDMA. The TD-CDMA can be deployed as a 5- or 10-MHz carrier.

The overall frame structure in TD-CDMA is also similar to WCDMA. However, in this case to allow TDD operation, the 15 time slots are partitioned into two sets, one for uplink and one for downlink transmissions. This uplink/downlink

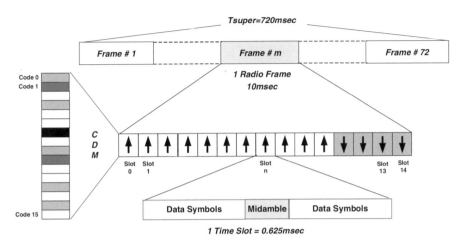

Figure 13.2   Frame structure of TD-CDMA.

partitioning is flexible and can be changed to accommodate asymmetric traffic loads. Typically, 1 time slot is needed for uplink access, 2 time slots for downlink control and paging, and the remaining 12 slots are used for uplink and downlink traffic channels.

The data transmissions in the TD-CDMA system are formatted in terms of one of several "bursts" and within each data burst a midamble is included. The midamble symbols are known to the receiver and can be used for channel estimation.

Also, each time slot can be code-multiplexed among multiple users by using orthogonal spreading codes of length 16 corresponding to Spreading Factor (SF) = 16 (see Fig. 13.2). In each time slot, multiple code channels can be allocated to the same user, resulting in a lower effective spreading factor and a higher data rate. Also, multiple slots can be aggregated to provide higher user data rates.

With different combinations of coding (i.e., 1/2 or 1/3) and spreading factor (i.e., 1–16) the data rate per time slot can be varied from 31.4 to 345.4 kbps. The actual user throughput will also depend on the number of allocated time slots. The peak user rate is about 1.5 Mbps in the downlink and 0.9 Mbps in the uplink. The overall sector throughput can be up to 1.4 Mbps in the downlink and about 750 kbps in the uplink.

Also similar to WCDMA, the data rates and throughput efficiency of TD-CDMA can be enhanced in the forward link by adding HSDPA capability to the air interface. This enhancement is expected to double the forward link peak rate and throughput.

Note that the TD-CDMA system at 5 MHz uses a 3.84-Mcps chip rate, but it can also be implemented as a 1.25-MHz carrier with a chip rate of 1.28 Mcps, in which case it is referred to as TD-SCDMA.

13.5 UMTS NETWORK ARCHITECTURE

The radio access network of WCDMA is designed to interwork with the legacy GSM/GPRS core networks with mobile application protocol (MAP) protocols as well as their evolution to an all-IP network. This is the same approach taken by 3GPP2 for the cdma2000 network.

Initial deployments of UMTS involve a common core network between the existing GSM/GPRS system and the new WCDMA or TD-CDMA networks.

While the circuit-switched traffic such as voice is routed to MSC/VLR/HLR, the packet-switched traffic is routed to the serving GPRS support node (SGSN) and gateway GPRS support node (GGSN). The SGSN and GGSN are responsible for IP routing and mobility management.

Figure 13.3 shows the UMTS network model for both WCDMA and TD-CDMA systems, which also interoperates with the evolved GSM/GPRS network. In this architecture, the key components of network are

- Radio Network Subsystem (RNS), which is a generalization of BSS functionality in legacy GSM networks and includes the following:
  - Node B is a logical node that includes radio transceivers and baseband signal processing such as coding and spreading for one or more cells. The

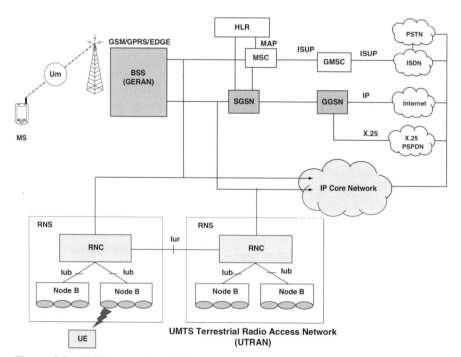

Figure 13.3   UMTS network model.

functions of a Node B can be compared with those of a BTS in the legacy GSM network.

– Radio Network Controller (RNC) controls a group of Node Bs connected to it and performs micromobility management for intra-RNC handoff. The functions of an RNC can be compared with a BSC in the GSM/GPRS network. The RNC routes both circuit-switched voice and packet data traffic.

• GPRS Support Nodes, including:

– SGSN, which is at the same hierarchical level as the MSC/VLR in the legacy GSM network and is responsible for the delivery of data packets from and to the mobile stations (MSs) within its service area. SGSN main functions include mobility management, authentication, ciphering, and routing by providing packet routing to and from the SGSN service area for all users in that service area.

– GGSN, which acts as a logical interface to the GPRS network and to external public data networks such as IP and X.25. It converts the GPRS packets coming from SGSN into appropriate format and sends them out on the corresponding packet data network (PDN).

Table 13.4 summarizes the mapping of some of the key network functions to different entities in the UMTS architecture.

TABLE 13.4   The Mapping of Functions to Network Elements in UMTS Networks

Function	Network Elements				
	MS (UE)	UTRAN	SGSN	GGSN	HLR
Network access control					
Registration					X
Authentication and authorization	X		X		X
Admission control	X	X	X		
Charging data collection			X	X	
Packet routing & transfer					
Relay	X	X	X	X	
Routing	X	X	X	X	
Address translation and mapping	X	X	X	X	
Encapsulation	X	X	X	X	
Tunneling		X	X	X	
Compression	X	X			
Ciphering	X	X			X
Mobility management	X		X	X	X
Radio resource management	X	X			

13.6 REFERENCES

1. 3GPP Technical Specification 25.211-214 V3.1.0 (UMTS-FDD)
2. 3GPP Technical Specification 25.221-224 V3.1.0 (UMTS-TDD)
3. *WCDMA for UMTS* by H. Holma and A. Toskala (Eds.), John Wiley & Sons, 2001

INDEX

f denotes figure, t denotes table.

CDMA2000® Evolution: System Concepts and Design Principles, by Kamran Etemad
ISBN: 0-471-46125-3 Copyright © 2004 John Wiley & Sons, Inc.

Date Due
